Local Gradient Theory for Dielectrics

Local Gradient Theory for Dielectrics

Fundamentals and Applications

Olha Hrytsyna
Vasyl Kondrat

Jenny Stanford
Publishing

Published by

Jenny Stanford Publishing Pte. Ltd.
Level 34, Centennial Tower
3 Temasek Avenue
Singapore 039190

Email: editorial@jennystanford.com
Web: www.jennystanford.com

British Library Cataloguing-in-Publication Data
A catalogue record for this book is available from the British Library.

Local Gradient Theory for Dielectrics: Fundamentals and Applications

ISBN 978-981-4800-62-4 (Hardcover)
ISBN 978-1-003-00686-2 (eBook)

To the brilliant memory of
Yaroslav Burak,
a famous Ukrainian scientist

Contents

Preface

Experimental studies in the second half of the twentieth century revealed some phenomena and effects that cannot be appropriately described within the classical theory. Among such problems, we can mention the near-surface and interface inhomogeneity of electromechanical fields, size effects of mechanical and electrical characteristics of a material, nonlinear dependence of the inverse capacitance of thin dielectric films on their thickness, high-frequency dispersion of longitudinal elastic waves, propagation of antiplane surface shear waves (SH waves) in isotropic solids, linear response of polarization of centrosymmetric cubic crystals to the temperature gradient (thermopolarization effect) as well as to the stress gradient (flexoelectric effect), the emergence of a bound electric charge on the free surfaces of dielectric bodies, etc. (Abazari et al. 2015; Axe et al. 1970; Boukai et al. 2008; Bullen and Bolt 1985; Catalan et al. 2004, 2005; Kraut 1971; Li et al. 2003; Ma and Cross 2001a; Mead 1961; Rafikov and Savinov 1994; Tagantsev 1987; Tang and Alici 2011a, 2011b; Zubko et al. 2007). Solutions to these problems call for the development of new generalized mathematical models of dielectrics that take into account the inhomogeneity of the state of physically small elements of the body and describe their physical properties more scrupulously and accurately.

An extension of the classical field theory toward the abovementioned mathematical models became possible due to an intensive development of new technologies, in particular, nanotechnologies. Here we should mention an extensive utilization and design of new composite and porous materials, including nanocomposite and nanoporous ones, the engineering of microscale/nanoscale structures, nanoelectromechanical devices, sensors, and actuators. In many cases, such theories allow for avoiding the singularities in solutions to problems with dislocations, cracks, line sources, point loads and charges, etc.

There are several approaches to constructing extended theories of thermoelastic polarized solids. One group of theories considers the additional degrees of freedom (i.e., microrotations, microdeformations, etc.) for material points in order to take into account the contribution of the microstructure changes to the macroscopic behavior of a body. In such a way, since the 1960s, the fundamentals of micromorphic, microstretch, micropolar continua theories of dielectrics have developed (Dixon and Eringen 1965a, 1965b; Demiray and Eringen 1973; Eringen 1999, 2004; Lee et al. 2004). Nonlocal and gradient-type theories compose another group of extended theories of dielectrics. The nonlocal field theory for piezoelectricity with functional constitutive relations was proposed by Eringen (1984, 2002) and Eringen and Kim (1977). The gradient-type theories of dielectrics were developed by allowing the stored energy density to depend on the gradient of some physical quantities, namely, the strain tensor gradient (Kogan 1964), the polarization gradient (Mindlin 1968), or the electric field gradient (Landau and Lifshitz 1984; Yang et al. 2004). Note that the latter theory is similar to the so-called theory with electric quadrupoles because the electric field gradient is a thermodynamic conjugate of the electric quadrupole (Kafadar 1971).

In 1987, Burak proposed a new continuum-thermodynamic approach to the construction of a gradient-type theory of thermoelastic solids. The mentioned approach is based on taking account of non-diffusive and non-convective mass fluxes associated with the changes in the material microstructure. These fluxes were related to the process referred to as the local mass displacement (Burak 1987). By employing this approach, papers (Burak et al. 2007, 2008; Hrytsyna 2017a; Hrytsyna and Kondrat 2018; Hrytsyna and Moroz 2019; Kondrat and Hrytsyna 2012a) present the foundations of a gradient-type theory of the deformation of electrothermoelastic non-ferromagnetic polarized medium. This theory was called the local gradient theory of dielectrics. It is based on the accounting for the local mass displacement and its effect on mechanical, heat, and electromagnetic fields. The present book is concerned with the mathematical and physical aspects of the local gradient theory of dielectrics and its applications.

The book consists of five chapters. A short overview of generalized continuum theories of dielectric media taking account of the nonlocal effects is given in the first chapter. This chapter contains a brief description of the well-known and the most common approaches to the development of such theories within the framework of the continuum description.

In the second chapter, the fundamental concepts and basic relations of the local gradient electrothermomechanics of non-ferromagnetic dielectric solid bodies are formulated. It is shown that a gradient-type theory of dielectrics can be formulated by considering the contribution of the mass fluxes caused by changes in the material microstructure. In order to describe the process of local mass displacement, we introduce the corresponding physical quantities and obtain the balance-type equation to which these quantities are subordinated. It is shown that due to the local mass displacement, the gradient-type constitutive relations are obtained. A complete set of equations that include the balance equations, respective physical and geometric relations, as well as the corresponding boundary and jump conditions are formulated. The connection of the constructed theory with some generalized theories of dielectrics is analyzed. It is shown that for the developed theory of dielectrics, the principle of conformity is fulfilled. It means that in the limiting case of neglecting the local mass displacement, the obtained equations coincide with the equations of the classical theory.

In the third chapter, the local gradient theory of dielectrics is generalized by taking into account (i) the tensor-like representation of the parameters related to the local mass displacement, (ii) the irreversibility and inertia of the polarization and local mass displacement, (iii) the rheological properties of a dielectric medium with fading memory, and (iv) the electric quadrupoles. This, in particular, enabled us to obtain a dynamically coupled set of equations of local gradient thermomechanics of polarized medium.

The mathematical models of local gradient electrothermomechanics of non-ferromagnetic polarized solids that are developed in the second and third chapters have become the basis for theoretical studies of near-surface inhomogeneity of coupled fields in dielectrics, for the description of size effects, wave processes, etc. The mentioned investigations compose the fourth and fifth chapters of the book. In these chapters, it is shown that by taking the local mass displacement, its irreversibility and inertia into consideration, the classical continuum theory of thermoelastic dielectrics is extended to accommodate electromechanical interaction in centrosymmetric materials. The theories of polarized solids generalized in such a way make it possible to study the transition modes of the formation of near-surface inhomogeneity of coupled fields in dielectric bodies as well as to investigate the perturbation of mechanical, thermal, and electromagnetic fields due to the effect of rapidly changeable loads. Within the linear approximation, these theories describe a number of experimentally observed phenomena, including the surface, size, flexoelectric, pyroelectric, and thermopolarization effects in isotropic media, anomalous dependence of the capacitance of thin dielectric films on their thickness, the dispersive properties of polarized media, etc. Note that the above phenomena are not explained within the framework of classical theory of dielectrics.

The book is based on the results obtained by the authors over the last 20 years. It should be noted that a certain part of the basic results of this book has been published in a series of papers (Burak et al. 2007, 2008; Burak and Hrytsyna 2011; Chapla et al. 2009; Hrytsyna 2008, 2010, 2011, 2013a, b, c, 2014, 2015, 2016, 2017a, b; Hrytsyna and Kondrat 2018; Hrytsyna and Moroz 2019; Kondrat and Hrytsyna 2009a, b, c, 2010a, b, c, 2011, 2012a, b, c, 2018) as well as presented at a number of international conferences.

In conclusion, we would like to express our gratitude to the people who supported and helped us throughout this project. We would like to express our appreciation to Prof. Yaroslav Burak, our mentor, who initiated these studies and persistently encouraged us to do above researches. He had a great influence on the formation of scientific judgments of both authors of this book.

We also thank Prof. Yuriy Povstenko, Prof. Vasyl Chekurin, and Prof. Yevhen Chaplya for discussions on a number of problems. We are grateful to Prof. Roman Kushnir for support of these investigations. Our special thanks go to Prof. Yuriy Tokovyy for his advice and great help throughout the long process of writing the book.

We wish especially to thank a number of our collaborators in the Centre of Mathematical Modelling and Pidstryhach Institute for Applied Problems of Mechanics and Mathematics, National Academy of Sciences of Ukraine, and Institute of Construction and Architecture, Slovak Academy of Sciences, who helped to carry out these researches by the valuable discussions, comments, and advice at seminars and in personal communications.

We are grateful to Orest Tsurkovsky for thorough proofreading of the manuscript.

We also thank Jenny Stanford Publishing and, personally, Stanford Chong for the suggestions regarding the preparation and publication of this book.

Last but not least, we would like to express our gratitude to our families and friends. Their encouragement, patience, help, and support have been very important to us.

Olha Hrytsyna

Vasyl Kondrat

Autumn 2019

Section I
EQUATIONS

Chapter 1

A Review of the Gradient and Nonlocal Theories of Electrothermoelastic Polarized Media

This chapter is devoted to the review of the integral- and gradient-type theories of an elastic, polarized medium. The nonlocal theories of dielectrics with functional constitutive relations of a spatial type are presented in Section 1.2. Section 1.3 contains a brief description of the gradient-type theories of dielectrics with classical and non-classical kinematics. Subsection 1.3.1 is devoted to a general analysis of the gradient-type theories of dielectrics. Within this subsection, a short description of the development of gradient theories of dielectrics is presented from the historical point of view. Fundamental equations of a micropolar electroelastic continuum are written in Subsection 1.3.2. In Subsection 1.3.3, the main equations of the microstretch theory of electroelastic media are briefly presented. In Subsection 1.3.4, we provide the essentials of the micromorphic continuum theory of electroelastic dielectrics. Subsection 1.3.5 is devoted to the strain gradient theories of dielectrics and their application to the investigation of direct and converse flexoelectric effects. In Subsection 1.3.6, we present, in a concise form, the fundamental equations of the gradient theory of piezoelectricity that includes a polarization gradient into the space of constitutive parameters. The theories of dielectrics with electric quadrupoles or an electric field gradient are discussed in Subsection 1.3.7. The concluding section

Local Gradient Theory for Dielectrics: Fundamentals and Applications
Olha Hrytsyna and Vasyl Kondrat

ISBN 978-981-4800-62-4 (Hardcover), 978-1-003-00686-2 (eBook)
www.jennystanford.com

of this chapter formulates the main aims the authors seek to achieve throughout this book.

1.1 Introduction

Many investigators have focused on the problem of an appropriate description and analysis of the regularities of electromechanical interaction in dielectric media. The brothers Pierre and Jacques Curie (1880, 1881) were the first to study the coupling between strain and electrical polarization. In the 1880s, they observed the emergence of positive and negative charges on the surfaces of quartz crystals, potassium sodium tartrate, and topaz, when compressing the above crystals in different directions. This phenomenon is referred to as *piezoelectric effect* originating from the Greek words "piezein" (which means "to squeeze or press") and "ēlektron" (which means "shining light"). Lippmann (1881), using the fundamental thermodynamics theory, mathematically deduced the possibility of deformation of crystals due to the action of an external electric field (*converse piezoelectric effect*). Pierre and Jacques Curie (1881) experimentally confirmed this phenomenon when they observed the vibrations of crystals due to the action of a fast-changing electric field.

The experiments, discovered by the Curie brothers, gave an impetus to the construction of mathematical models for describing the interconnection between electromagnetic, deformation, and thermal processes in crystalline bodies. The phenomenological theory of piezoelectric phenomena was first proposed by Voigt, who in 1884 established a connection between piezoelectric properties and the crystalline structure of bodies. Later he generalized the available data about piezoelectricity early in the twentieth century in his book "Lehrbuch der Kristall-Physik" (Voigt 1910).

Direct and converse piezoelectric effects play an essential role in many electrical engineering applications. For example, the above phenomena were used to develop filters and regulating devices. Their first practical use was related to electroacoustics and measuring techniques. During the years of the First World War, using piezoelectric materials, Langevin (1918) created submarine sonars for sound location (abbreviation originating from the words SOund NAvigation and Ranging). In 1920, using a plate of natural quartz, a quartz resonator was invented by Cady (1922, 1946). In 1925, Pierce (1925) invented an acoustic interferometer. Powerful ultrasonic vibrations were effectively used by Wood and Loomis (1927). With the use of piezoelectric crystal, a gramophone pickup was developed, becoming the first sound reproducing device invented. The discovery of crystals with ferroelectric properties, as well as of piezoceramics and piezopolymers, continued the expansion of the practical application of the piezoelectric effect and stimulated the relevant theoretical and experimental investigations that are widely represented in the scientific literature.

The solutions to a wide range of practically important problems were found based on the relations of the classical theory of dielectrics. However, since the middle of the twentieth century, a number of experimental results have been accumulated in the scientific literature that cannot be appropriately described using the classical macroscopic theory of piezoelasticity. For example, the classical theory of elastic dielectric continuum fails to capture the linear response of polarization of crystals (including centrosymmetric crystals and isotropic materials) to an non-uniform deformation or a temperature gradient, which are known as flexoelectric and thermopolarization effects, respectively (Catalan et al. 2004, 2005; Ma and Cross 2001a; Rafikov and Savinov 1994; Tagantsev 1987; Zubko et al. 2007). Mead (1961) showed that the experimentally obtained electrical capacitance of a thin dielectric film is smaller than the prediction of the classical theory. This phenomenon is known in scientific literature as Mead's anomaly. The rotation of the direction of mechanical displacement along the path of a transverse elastic wave and the rotation of a plane of light polarization caused by passing through a polarized crystal of an alpha quartz (the so-called acoustical and optical activity in alpha quartz, respectively) (Joffrin and Levelut 1970; Pine 1970) as well as the flexoelectric rotation of polarization in thin films (Catalan et al. 2011) and high-frequency dispersion of elastic waves (Axe et al. 1970; Taucher and Guzelsu 1972) are absent from the classical theory but have been observed in the laboratory. The classical theory is incapable of describing a size-dependent behavior of the mechanical, thermal, and electromagnetic properties (e.g., elastic moduli, hardness, surface tension, stiffness, and dielectric permeability) of piezoelectric materials at nanoscale (Abazari et al. 2015; Angadi and Thanigaimani 1994; Gharbi et al. 2009; Han and Vlassak 2009; Jurov et al. 2011; Kim et al. 1995;

Robinson et al. 2012; Tang and Alici 2011a, 2011b; Yang et al. 2006). For example, Abazari et al. (2015) showed that the Young modulus of crystalline materials measured in bending stiffness experiments increases with thickness reduction. Boukai et al. (2008) reported that by reducing the size of a device, the thermal conductivity can be different from its bulk values.

The description of these effects required a development of new generalized theories of dielectrics, which at a continuum level would take into account the long-range effects, the influence of microstructure on physical and mechanical properties of a material, etc. There exist various ways of constructing such generalized continuum theories. One such way is to modify the constitutive equations in classical continuum theories of dielectrics from local to gradient-type or nonlocal (see Scheme 1.1).

The theories where the functional link between the conjugate constitutive parameters is predetermined (Eringen and Kim 1977; Eringen 1984) are called "*nonlocal*" or "*strongly nonlocal*" theories.

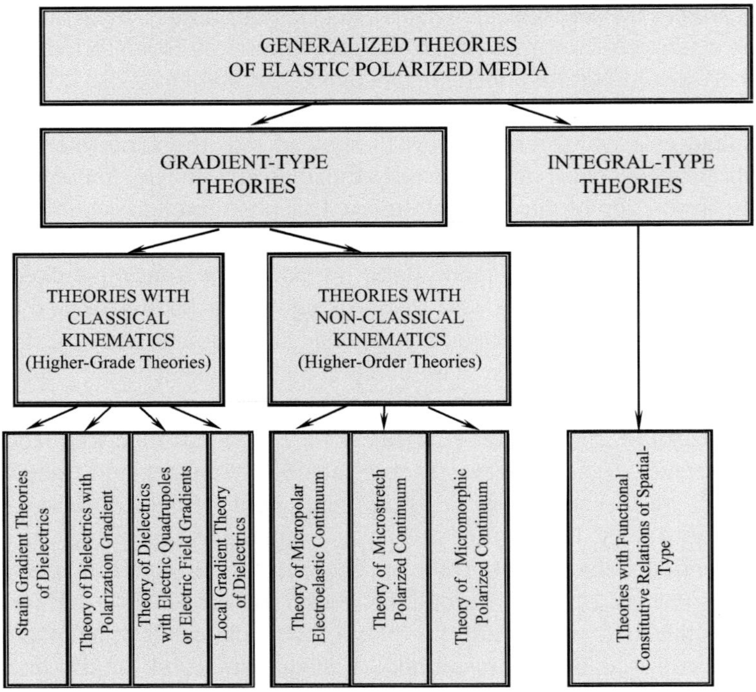

Scheme 1.1

Another way consists in formulating gradient-type constitutive equations. Within the scope of a continuum description, the gradient-type theories of dielectrics are constructed by introducing: (i) the strain gradients (Toupin 1962; Kogan 1964; Mindlin 1965; Tagantsev 1986), (ii) the polarization gradient (Mindlin 1968; Chowdhury and Glockner 1977a), (iii) the electric field gradient or higher electric quadrupoles (Kafadar 1971; Demiray and Eringen 1973; Landau and Lifshitz 1984; Yang et al. 2004), and (iv) the additional degrees of freedom (Eringen 1999) into the space of constitutive parameters. These theories are referred to as *gradient* or *weakly nonlocal* theories (Kunin 1975; Maugin 1979; Ván 2003).

Here we propose a short list of well-established nonlocal and gradient-type theories of an elastic dielectric medium developed within the framework of continuum description. The number of works published in this field is so large that it is practically infeasible to do justice to all investigations. Therefore, here we mainly focus on conceptual works that contain the fundamentals of a mathematical description of nonlocal and gradient-type theories, including generalized theories of polarized media with additional degrees of freedom. At the same time, we do not focus on applied research implemented on the basis of these theories, nor on the development of appropriate numerical methods. In view of the large amount of relevant investigations, such a review requires a separate analysis. We only note that integral and gradient-type theories, in a number of cases, make it possible to avoid a singularity of solutions to the problems with point charges, concentrated forces, defects (e.g., cracks, dislocations), etc. These theories describe nonlocal effects, such as the size and surface phenomena. The above theories have been widely used to study the mechanical properties of piezoelectric nanomaterials as well as to investigate the shock, surface, and acceleration waves. Generalized theories of dielectrics predict that short plane acoustic waves are dispersive, which well agrees with experiments and lattice dynamics. Mindlin (1972a) and later Eringen and Kim (1977) showed that integral- and gradient-type theories are closer to the lattice dynamics than the classical continuum theory.

Note that the gradient and nonlocal theories of dielectrics predict the solutions that differ from the classical solutions in the near-surface and interface regions, in transition modes of shock waves, where the effect of the fields' gradientality is significant (Maugin 1988). Outside these regions, the solutions to the boundary-value problems based on the relations of generalized and classical theories of dielectrics practically coincide (Kunin 1975).

In the present book, we do not consider the theory of polarized media developed with the use of statistical methods. We also do not aim to review the generalized theories of dielectrics built on the approaches and methods of crystal lattice theory and molecular dynamics. An overview of these works can be the subject of a separate study. Our focus is on the mathematical models of continuum mechanics where we restrict ourselves to considering the extended mathematical models of dielectric media constructed within the framework of a one-continuum approach. Readers interested in considering a generalized continuum model for diatomic solids and compound continua can refer to the works by Mindlin (1972b, 1972c), Demiray (1977), Demiray and Dost (1988), Kalpakides (1996), Hadjigeorgiou et al. (1999a), and Khoroshun (2006).

Let us note a number of review publications on different approaches to the construction of generalized theories of polarized media. Trends in the development of the nonlocal and gradient theories of dielectrics are highlighted in the monographs by Nowacki (1983), Maugin (1988), and Eringen (2002) and in review articles by Yang (2006), Kondrat and Hrytsyna (2009s), and Yan and Jiang (2017). A detailed description of a micromorphic, microstretch, and micropolar continuum may be found in Eringen (1999). Reviews by Wang (2016) and Yan and Jiang (2017) outline the recent progress and achievements in research related to the continuum mechanics modelling of the size-dependent mechanical and physical properties of piezoelectric nanomaterials. For a broader picture of flexoelectricity and its practical applications (in sensors, actuators, energy harvesters, nanocomposites, etc.), the readers are referred to review articles by Jiang et al. (2013), Nguyen et al. (2013), Yudin and Tagantsev (2013), Zubko et al. (2013), Krichen and Sharma (2016), and references therein. The various formats of gradient-type theories in continuum mechanics were discussed by Jirásek (2004), Askes and Aifantis (2011), dell'Isola et al. (2017). The higher grade material theories with internal variables or additional degrees of freedom were compared by Papenfuss and Forest (2006).

A separate branch of literature defines additional material constants related to nonlocal effects. We do not analyze these works here because the list would be too extensive. We only note that a number of authors (Kröner 1963; Askar 1972; Lakes 1988, 1995; Miller and Shenoy 2000; Liebold and Müller 2015) use experimental data to find additional characteristics of polarized matter. A brief review of some experimental works related to defining elastic moduli within the framework of nonlocal theories of mechanics is given by Lakes (1995). Theoretical methods (i.e., discrete atomistic methods such as molecular dynamics simulations, crystal lattice dynamics) for estimating the flexoelectric properties of crystalline dielectrics and additional constants in the second strain gradient elasticity were used by Mindlin (1967, 1972a), Askar et al. (1970), Maranganti and Sharma (2009), Naumov et al. (2009), Polyzos and Fotiadis (2012), Ojaghnezhad and Shodja (2013), et al.

1.2 Nonlocal Theories of Dielectrics

The nonlocal theory of elasticity was developed by Edelen (1969), Edelen and Laws (1971), Eringen and Edelen (1972), et al. This theory was generalized by Eringen and Kim (1977), Eringen (1984, 2002), and Eringen and Maugin (1990) for the case of piezoelectric bodies. In order to describe nonlocal effects in materials that possess piezoelectric properties, they used constitutive equations of integral type. As a rule, kernels in such relations depend on the difference between the position vector of the current material point and the position vectors of the neighboring body points. In such a way, these relations take the long-range effects into account. For electroelastic media, the nonlocal constitutive equations can be expressed as follows:

$$\hat{\boldsymbol{\sigma}}(\mathbf{r}) = \int\limits_{(V)} \left[\hat{\mathbf{C}}^{(4)}(\mathbf{r},\mathbf{r}') : \hat{e}(\mathbf{r}') - \hat{\mathbf{h}}^{(3)}(\mathbf{r},\mathbf{r}') \cdot \mathbf{E}(\mathbf{r}') \right] dV(\mathbf{r}'), \tag{1.1}$$

$$\boldsymbol{\Pi}_e(\mathbf{r}) = \int\limits_{(V)} \left[\hat{\boldsymbol{\varepsilon}}(\mathbf{r},\mathbf{r}') \cdot \mathbf{E}(\mathbf{r}') + \hat{\mathbf{h}}^{(3)}(\mathbf{r},\mathbf{r}') : \hat{e}(\mathbf{r}') \right] dV(\mathbf{r}'). \tag{1.2}$$

Here, $\mathbf{r}$ is the position vector; $\hat{\boldsymbol{e}}$ and $\hat{\boldsymbol{\sigma}}$ are the strain and stress tensors, respectively; $\mathbf{E}$ is the electric field vector; $\boldsymbol{\Pi}_e$ denotes the polarization vector; $\hat{\mathbf{C}}^{(4)}$, $\hat{\mathbf{h}}^{(3)}$, and $\hat{\boldsymbol{\varepsilon}}$ are tensors of the fourth, third, and second orders whose components are the material characteristics; the dot denotes the convolution of the corresponding indices. Note that here and in what follows, bold symbols stand for vector quantities, bold symbols with caps denote tensor quantities, and the superscript in parentheses indicates the order of the tensor quantity, but usually we omit these superscripts for tensors of the second order and retain them for tensors of the third and higher orders.

The relations (1.1) and (1.2) take into consideration that the stress tensor $\hat{\boldsymbol{\sigma}}$ and the polarization vector $\boldsymbol{\Pi}_e$ at a current point $\mathbf{r}$ depend not only on the values of the strain tensor and on the electric field at the same point but also on the values of the mentioned functions at all other points of the body.

Eringen (2002) showed that the integral relations (1.1) and (1.2) can be rewritten in a differential form, that is,

$$\left[1-(ae_0)^2\Delta\right]\hat{\boldsymbol{\sigma}} = \hat{\mathbf{C}}^{(4)} : \hat{\boldsymbol{e}} - \hat{\mathbf{h}}^{(3)} \cdot \mathbf{E},$$

$$\left[1-(ae_0)^2\Delta\right]\boldsymbol{\Pi}_e = \hat{\boldsymbol{\varepsilon}} \cdot \mathbf{E} + \hat{\mathbf{h}}^{(3)} : \hat{\boldsymbol{e}},$$

where a is an internal characteristic length (for example, lattice parameter), e_0 is a material constant determined by experimental results or by methods of molecular dynamics or lattice dynamics, and Δ is the Laplacian.

Eringen and Kim (1977) analyzed the link between the nonlocal theory of dielectrics and the lattice dynamics.

The nonlocal theory of dielectrics was used to explain a number of phenomena that cannot be justified by classical theories of electroelastic media. In particular, Eringen (1984) studied the dispersion of short plane waves and the potential field of a point electric charge. In order to justify the nonlinear dependence of the capacitance of thin dielectric films on their thickness, Yang (1997) used the nonlocal law for dielectric polarization

$$\Pi_e = \frac{\varepsilon_0 \chi}{2\alpha} \int_0^h E(x')\mathrm{e}^{-|x'-x|/\alpha}\, dx'. \tag{1.3}$$

Here, Π_e and E are components of polarization vector and electric field vector, respectively; α is a characteristic length of microscopic nonlocal interactions; χ is the dielectric susceptibility; and ε_0 is the electric constant. Yang showed a good agreement between the nonlocal solution for thin film capacitance and Mindlin's (1969) theoretical investigations as well as Mead's (1961) experimental results.

The nonlocal theory of polarized solids was used effectively to analyze free vibrations of elastic and thermoelastic piezoelectric nanobeams and nanoplates (Ke and Wang 2012, 2014; Liu et al. 2013). This theory satisfactorily explains the high-frequency vibration and wave dispersion. Therefore, nonlocal theory is sometimes referred to as "*nonlocal models of media with spatial dispersion*."

Note that the nonlocal theory of electrothermoelastic polarized solids results in complicated integro-differential equations that are usually difficult to solve.

1.3 Gradient-Type Theories of Dielectrics

1.3.1 General Characteristics

The goal of all generalized gradient-type theories of elastic media is to enclose the mechanical properties at a microscopic level into a macroscopic description. Historically, the first publication on the theory of generalized elastic continuum was the work by the Cosserat (1909) brothers that proposed the theory of polarized elastic solids in 1909. Two decades later, a gradient theory of elastic media was proposed by Jaramillo (1929). The next significant step in the development of generalized theories of elasticity was made in the 1950–60s (Grad 1952; Günther 1958; Pidstryhach 1967; Schaëfer 1967). In this period, the foundations of the Toupin couple–stress theory (Toupin 1962), the Mindlin gradient theory of elastic media with microstructure (Mindlin 1964), and Eringen–Sukhubi theory of micromorphic continuum (Eringen and Suhubi 1964; Suhubi and Eringen 1964) were made.

A significant contribution to the development of the theory of media with microstructure belongs to Eringen, who formulated and developed the theory of micropolar fluids (Eringen 1966b), the linear (Eringen 1966a) and nonlinear (Eringen 1970) theories of micropolar elastic continuum, the micromorphic theory (Eringen 1964, 1968; Eringen and Suhubi 1964; Suhubi and Eringen 1964), as well as the theories of both the fluid (Eringen 1969) and elastic (Eringen 1999) microstretch continuum. These results became the basis for a number of monographs (Eringen 1967, 1976, 1999, 2002; Eringen and Maugin 1990).

In the 1960s, the aforementioned generalized theories of elasticity in a natural manner were extended to dielectric media. During the 1960–70s, the foundations of micropolar, microstretch, and micromorphic theories of polarized continuum as well as the strain gradient theories of dielectrics were formulated. Mindlin (1968) proposed a gradient-type theory of ionic crystals that takes account of the dependence of internal energy on the polarization gradient. The period of 1970–80s was marked by the development of an extended theory of dielectrics with a gradient of the electric field vector that is equivalent to the theory of elastic dielectrics with electric quadrupoles (Kafadar 1971; Landau and Lifshitz 1984).

The mentioned gradient-type theories of dielectrics can be conventionally subdivided into two groups (see Scheme 1.1, p. 6). One group of theories uses the classical kinematics. In such theories, similar to the classical models, there is only one kinematic characteristic, namely, the displacement vector $\mathbf{u}(\mathbf{x},t)$. These theories consider elastic media as a continuous collection of infinitesimally small particles represented by geometrical points—particle centroids. However, unlike the local (i.e., macroscopic) theories, the gradient-type theories take the long-range effects into account. Such group of theories is sometimes called the *gradient theories of materials of a simple structure* (Kunin, 1975), or the *higher-grade theories* (Tekoğlu and Onck, 2008). This group of theories includes the generalized theories of dielectrics, whose space of constitutive parameters is expanded: (i) by the strain gradients, (ii) by the polarization gradient, (iii) by the electric field gradients, or by the higher-order electric moments (i.e., quadrupole, octupole, etc.).

Another group of extended theories of dielectrics includes the theories that consider the additional degrees of freedom (i.e., microrotations, microdeformations, etc.) for material points in order to incorporate the contribution of the microstructure changes to the macroscopic behavior of the body. Such theories are sometimes called the *gradient theories of materials of complicated structure* (Kunin 1975), the *theories of media with microstructure*, the *theories of media with additional degrees of freedom*, or the *higher-order theories* (Tekoğlu and Onck, 2008). Within this group of theories, a medium is considered to be a continuous collection of a large number of deformable particles (i.e., macroelements), each of them being characterized by a finite size, structure, or orientation. Macroelements are considered to be a system of interacting particles (i.e., microelements) that can be deformed and can rotate relative to the macroelement mass center. Such a deformation and rotations of microelements can eventually affect the macroscopic behavior of the body. Thus, in micromorphic theories, along with the convective (i.e., translational) motion of a macroelement, the rotational motion of its microparticles or their deformation is considered as well. Different variants of theories of continua with a microstructure can be proposed depending on the particular forms of motion of the microelements taken into account (i.e., translational or rotational). Note that within the micropolar continua, the macroelement is considered to be a rigid oriented body, while within the micromorphic continua, the macroelement is considered to be a solid deformable body. In general, this group of mathematical theories contains the micropolar theories of electromagnetic elastic continua (Eringen 1999; Chen 2013), the micromorphic theory of electromagnetic elastic and thermoelastic continua (Eringen 2003; Lee et al. 2004; Romeo 2011), the theory of microstretch continua (Eringen 1999; Galeş 2011), as well as various modifications of the above theories. Among the theories of dielectrics with non-classical kinematics, the most general is the micromorphic theory. Under some assumptions, the mathematical models of microstretch, polar, and micropolar continua may be obtained from the micromorphic theory.

Herein below we present a brief description of some gradient-type mathematical models of the elastic polarized media.

1.3.2 Polar and Micropolar Electroelastic Continuum

The classical theory of elasticity envisions a solid body as a continuum of material points, each with infinitesimal size and no inner structure. The motion of such points is described only by one kinematic characteristic, namely, the displacement vector. Hence, the material point has three degrees of freedom in the classical theory. The *Cosserat continuum theory* (or otherwise the theory of *polar continuum*) supplies additional degrees of freedom to a material point.

The theory of polar continuum was developed by E. and F. Cosserat in 1909 (Cosserat and Cosserat 1909). The Cosserat brothers incorporated a local rotation of material points in addition to the translation assumed in a classical continuum theory of elasticity.

Hence, the kinematic properties of each point of the Cosserat continuum are described by two independent vector quantities, namely, by the displacement vector $\mathbf{u}$ and by the rotation vector $\boldsymbol{\omega}$. In the Cosserat theory, in contrast to the classical (i.e., symmetric) theory of elasticity, the stress state is described by a non-symmetric stress tensor. Therefore, the Cosserat theory is often referred to as the theory of *asymmetric elasticity*. Note that within the framework of this theory, the six elastic moduli have been introduced to describe the mechanical behavior of an isotropic elastic material.

In the polar continuum theory, the rotation vector is taken independent of the displacement vector. Nowacki (1986) proposed another linear variant of generalized theory of elasticity where the rotation vector $\boldsymbol{\omega}$ is fully described by the displacement vector $\mathbf{u}$ through the formula

$$\boldsymbol{\omega} = \frac{1}{2}\nabla \times \mathbf{u}.$$

Such a theory was referred to as the *Cosserat pseudocontinuum theory*. In this variant of the extended theory of elasticity, for an isotropic body, the number of elastic constants is reduced from six to four.

Note that the Cosserat continuum and Cosserat pseudocontinuum theories were effectively used to study the mechanical behavior of materials that allow significant changes in the orientation of a microstructure such as liquid crystals and ferroelectrics (Chen et al. 2004).

In the 1960s and 1970s, Suhubi and Eringen (1964), Eringen and Suhubi (1964), and Eringen (1966a) proposed the modern formulation of the Cosserat continuum theory. In order to describe the behavior of materials possessing a microstructure, they assumed that a material continuum is a collection of small rigid bodies (i.e., particles) undergoing both translational and rotational motions. Such a theory they originally called the *theory of simple micro-elastic solids*. At present, it is known as the *micropolar* (or *asymmetric*, or *moment*) theory of elasticity. Note that assuming a constant microinertia, the Eringen micropolar theory may be reduced to the Cosserat theory.

Eringen (1999) studied the micropolar materials that possess piezoelectric properties, that is, the electric field can influence the mechanical behavior of such materials. He established the equations governing the motion of a micropolar electrothermoelastic continuum. Such a theory was obtained by introducing additional terms into the energy balance equations and into thermomechanical balance laws as well as by adding the Maxwell equations and corresponding constitutive relations to the vectors of electromagnetic field.

In the reference configuration, a material point *P* of the micropolar media is identified by a position vector **X**, and by three rigid directors Ξ attached to this point. The motion of this point can be expressed by the formulae

$$x_k = x_k(\mathbf{X}, t), \quad \xi_k = \chi_{kK}(\mathbf{X}, t)\Xi_K \tag{1.4}$$

The inverse motion for a material point of micropolar continuum can be written in the form

$$X_K = X_K(\mathbf{x}, t), \quad \Xi_K = \bar{\chi}_{Kk}(\mathbf{x}, t)\xi_k, \tag{1.5}$$

where $\chi_{kK}\bar{\chi}_{Kl} = \delta_{kl}$ and $\bar{\chi}_{Kk}\chi_{kL} = \delta_{KL}$.

In view of formulae (1.4) and (1.5), the two measures of deformation $\widehat{\mathbf{A}}^{(2)} = \{A_{KL}\}$ and $\widehat{\mathbf{\Gamma}}^{(2)} = \{\Gamma_{KL}\}$ can be introduced for the micropolar media, where

$$A_{KL} \equiv x_{k,K}\chi_{kL}, \quad \Gamma_{KL} \equiv \frac{1}{2}\in_{KMN} \chi_{kM,L}\chi_{kN}. \tag{1.6}$$

Here, $\widehat{\mathbf{A}}^{(2)}$ and $\widehat{\mathbf{\Gamma}}^{(2)}$ are the Cosserat deformation tensor and Wryness tensor (Chen 2013), respectively; the summation over repeated subscripts is implied; subscripts preceded by a comma denote a partial differentiation with respect to the corresponding spatial coordinate; and $\widehat{\boldsymbol{\in}}^{(3)} = \{\in_{KMN}\}$ is the permutation symbol (the Levi–Civita symbol).

For the case of isothermal approximation, the nonlinear dynamic behavior of elastic non-ferromagnetic polarized media was characterized by geometric relations (1.4)–(1.6), as well as by the conservation law of mass

$$\frac{\partial \rho}{\partial t} + \nabla \cdot (\rho \mathbf{v}) = 0, \tag{1.7}$$

by the conservation law of linear momentum

$$\nabla \cdot \hat{\mathbf{t}}^{(2)} + \rho \mathbf{F} + \mathbf{F}_e = \rho \frac{d\mathbf{v}}{dt}, \tag{1.8}$$

by the conservation law of angular momentum

$$\nabla \cdot \widehat{\mathbf{m}}^{(2)} + \widehat{\in}^{(3)} : \hat{\mathbf{t}}^{(2)} + \rho\mathbf{M} + \mathbf{M}_e = \rho j \frac{d\boldsymbol{\varphi}}{dt}, \tag{1.9}$$

by the Maxwell equations

$$\nabla \cdot \mathbf{D} = \rho_e, \quad \nabla \cdot \mathbf{B} = 0, \tag{1.10}$$

$$\nabla \times \mathbf{E} = -\frac{\partial \mathbf{B}}{\partial t}, \quad \nabla \times \mathbf{H} = \mathbf{J}_e - \frac{\partial \mathbf{D}}{\partial t}, \tag{1.11}$$

by the law of conservation of charge

$$\nabla \cdot \mathbf{J}_e + \frac{\partial \rho_e}{\partial t} = 0, \tag{1.12}$$

by the constitutive relations for electromagnetic field vectors

$$\mathbf{D} = \varepsilon_0 \mathbf{E} + \boldsymbol{\Pi}_e, \ \mathbf{B} = \mu_0 \mathbf{H}, \tag{1.13}$$

by the constitutive relations

$$\widehat{\mathbf{T}}^{(2)} = \rho \frac{\partial f}{\partial \widehat{\mathbf{A}}^{(2)}}, \ \widehat{\mathbf{M}}^{(2)} = \rho \frac{\partial f}{\partial \widehat{\boldsymbol{\Gamma}}^{(2)}}, \ \boldsymbol{\Pi}_e = -\rho \frac{\partial f}{\partial \mathbf{E}_*}. \tag{1.14}$$

Here, $\hat{\mathbf{t}}^{(2)}$ and $\widehat{\mathbf{m}}^{(2)}$ are the stress and couple stress tensors, respectively; $\mathbf{F}$ is the body force density; $\mathbf{F}_e$ is the ponderomotive force; ρ denotes the mass density; $\mathbf{v}$ is the velocity vector; $\boldsymbol{\varphi}$ is the gyration vector; $\mathbf{M}$ is the body couple vector; j is microinertia; $\mathbf{E}$ and $\mathbf{H}$ are the vectors denoting the electric and magnetic fields, respectively; $\mathbf{D}$ and $\mathbf{B}$ are the electric and magnetic induction vectors; $\mathbf{J}_e$ denotes the density of electric current related to the displacement of free charges; $\boldsymbol{\Pi}_e$ is the polarization vector; ρ_e is the density of free electric charges; t is the time; ε_0 and μ_0 are the electric and magnetic constants; «×» denotes the vector product; and $f = f\left(\widehat{\mathbf{A}}^{(2)}, \widehat{\boldsymbol{\Gamma}}^{(2)}, \mathbf{E}_*\right)$ is the generalized Helmholtz free energy density. For non-ferromagnetic solids, vector $\mathbf{M}_e$ is defined by the formula (Chen 2013)

$$\mathbf{M}_e = \widehat{\in}^{(3)} : (\boldsymbol{\Pi}_e \otimes \mathbf{E}_*),$$

where $\mathbf{E}_*$ is the electric field vector in the co-moving frame and «⊗» denotes the dyadic product.

Note that tensors $\hat{\mathbf{t}}^{(2)} = \{t_{kl}\}$ and $\widehat{\mathbf{T}}^{(2)} = \{T_{KL}\}$ as well as $\widehat{\mathbf{m}}^{(2)} = \{m_{kl}\}$ and $\widehat{\mathbf{M}}^{(2)} = \{M_{KL}\}$ are related through the formulae

$$T_{kl} = j^{-1} T_{KL} x_{k,K} \chi_{lL},$$

$$m_{kl} = j^{-1} M_{KL} x_{k,K} \chi_{lL}.$$

Note that the micropolar theory of electroelastic continua was effectively used for investigation of the molecular crystals, granular materials, and some types of piezoelectric composites.

1.3.3 Microstretch Continuum

The theory of *microstretch elastic continuum* was introduced by Eringen (1971) as a generalization of the micropolar theory. The material particle in a microstretch continuum can perform both microrotation and volumetric extension in addition to the bulk deformation. There are four additional degrees of freedom in such a continuum. In total, each point of a microstretch elastic solid possesses seven degrees of freedom, namely, three degrees for translation, three degrees for microrotations, and one degree for microstretch.

Some years later, Eringen (2004) introduced the electromagnetic theory of microstretch elastic solids where he obtained the constitutive relations and field equations for thermomicrostretch isotropic solids subjected to electromagnetic fields. The linear theory of microstretch piezoelectric and thermopiezoelectric media were developed by Ieşan (2006, 2008), Ieşan and Quintanilla (2007), and Quintanilla (2008). Here, we briefly present the main equations of the microstretch theory for ideal dielectric media.

In a linear approximation, the basic relations governing the processes in microstretch electroelastic continuum consist of the following (Galeş 2011):

The equations of motion

$$\nabla \cdot \hat{\mathbf{t}}^{(2)} + \rho_0 \mathbf{F} = \rho_0 \frac{\partial^2 \mathbf{u}}{\partial t^2},$$

$$\nabla \cdot \widehat{\mathbf{m}}^{(2)} + \widehat{\in}^{(3)} : \hat{\mathbf{t}}^{(2)} + \rho_0 \mathbf{M} = \rho_0 \hat{\mathbf{j}}^{(2)} \cdot \frac{\partial^2 \boldsymbol{\omega}}{\partial t^2},$$

$$\nabla \cdot \boldsymbol{\pi} - \sigma + \rho_0 M = \rho_0 j_0 \frac{\partial^2 \varphi}{\partial t^2}.$$

The Maxwell equations

$$\nabla \cdot \mathbf{D} = 0, \quad \nabla \cdot \mathbf{B} = 0, \tag{1.15}$$

$$\nabla \times \mathbf{E} = -\frac{\partial \mathbf{B}}{\partial t}, \quad \nabla \times \mathbf{H} = -\frac{\partial \mathbf{D}}{\partial t}. \tag{1.16}$$

The geometrical equations

$$\hat{e}^{(2)} = \nabla \otimes \mathbf{u} + \widehat{\in}^{(3)} \cdot \boldsymbol{\omega}, \quad \hat{\kappa}^{(2)} = \nabla \otimes \boldsymbol{\omega}, \ \boldsymbol{\zeta} = \nabla \varphi .$$

The constitutive relations

$$\hat{\mathbf{t}}^{(2)} = \rho_0 \frac{\partial f}{\partial \hat{\mathbf{e}}^{(2)}}, \ \widehat{\mathbf{m}}^{(2)} = \rho_0 \frac{\partial f}{\partial \hat{\boldsymbol{\kappa}}^{(2)}}, \ \sigma = \rho_0 \frac{\partial f}{\partial \varphi}, \ \boldsymbol{\pi} = \rho_0 \frac{\partial f}{\partial \zeta}, \ \mathbf{D} = -\rho_0 \frac{\partial f}{\partial \mathbf{E}}.$$

Here, $\hat{\mathbf{t}}^{(2)}$, $\widehat{\mathbf{m}}^{(2)}$, and $\tilde{\mathbf{j}}^{(2)}$ are the stress, couple stress, and microinertia tensors, respectively; ρ_0 denotes the reference mass density; $\mathbf{u}$ and $\boldsymbol{\omega}$ are the mechanical displacement and microrotation vectors, respectively; $\mathbf{M}$ denotes the microstretch body force; $\boldsymbol{\pi}$ is the microstretch stress vector; σ is the microstress function; j_0 is the microstretch inertia; φ is the microstretch function; $\hat{\boldsymbol{e}}^{(2)}$, $\hat{\boldsymbol{\kappa}}^{(2)}$, and $\boldsymbol{\zeta}$ are kinematic strain measures; and $f = f\left(\hat{\boldsymbol{e}}^{(2)}, \hat{\boldsymbol{\kappa}}^{(2)}, \varphi, \boldsymbol{\zeta}, \mathbf{E}\right)$ is the generalized free energy density.

It is worth noting that Maxwell's equations (1.15) and (1.16) correspond to ideal dielectrics where there are no free charges.

1.3.4 Micromorphic Continuum

The micromorphic theory envisions a material body as a continuum of physically small deformable particles of a finite size (i.e., macroelements) possessing an inner structure. The above macrovolume elements are constructed with microelements. The motion of the macroelement is caused by its convective translation, rotation around the center of mass and deformation. The deformation of micromorphic continua is expressed as the sum of the macroscopic continuous deformation (*macro-strains*) and the microscopic internal deformation of microvolume elements (*microscopic internal strains*).

Within the framework of the micromorphic theory of first-grade, each particle of the body has nine independent degrees of freedom describing its stretches and rotations, in addition to the three translational degrees of freedom of its geometric center. Note that in the micromorphic continuum theory of second-grade, each point is characterized by 24 additional degrees of freedom.

The micromorphic theory of elastic continua was proposed by Eringen and Suhubi (1964) and Suhubi and Eringen (1964). Mindlin (1964) developed the linear variant of the theory of micromorphic media.

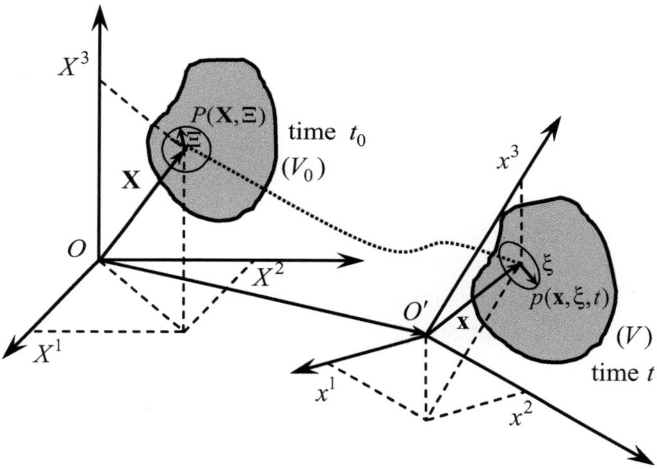

Figure 1.1 Macro- and micromotion of a material particle within micromorphic continua.

Eringen (2003) and Lee et al. (2004) exploited a purely mechanical micromorphic continuum theory of elasticity to derive linear constitutive relations and governing equations for micromorphic thermoelastic solids subjected to electromagnetic interactions.

Herein below, we briefly present the basic ideas and fundamental equations that describe the coupled fields within the framework of the first-grade micromorphic theory of dielectric media.

Lee et al. (2004) assumed that the small deformable particle P can be represented by its position vector $\mathbf{X}$ and by a vector Ξ attached to P representing its microstructure in the reference state at time $t = t_0$ (Fig. 1.1). The motions that carry $P(\mathbf{X},\Xi)$ to $p(\mathbf{x},\xi,t)$ in a spatial configuration (i.e., deformed state) at time t can be expressed as follows:

$$x_k = x_k(\mathbf{X}, t), \quad \xi_k = \chi_{kK}(\mathbf{X}, t)\Xi_K. \tag{1.17}$$

Note that the first equation from (1.17) describes the motion of the centroid of the particle (i.e., its macromotion), while the second equation specifies the change in orientation and deformation for the inner structures of the particle (i.e., its micromotion). The inverse motions can be expressed by the formulae $X_K = X_K(\mathbf{x}, t)$ and $\Xi_K = \bar{\chi}_{Kk}(\mathbf{x}, t)\xi_k$, where $\chi_{kK}\bar{\chi}_{Kl} = \delta_{kl}$ and $\bar{\chi}_{Kk}\chi_{kL} = \delta_{KL}$.

The generalized Lagrangian strain tensors of the micromorphic theory are defined as follows (Lee et al. 2004):

$$\alpha_{KL} \equiv x_{k,K}\bar{\chi}_{Lk} - \delta_{KL},$$

$$\beta_{KL} \equiv \chi_{kK}\chi_{kL} - \delta_{KL} = \beta_{LK},$$

$$\gamma_{KLM} \equiv \bar{\chi}_{Kk}\chi_{kL,M}\,.$$

The balance laws of micromorphic polarized media consist of thermomechanical equations and equations of electromagnetic field.

The electromagnetic balance laws include the Maxwell equations (1.10), (1.11), and the law of conservation of charge (1.12) as well as the constitutive relations (1.13).

The thermomechanical balance laws for micromorphic continua were obtained by Eringen and Suhubi (1964). Here, for simplicity, we consider an isothermal approximation. For such a case, the nonlinear balance laws of micromorphic polarized continuum are as follows (Lee et al. 2004):

$$\frac{\partial\rho}{\partial t} + \nabla\cdot(\rho\mathbf{v}) = 0,$$

$$\frac{d\hat{\mathbf{i}}^{(2)}}{dt} = \hat{\mathbf{i}}^{(2)}\cdot\widehat{\boldsymbol{\omega}}^{(2)\mathrm{T}} + \widehat{\boldsymbol{\omega}}^{(2)}\cdot\hat{\mathbf{i}}^{(2)},$$

$$\nabla\cdot\hat{\mathbf{t}}^{(2)} + \rho\mathbf{F} + \mathbf{F}_e = \rho\frac{d\mathbf{v}}{dt},$$

$$\nabla\cdot\widehat{\mathbf{m}}^{(3)} + \hat{\mathbf{t}}^{(2)\mathrm{T}} - \hat{\mathbf{s}}^{(2)\mathrm{T}} + \rho\widehat{\mathbf{M}}^{(2)} + \widehat{\mathbf{M}}_e^{(2)} = \rho\widehat{\mathbf{W}}^{(2)},$$

where $\hat{\mathbf{i}}^{(2)} = \{i_{km}\}$ and $\widehat{\boldsymbol{\omega}}^{(2)} = \{\omega_{kl}\}$ are the microinertia and microgyration tensors; $\hat{\mathbf{t}}^{(2)} = \{t_{kl}\}$, $\hat{\mathbf{s}}^{(2)} = \{s_{kl}\}$, and $\widehat{\mathbf{m}}^{(3)} = \{m_{kml}\}$ are the macrostresses, microstresses, and moment stresses, respectively; $\widehat{\mathbf{M}}^{(2)} = \{M_{kl}\}$ is the external body moment; superscript T denotes the transpose tensor. The spin inertia $\widehat{\mathbf{W}}^{(2)} = \{W_{kl}\}$ and component ω_{kl} of the microgyration tensor are defined as follows:

$$W_{kl} = i_{ml}\left(\frac{d\omega_{km}}{dt} + \omega_{kn}\omega_{nm}\right),$$

$$\omega_{kl} = \frac{d\chi_{kK}}{dt}\bar{\chi}_{Kl}\,.$$

For non-ferromagnetic solids, tensor $\widehat{\mathbf{M}}_e^{(2)}$ is given by the formula

$$\widehat{\mathbf{M}}_e^{(2)} = \mathbf{\Pi}_e \otimes \mathbf{E}_*\,.$$

The constitutive equations for a micromorphic continuum with electromagnetic interactions can be expressed as (Lee et al. 2004):

$$\hat{\mathbf{T}}_m^{(2)} = \rho_0 \frac{\partial f}{\partial \hat{\boldsymbol{\alpha}}^{(2)}},\ \hat{\mathbf{S}}^{(2)} = \rho_0 \frac{\partial f}{\partial \hat{\boldsymbol{\beta}}^{(2)}},\ \hat{\mathbf{\Gamma}}^{(3)} = \rho_0 \frac{\partial f}{\partial \hat{\boldsymbol{\gamma}}^{(3)}},\ \mathbf{\Pi}_e = -\rho_0 \frac{\partial f}{\partial \mathbf{E}_*},$$

where f is the generalized Helmholtz free energy; $\hat{\mathbf{T}}_m^{(2)} = \{T_{KL}^m\}$, $\hat{\mathbf{S}}^{(2)} = \{S_{KL}\}$, and $\hat{\mathbf{\Gamma}}^{(3)} = \{\Gamma_{KLM}\}$ are the generalized second-order Piola–Kirchhoff stress tensors of the micromorphic theory (i.e., the generalized stresses in Lagrangian description). These tensors are defined as

$$T_{KL}^m \equiv j t_{kl}^m X_{K,k} \chi_{lL},$$

$$S_{KL} \equiv \frac{1}{2} j s_{kl} \bar{\chi}_{Kk} \bar{\chi}_{Ll},$$

$$\Gamma_{KLM} \equiv j m_{mkl} X_{M,m} \chi_{kK} \bar{\chi}_{Ll},$$

where $j \equiv \det(x_{k,K})$ is the Jacobian of the deformation gradient, $t_{kl}^m \equiv t_{kl} + \Pi_{ek} E_{*l}$, and the superscript "$m$" refers to the "mechanical" parts of the stress components.

The micromorphic elastic continuum model was exploited by Romeo (2011) to derive the nonlinear governing equations for polarizable media with the dipole and quadrupole moments. Besides the micromechanical fields, Romeo defined electric dipole and quadrupole densities and derived electromagnetic force, couple stress, and power in terms of these densities. Romeo (2011) replaced the balance equation for the spin inertia by suitable balance equations for dipole and quadrupole per unit mass. Romeo (2012) also generalized the micromorphic continuum model for ferroelectric materials. In contrast with the Eringen (1999) micromorphic theory, Romeo introduced polarization as a microfield arising from a charge microdensity. He showed that polarization and quadrupole densities depend on microstrain.

Continuum theories of micromorphic polarized media are employed to study various problems, including the wave propagation (for example, see Lee and Chen (2004), Eringen (2006)). Cao et al. (2014) used the linear micromorphic electroelastic theory in order to solve the bending problems of a piezoelectric micro-beam. They found that the predictions from the micromorphic electroelastic theory are remarkably different from those from the classical theory when the beam thickness is the same as the characteristic length scale parameter. Wang and Lee (2010) discussed the applications of micromorphic theory in microscale/nanoscale.

It should be noted that the micromorphic theory is useful for the investigation of covalent, molecular, and ionic crystals, nanocomposites, as well as the porous elastic bodies (such as bones, ceramics, and some synthetic materials) (Chen et al. 2004). A disadvantage of this theory is a large number of material constants. For example, in the linear theory of the elastic continua, there are 903 elastic moduli (Mindlin 1964). Chen and Lee (2003) and later Zeng et al. (2006) discussed some algorithms to determine the material constants for micromorphic solids.

The micromorphic theory is the most general gradient-type theory of the media with microstructure. As a partial case, it contains all other mathematical models of the media with additional degrees of freedom. Depending on which exact types of inner motion are dominant, based on the micromorphic theory, different variants of gradient-type models of mechanics can be constructed. If the microstructure of a material is considered to be rigid (i.e., the inner motion of the microstructure within the small body element should be neglected), the micromorphic theory can be reduced to the micropolar theory. From the micromorphic theory, there can be developed a Mindlin strain gradient theory assuming that the particles are deformed in the same way as the entire continuum. Upon the assumption that microrotation and macrorotation coincide, one can obtain the couple stress theory. In case the particles are converged to geometric points endowed with mass, all theories of elastic polarized media with a microstructure will be reduced to the classical theory of dielectrics.

1.3.5 Strain Gradient Theories of Dielectrics: Flexoelectric Effect

Piezoelectric effect is not characteristic of isotropic and centrosymmetric dielectrics. However, such materials can polarize under a non-uniform mechanical deformation (Fig. 1.2). Maskevich and Tolpygo (1957) were the first to predict the existence of a coupling between the mechanical and electric fields in centrosymmetric crystals and in isotropic materials.

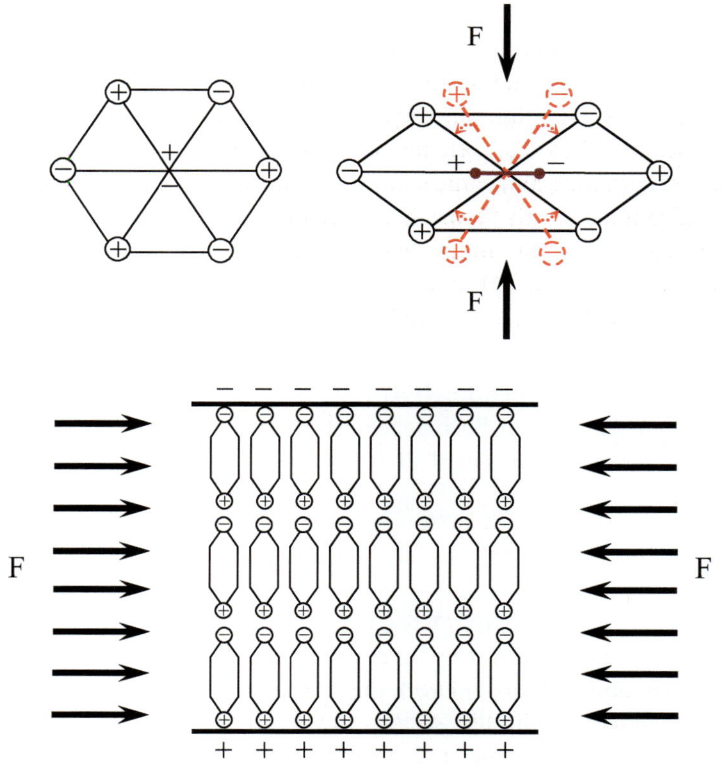

Figure 1.2 Flexoelectric effect.

The polarization of solid crystals, including centrosymmetric ones, due to their non-uniform deformation, was called by Indenbom et al. (1981) the *flexoelectric effect*, by analogy to a similar phenomenon in liquid crystals (De Gennes 1974). The *converse flexoelectric effect* describes the body deformation and the generation of an elastic strain by an electric field gradient.

Zholudev (1966) was the first to carry out the laboratory studies of the flexoelectric effect in solid crystalline materials. Bursian and Trunov (1974) observed the flexoelectric phenomena during the bending of crystal plates. Catalan et al. (2004, 2005) investigated the effect of flexoelectricity on the dielectric characteristics of inhomogeneously deformed thin films. Dumitrică et al. (2002) studied the normal polarization in carbon nanoshells due to the bending deformations. Ma and Cross (2001a, 2001b, 2003) investigated the connection between an elastic strain gradient and electric polarization in various ceramic materials.

A phenomenological description of the flexoelectric effect was proposed by Kogan (1964) more than 50 years ago. In order to describe such an effect within the framework of the linear theory, he proposed to take into account the dependence of the polarization vector not only on the strain tensor $\hat{\boldsymbol{e}}$, but also on its gradient $\boldsymbol{\nabla}\otimes\hat{\boldsymbol{e}}$.

Based on the strain gradient theories of elasticity developed by Toupin (1962) and Mindlin (1965), an extension of a mathematical model to elastic polarized materials, including strain gradient effects, was proposed in order to describe the flexoelectric effect. In such models, the strain gradient is included into the internal energy density function: $U = U(\hat{\boldsymbol{e}}, \boldsymbol{\nabla}\otimes\hat{\boldsymbol{e}}, \boldsymbol{\Pi}_e)$. For dielectrics with the incorporation of flexoelectricity, the constitutive equations are as follows (Enakoutsa et al. 2016):

$$\hat{\boldsymbol{\sigma}} = \frac{\partial U}{\partial \hat{\boldsymbol{e}}}, \quad \hat{\boldsymbol{\tau}}^{(3)} = \frac{\partial U}{\partial \hat{\boldsymbol{e}}^{(3)}}, \quad \mathbf{E} = \frac{\partial U}{\partial \boldsymbol{\Pi}_e},$$

where

$$\hat{\boldsymbol{e}} = \frac{1}{2}\left[\boldsymbol{\nabla}\otimes\mathbf{u} + (\boldsymbol{\nabla}\otimes\mathbf{u})^{\mathrm{T}}\right], \quad \hat{\boldsymbol{e}}^{(3)} = \boldsymbol{\nabla}\otimes\boldsymbol{\nabla}\otimes\mathbf{u}. \tag{1.18}$$

Here, $\hat{\boldsymbol{\sigma}}$ and $\hat{\boldsymbol{\tau}}^{(3)}$ are the ordinary and higher-order stress tensors, respectively.

For a linear approximation, the consideration of the electric field–strain gradient coupling leads to the following constitutive equations:

$$\hat{\boldsymbol{\sigma}} = \hat{\mathbf{C}}^{(4)} : \hat{\boldsymbol{e}} + \hat{\boldsymbol{\varepsilon}}^{(5)} \overset{(3)}{\cdot} (\boldsymbol{\nabla} \otimes \hat{\boldsymbol{e}}) + \hat{\mathbf{h}}^{(3)} \cdot \boldsymbol{\Pi}_e, \tag{1.19}$$

$$\hat{\boldsymbol{\tau}}^{(3)} = \hat{\boldsymbol{e}} : \hat{\boldsymbol{\varepsilon}}^{(5)} + \hat{\mathbf{g}}^{(6)} \overset{(3)}{\cdot} (\boldsymbol{\nabla} \otimes \hat{\boldsymbol{e}}) + \hat{\mathbf{f}}^{(4)} \cdot \boldsymbol{\Pi}_e, \tag{1.20}$$

$$\mathbf{E} = \hat{\boldsymbol{\chi}} \cdot \boldsymbol{\Pi}_e + \hat{\boldsymbol{e}} : \hat{\mathbf{h}}^{(3)} + (\boldsymbol{\nabla} \otimes \hat{\boldsymbol{e}}) \overset{(3)}{\cdot} \hat{\mathbf{f}}^{(4)}. \tag{1.21}$$

Here, $\hat{\mathbf{C}}^{(4)}$ is the fourth-order tensor of the elastic constants; $\hat{\mathbf{h}}^{(3)}$ and $\widehat{\boldsymbol{f}}^{(4)}$ are the third-order and fourth-order tensors of piezoelectric and flexoelectric constants, respectively; $\hat{\boldsymbol{\chi}}$ is the second-order reciprocal dielectric susceptibility tensor; $\hat{\boldsymbol{\varepsilon}}^{(5)}$ and $\hat{\mathbf{g}}^{(6)}$ denote the fifth-order and sixth-order generalized second-gradient elastic constants that represent the "strain–strain gradient" and "strain gradient–strain gradient" coupling, respectively; the dot denotes the convolution of the corresponding indices, while the number in parentheses above the dot indicates the number of the index convolution.

For centrosymmetric materials, the constitutive equations (1.19)–(1.21) reduce to

$$\hat{\boldsymbol{\sigma}} = \hat{\mathbf{C}}^{(4)} : \hat{\boldsymbol{e}}, \tag{1.22}$$

$$\hat{\boldsymbol{\tau}}^{(3)} = \hat{\mathbf{g}}^{(6)} \overset{(3)}{\cdot} (\boldsymbol{\nabla} \otimes \hat{\boldsymbol{e}}) + \hat{\mathbf{f}}^{(4)} \cdot \boldsymbol{\Pi}_e, \tag{1.23}$$

$$\mathbf{E} = \hat{\boldsymbol{\chi}} \cdot \boldsymbol{\Pi}_e + (\boldsymbol{\nabla} \otimes \hat{\boldsymbol{e}}) \overset{(3)}{\cdot} \hat{\mathbf{f}}^{(4)}, \tag{1.24}$$

where

$$C_{ijkl} = \underline{\lambda}\delta_{ij}\delta_{kl} + \underline{\mu}\left(\delta_{jk}\delta_{il} + \delta_{ik}\delta_{jl}\right),$$

$$f_{ijkl} = f_{12}\delta_{ij}\delta_{kl} + f_{44}\left(\delta_{jk}\delta_{il} + \delta_{ik}\delta_{jl}\right),$$

$$\chi_{ij} = \chi\delta_{ij},$$

$$\begin{aligned} g_{ijklpq} = {} & g_1\left(\delta_{ij}\delta_{kl}\delta_{pq} + \delta_{ij}\delta_{kp}\delta_{lq} + \delta_{ik}\delta_{jq}\delta_{lp} + \delta_{iq}\delta_{jk}\delta_{lp}\right) \\ & + g_2\delta_{ij}\delta_{kq}\delta_{lp} + g_3\left(\delta_{ik}\delta_{jl}\delta_{pq} + \delta_{ik}\delta_{jp}\delta_{lq} + \delta_{il}\delta_{jk}\delta_{pq} + \delta_{ip}\delta_{jk}\delta_{lq}\right) \\ & + g_4\left(\delta_{il}\delta_{jp}\delta_{kq} + \delta_{ip}\delta_{jl}\delta_{kq}\right) \\ & + g_5\left(\delta_{il}\delta_{jq}\delta_{kp} + \delta_{ip}\delta_{jq}\delta_{kl} + \delta_{iq}\delta_{jl}\delta_{kp} + \delta_{iq}\delta_{jp}\delta_{kl}\right). \end{aligned}$$

Here, $\underline{\lambda}$ and $\underline{\mu}$ are the Lamé moduli; $f_{12}, f_{44}, \chi, g_i$ $(i = \overline{1,5})$ are material constants; and δ_{ij} denotes the Kronecker delta.

The constitutive equations (1.22)–(1.24) describe a linear response of polarization to a strain gradient $\nabla \otimes \hat{e}$ in the absence of electric field for isotropic materials.

The extended theory of crystalline dielectrics with piezoelectric, flexoelectric, and thermopolarization effects was developed by Tagancev (1986, 1991) and Gurevich and Tagantsev (1982). Wang et al. (2004), considering the rotation gradient effect, generalized the Toupin couple–stress theory of elasticity for piezoelectric materials. They related the electric polarization vector to the macroscopic rotation gradient. Using the variational principle, Shen and Hu (2010) developed a more comprehensive theory of nanosized dielectrics with the flexoelectric effect, the surface effects, and the electrostatic force. Hadjesfandiari (2013) proposed a variant of size-dependent theory of piezoelectricity where he showed that electric polarization can be generated as a result of coupling to the mean curvature tensor.

Maranganti et al. (2006) showed that flexoelectric effect noticeably manifests itself in nanoscale systems, around interfaces or in general in the vicinity of high field gradients. Majdoub et al. (2008a, 2008b, 2009), Majdoub (2010), and Liang et al. (2017) also examined the influence of strain gradients on the mechanical and electric properties of polarized solids and showed that flexoelectricity is a size-dependent effect, which becomes more significant in nanoscale systems. For example, Majdoub et al. (2008a), who investigated the flexoelectric effect in a nanocantilever beam, reported on the dramatic enhancement in energy harvesting in such piezoelectric nanostructures due to flexoelectricity. They pointed to the extreme dependence of the energy conversion factor on the nanobeam thickness.

The theory of flexoelectricity with strain gradients was effectively used in designing a new class of piezoelectric materials (Zhu et al. 2006; Fousek et al. 1999). Sharma et al. (2007) used this theory to study the possibility of creating piezoelectric nanocomposites without using piezoelectric materials. They showed that the improvement in piezoelectric properties of composites can be achieved by reducing the size of inclusions and by choosing the non-centrosymmetric shape of these inclusions. This effect can also be enhanced by an optimal selection of materials both for a matrix and for inclusions.

In order to evaluate the flexoelectric effect, one needs to ascertain the magnitude of the flexoelectric coefficients. Therefore, numerous works, both theoretical and experimental, are devoted to laboratory investigations of the flexoelectric effect, which relate to defining the flexoelectric modules for various materials (for example, see Askar and Lee 1974; Cross 2006; Fu et al. 2006; Gharbi et al. 2009; Hong et al. 2010; Kalinin and Meunier 2008; Ma and Cross 2001a, 2001b, 2003, 2006; Maranganti and Sharma 2009; Robinson et al. 2012; Zubko et al. 2007). A review of some experimental methods for quantifying the flexoelectric constants can be found in Mao (2016) and Zubko et al. (2013).

Nowadays, the strain gradient theory of polarized solids is effectively used for constructing a generalized theory of beams, plates, shells, etc. (Yang et al. 2015, Sladek et al. 2017, Zhang et al. 2016). For example, Yan and Jiang (2013a, 2013b) and Zhang et al. (2014) developed modified piezoelectric nanobeam and nanoplate models with the flexoelectric effect. Based on these theories, they investigated the static and dynamic behaviors of piezoelectric nanomaterials. Sladek et al. (2017) have showed that for plates described by the nonlocal theory, the size effect appears in the governing equations at dynamic terms or when the Laplacian of the in-plane and transversal loads is nonzero.

For studying the fundamentals of and recent advances in flexoelectricity, readers can refer to the following reviews: Jiang et al. 2013; Krichen and Sharma 2016; Mao 2016; Nguyen et al. 2013; Yan and Jiang 2017; Yudin and Tagantsev 2013; Zubko et al. 2013. A short review of experimental works that demonstrate the effect of flexoelectricity in novel crystalline materials can be found in Sharma et al. (2007).

1.3.6 Polarization Gradient Theory of Dielectrics

Burak (1966) was the first to introduce the polarization gradient into the space of constitutive parameters of polarized media. He assumed that the thermodynamic behavior of a thermoelastic dielectric is characterized by the mechanical stress $\hat{\boldsymbol{\sigma}}$ and strain tensor $\hat{\boldsymbol{e}}$, by the specific entropy s and absolute temperature T, as well as by the electric stresses tensor $\hat{\mathbf{T}}$, and by the specific density of polarization (i.e., bound) charge $\hat{\mathbf{\Omega}}$, where

$$\hat{\mathbf{\Omega}} = -\frac{1}{2}\left[\nabla \otimes \boldsymbol{\pi}_e + (\nabla \otimes \boldsymbol{\pi}_e)^{\mathrm{T}}\right].$$

Tensor $\hat{\mathbf{T}}$ characterizes the power effect of the electric field on the bound charges. Here, $\boldsymbol{\pi}_e = \boldsymbol{\Pi}_e/\rho$ is the polarization per unit mass; $\boldsymbol{\pi}_e = \mathbf{w}\omega_0/\rho_0$, where ω_0 and ρ_0 are initial values of the density of positive charges and mass density, respectively; $\mathbf{w} = \mathbf{u}_+ - \mathbf{u}_-$ is the difference between the displacement vectors of positive and negative charges.

Burak obtained the following Gibbs equation:

$$df = -sdT + \frac{1}{\rho_0}\hat{\boldsymbol{\sigma}} : d\hat{\boldsymbol{e}} + \frac{1}{\omega_0}\hat{\mathbf{T}} : d\hat{\mathbf{\Omega}}, \qquad (1.25)$$

where $f = f\left(T, \hat{\boldsymbol{e}}, \hat{\mathbf{\Omega}}\right)$ is the generalized free energy density.

The Gibbs equation (1.25) is the basis for formulating constitutive relations. Assuming the isothermal approximation, for linear isotropic media, these relations are as follows:

$$\hat{\boldsymbol{\sigma}} = 2G\hat{\boldsymbol{e}} + \rho_0 R\hat{\mathbf{\Omega}} + \left[\left(K - \frac{2}{3}G\right)e + \rho_0\left(N - \frac{1}{3}R\right)\Omega\right]\hat{\mathbf{I}}, \qquad (1.26)$$

$$\hat{\mathbf{T}} = 2L\hat{\mathbf{\Omega}} + \omega_0 R\hat{\boldsymbol{e}} + \left[\left(M - \frac{2}{3}L\right)\Omega + \omega_0\left(N - \frac{1}{3}R\right)e\right]\hat{\mathbf{I}}. \qquad (1.27)$$

Here, $\Omega = \nabla \cdot \boldsymbol{\pi}_e$ is the first invariant of tensor $\hat{\mathbf{\Omega}}$; K and G are the isothermal modulus of dilatation and the shear modulus, respectively; M and L are the moduli of volumetric and deviatoric polarization, respectively; R and N are the coupling factors between the processes of polarization and deformation; and $\hat{\mathbf{I}}$ denotes the unit tensor.

Note that the constitutive equations (1.26) and (1.27) take into account the coupling between mechanical and electric fields in isotropic materials. However, the gradient theory of dielectrics proposed by Burak (1966) has not been further developed in the scientific literature.

Another gradient-type theory of dielectrics was proposed by Mindlin (1968). He assumed that the internal energy U^L depends not only on the deformation tensor $\hat{\boldsymbol{e}}$ and on the polarization vector $\boldsymbol{\Pi}_e$, as the Toupin theory of piezoelectrics suggests, but also on the

gradient of the polarization vector, namely, $U^L(\hat{e}, \mathbf{\Pi}_e, \mathbf{\nabla}\otimes\mathbf{\Pi}_e)$. In order to determine the coupled mechanical and electric fields in an elastic dielectric body that occupies the region (V), Mindlin (1968) developed a set of equations that includes:
the mechanical equations of motion

$$\mathbf{\nabla}\cdot\hat{\boldsymbol{\sigma}}+\mathbf{F}_v=\rho\frac{\partial^2\mathbf{u}}{\partial t^2}\quad\forall\mathbf{r}\in(V),\tag{1.28}$$

the equations of electrostatic field for a medium and for a vacuum

$$-\varepsilon_0\mathbf{\nabla}^2\varphi_e+\mathbf{\nabla}\cdot\mathbf{\Pi}_e=0\ \forall\mathbf{r}\in(V),\tag{1.29}$$

$$\mathbf{\nabla}^2\varphi_{ev}=0\ \forall\mathbf{r}\in(V_v),\tag{1.30}$$

the constitutive equations

$$\hat{\boldsymbol{\sigma}}=\frac{\partial U^L}{\partial\hat{e}},\ \mathbf{E}^L=-\frac{\partial U^L}{\partial\mathbf{\Pi}_e},\ \hat{\mathbf{E}}=\frac{\partial U^L}{\partial(\mathbf{\nabla}\otimes\mathbf{\Pi}_e)},\tag{1.31}$$

the expression

$$\mathbf{\nabla}\cdot\hat{\mathbf{E}}+\mathbf{E}^L-\mathbf{\nabla}\varphi_e+\mathbf{E}^0=0\ \forall\mathbf{r}\in(V),\tag{1.32}$$

the boundary condition

$$\forall\mathbf{r}\in(\Sigma):\ \hat{\boldsymbol{\sigma}}\cdot\mathbf{n}=\boldsymbol{\sigma}_a,\ \hat{\mathbf{E}}\cdot\mathbf{n}=0,\ \left(-\varepsilon_0\left|\mathbf{\nabla}\varphi_e\right|+\mathbf{\Pi}_e\right)\cdot\mathbf{n}=0.\tag{1.33}$$

Here, ρ is the mass density; $\hat{\boldsymbol{\sigma}}$ denotes the stress tensor; $\mathbf{E}^L$ and $\mathbf{E}^0$ are the local and external electric field vectors, respectively; $\hat{\mathbf{E}}=\left\{E_{ij}\right\}$ is the higher-order electric field; φ_e is the electric potential; φ_{ev} is the electric potential of vacuum; $\left|\mathbf{\nabla}\varphi_e\right|$ denotes the jump in the gradient of the electric potential across the body surface (Σ); $\mathbf{F}_v$ is the external body force; σ_a denotes the surface traction; and $\mathbf{n}$ is the unit vector normal to the material surface (Σ).

Note that Eq. (1.32) is referred to as the "*equation of intramolecular force balance*" (Maugin 1977, 1988; Mindlin 1972a). Assuming $\hat{\mathbf{E}}=0$ and $\mathbf{E}^L = -\mathbf{E}$, in the absence of the external electric field, Eq. (1.32) yields $\mathbf{E} = -\mathbf{\nabla}\varphi_e$, where $\mathbf{E}$ is the macroscopic electric field vector.

A nonlinear gradient theory of piezoelectricity was proposed by Suhubi (1969). In the mid-1970s, Chowdhury et al. proposed and developed the linear (Chowdhury and Glockner 1977a) and nonlinear (Chowdhury and Glockner 1976; Chowdhury et al. 1979) theories of thermoelastic media with polarization gradient.

Based on this theory, Mindlin (1968) obtained a mathematical representation for a surface energy of polarization and deformation, which is absent from the classical theory. The Mindlin gradient theory of dielectrics made it possible to describe the coupling of mechanical and electromagnetic fields in centrosymmetric (including isotropic) materials.

Mindlin (1973) showed that within the gradient theory of dielectrics, the potential of an electric field caused by a point charge is a bounded function.

The gradient theory of piezoelectrics made it possible to accommodate a number of phenomena that do not cover the classical theory of electromechanical interaction in elastic dielectric media, namely:

- Near-surface inhomogeneity of electromechanical fields and size phenomena,
- Nonlinear dependence of the electrical capacitance of thin dielectric films on its thickness (the so-called Mead's anomaly) (Mindlin 1969, 1972a),
- Acoustical activity (i.e., the rotation of the direction of mechanical displacement along the path of a transverse elastic wave) and optical activity (i.e., the rotation of the plane of light polarization caused by passing through a polarized crystal) in an alpha quartz due to the action of electromagnetic field (Mindlin and Toupin 1971),
- Electromagnetic radiation from a vibrating quartz plate and sphere as well as electromechanical vibrations of centrosymmetric cubic crystal plates (Mindlin 1971, 1972a, 1972b, 1974).

Mindlin (1969, 1972a) justified the gradient theory of dielectrics using the microscopic theory of lattice. He showed that in terms of the lattice theory, Eq. (1.32) corresponds to the condition of equilibrium of the atom shell (i.e., the outer electrons) under the action of the core of the same atom, of the adjacent atoms on the shell, and of the surrounding Maxwell field. Herewith, the vector $\mathbf{E}^L$ and tensor $\hat{\mathbf{E}}$ quantities characterize the "core"–"shell" and "shell"–"shell" interactions, respectively.

Mindlin also introduced the fundamental length scale characteristics l_1 and l_2 that characterize the material structure.

Note that such parameters are absent in classical theories. Mindlin (1972a), and later Maugin (1988), showed that l_1 and l_2 are quantities of the order of magnitude of interatomic distance. Askar et al. (1970) obtained the values of these material coefficients using the theory of lattice dynamics. They established that these characteristic lengths are $l_1 \approx 7.3$ nm and $l_2 \approx 19.9$ nm for sodium chloride and $l_1 \approx 9.3$ nm and $l_2 \approx 22.2$ nm for potassium chloride.

The relations of gradient theory of piezoelectrics have been widely used to study:

- Different types of waves, including shock waves and acceleration waves (Chowdhury and Glockner 1977a; Collet 1981, 1982; Dost 1983; Dost et al. 1984; Majorkowska-Knap and Lenz 1989; Nowacki 1983; Nowacki and Glockner 1981; Romeo 2010).
- Stress-state behavior of the bodies with cracks and defects (Askar et al. 1970, 1971; Gou 1971; Nowacki 2004, 2006; Nowacki and Hsieh 1986).
- Fundamental solutions for the problems of concentrated body forces, electric forces, point charges, etc. (Chowdhury 1982; Chowdhury and Glockner 1974, 1977b, 1980, 1981; Schwartz 1969).

Askar et al. (1970, 1971) showed that within the framework of Mindlin's gradient theory of dielectrics, the surface energy density of the linear crack is bounded in ion crystals.

The theory for elastic dielectrics with polarization gradient is also discussed in the works by Sahin and Dost (1988), Majdoub et al. (2008a), Shen and Hu (2010), Yue et al. (2014), and Wang (2016). The mentioned authors proposed more general continuum theories of dielectrics with strain and electric gradient effects. They postulated the dependence of internal energy U^L both on polarization gradient and on strain gradient, namely, $U^L = U^L(\hat{\boldsymbol{e}}, \boldsymbol{\Pi}_e, \boldsymbol{\nabla} \otimes \boldsymbol{\Pi}_e, \boldsymbol{\nabla} \otimes \boldsymbol{\nabla} \otimes \mathbf{u})$. In such a way, they incorporated the flexoelectric effect into constitutive equations and wrote them as follows:

$$\hat{\boldsymbol{\sigma}} = \frac{\partial U^L}{\partial \hat{\boldsymbol{e}}}, \ \hat{\boldsymbol{\tau}}^{(3)} = \frac{\partial U^L}{\partial (\boldsymbol{\nabla} \otimes \boldsymbol{\nabla} \otimes \mathbf{u})}, \ \mathbf{E}^L = -\frac{\partial U^L}{\partial \boldsymbol{\Pi}_e}, \ \hat{\mathbf{E}} = \frac{\partial U^L}{\partial (\boldsymbol{\nabla} \otimes \boldsymbol{\Pi}_e)}.$$

Here, $\hat{\boldsymbol{\sigma}}$ and $\hat{\boldsymbol{\tau}}^{(3)}$ are ordinary and higher-order stress tensors.

These equations can be written in the explicit form (Shen and Hu 2010):

$$\hat{\boldsymbol{\sigma}} = \hat{\mathbf{C}}^{(4)} : \hat{\boldsymbol{e}} + \hat{\mathbf{h}}^{(3)} \cdot \boldsymbol{\Pi}_e + \hat{\boldsymbol{\gamma}}^{(4)} : (\boldsymbol{\nabla} \otimes \boldsymbol{\Pi}_e) + \hat{\boldsymbol{\varepsilon}}^{(5)} \overset{(3)}{\cdot} (\boldsymbol{\nabla} \otimes \hat{\boldsymbol{e}}),$$

$$\hat{\boldsymbol{\tau}}^{(3)} = \hat{\boldsymbol{e}} : \hat{\boldsymbol{\varepsilon}}^{(5)} + \hat{\mathbf{f}}^{(4)} \cdot \boldsymbol{\Pi}_e + (\boldsymbol{\nabla} \otimes \boldsymbol{\Pi}_e) : \hat{\boldsymbol{\eta}}^{(5)} + \hat{\mathbf{g}}^{(6)} \overset{(3)}{\cdot} (\boldsymbol{\nabla} \otimes \hat{\boldsymbol{e}}),$$

$$\mathbf{E} = \hat{\boldsymbol{e}} : \hat{\mathbf{h}}^{(3)} + \hat{\boldsymbol{\chi}} \cdot \boldsymbol{\Pi}_e + \hat{\mathbf{b}}^{(3)} \cdot (\boldsymbol{\nabla} \otimes \boldsymbol{\Pi}_e) + (\boldsymbol{\nabla} \otimes \hat{\boldsymbol{e}}) \overset{(3)}{\cdot} \hat{\mathbf{f}}^{(4)},$$

$$\hat{\mathbf{E}} = \hat{\boldsymbol{e}} : \hat{\boldsymbol{\gamma}}^{(4)} + \boldsymbol{\Pi}_e \cdot \hat{\mathbf{b}}^{(3)} + \hat{\boldsymbol{\beta}}^{(4)} : (\boldsymbol{\nabla} \otimes \boldsymbol{\Pi}_e) + \hat{\boldsymbol{\eta}}^{(5)} \overset{(3)}{\cdot} (\boldsymbol{\nabla} \otimes \hat{\boldsymbol{e}}).$$

Here, $\hat{\boldsymbol{e}}$ is the strain tensor; $\hat{\mathbf{C}}^{(4)}$ and $\hat{\mathbf{h}}^{(3)}$ are the elastic and piezoelectric coefficients; $\hat{\boldsymbol{\chi}}$ is the tensor of dielectric susceptibility; $\hat{\mathbf{f}}^{(4)}$ and $\hat{\boldsymbol{\gamma}}^{(4)}$ are the direct and the converse flexoelectric coefficients that represent "the strain gradient $\boldsymbol{\nabla} \otimes \hat{\boldsymbol{e}}$ and the polarization vector $\boldsymbol{\Pi}_e$" and "the polarization gradient $\boldsymbol{\nabla} \otimes \boldsymbol{\Pi}_e$ and the elastic strain tensor $\hat{\boldsymbol{e}}$" coupling, respectively. Tensors $\hat{\mathbf{b}}^{(3)}$, $\hat{\boldsymbol{\beta}}^{(4)}$, and $\hat{\boldsymbol{\eta}}^{(5)}$ represent "the polarization $\boldsymbol{\Pi}_e$ and the polarization gradient $\boldsymbol{\nabla} \otimes \boldsymbol{\Pi}_e$", "the polarization gradient $\boldsymbol{\nabla} \otimes \boldsymbol{\Pi}_e$ and the polarization gradient $\boldsymbol{\nabla} \otimes \boldsymbol{\Pi}_e$", and "the polarization gradient $\boldsymbol{\nabla} \otimes \boldsymbol{\Pi}_e$ and the strain gradient $\boldsymbol{\nabla} \otimes \hat{\boldsymbol{e}}$" coupling, respectively. Tensors $\hat{\boldsymbol{\varepsilon}}^{(5)}$ and $\hat{\mathbf{g}}^{(6)}$ represent "the strain $\hat{\boldsymbol{e}}$ and the strain gradient $\boldsymbol{\nabla} \otimes \hat{\boldsymbol{e}}$" and "the strain gradient $\boldsymbol{\nabla} \otimes \hat{\boldsymbol{e}}$ and the strain gradient $\boldsymbol{\nabla} \otimes \hat{\boldsymbol{e}}$" coupling, respectively.

Some authors enriched the plate bending theory replacing the classical theory by the gradient theory of dielectrics. For example, based on the Kirchhoff plate model and on the Mindlin linear gradient theory of dielectrics, Wang (2016) developed a modified continuum model of plates and used it to study the size-dependent flexoelectric effect in a cantilevered piezoelectric nanoplate.

A complete theory of elastic dielectrics with a gradient of polarization and polarization inertia was proposed by Tiersten and Tsai (1972), Maugin and Pouget (1980), Dost and Sahin (1986), and Hadjigeorgiou et al. (1999a, 1999b). One can find an extensive account of this generalized theory of dielectrics in the books by Maugin (1988) and Maugin et al. (1992). Maugin and Pouget (1980) and Pouget and Maugin (1980, 1981a, 1981b) used this theory to study acousto-optical and surface waves, including the Rayleigh waves. Collet (1984) analyzed the effect of shock waves on deformable ferroelectrics. Sahin and Dost (1988), along with the polarization gradient and the polarization inertia, considered the strain gradient as well. Pouget et al. (1986a, 1986b) and Askar et al. (1984) showed that the thus generalized theory of dielectrics has a support from lattice dynamics.

1.3.7 Theory of Dielectrics with Electric Quadrupole or Electric Field Gradient

Another approach to the construction of gradient-type theories of polarized medium was proposed by Kafadar (1971) who demonstrated that the electric multipoles have a significant impact on the piezoelectric and elastic behavior of dielectrics. Within the framework of continuum description, he represented the distribution of bound charges of an arbitrary system of particles by the following formulae:

$$Q(\mathbf{r}) = \int_{(V')} q(\mathbf{r}')dV', \ldots$$

$$Q^{k_1\ldots k_n}(\mathbf{r}) = \frac{1}{n!}\int_{(V')} q(\mathbf{r}')r^{k_1}\ldots r^{k_n}dV' = Q^{[k_1\ldots k_n]}(\mathbf{r}),$$

where q is the free charge volume density; Q is the charge magnitude of a system of particles within a material volume (V'); $Q^{k_1\ldots k_n}$ denotes the electric multipole of the order n (i.e., Q^{k_1} is the polarization vector, $Q^{k_1k_2}$ is the quadrupole tensor, etc.); k_j $(j=\overline{1,n})$ are the indices that acquire the values 1, 2, and 3; $r^{k_j} = \mathbf{i}^{k_j}(\mathbf{r}_0)\cdot(\mathbf{r}-\mathbf{r}')$; $\mathbf{i}^{k_j}$ is the base vector; $\mathbf{r}_0$ denotes the position vector of a reference point; and the parentheses enclosing a set of indices denote symmetrization.

Kafadar assumed that at the electrostatic approximation, the Lagrangian is taken to be a function of the electric field vector $\mathbf{E}$ as well as its gradients of the first and higher orders, i.e., $\nabla\otimes\mathbf{E}$, $\nabla\otimes\nabla\otimes\mathbf{E}$, ... (or its components E_{k_1}, E_{k_1,k_2}, E_{k_1,k_2k_3} ...; here, the comma indicates differentiation with respect to the corresponding spatial variables). Hence, he obtained the following constitutive relations:

$$Q^{k_1 \ldots k_n} = \rho \frac{\partial L}{\partial E_{k_1, k_2 \ldots k_n}} \tag{1.34}$$

for the components of electric multipoles.

Kafadar also pointed out the possibility of extending the proposed approach to deformable media. To this end, the dependence of a Lagrangian of an electroelastic continuum on the strain tensor, on the electric field vector, and on their gradients should be taken into account.

Similarly, he also investigated material systems whose behavior is determined both by electric and magnetic multipoles. Kafadar received the linear constitutive equations for an undeformed body, restricting himself to the dipole and quadrupole multipoles only. Using the obtained relations, he solved a number of problems of electrostatics and showed that due to the consideration of electric quadrupole, the potential and electric fields generated by a point charge in an infinite dielectric medium become the bounded functions. He showed that unlike the classical theory, the capacitance of a conducting sphere embedded in an infinite dielectric medium is a nonlinear function of its radius, while the index of refraction for such a dielectric depends on frequency.

The theory of dielectrics with electric quadrupoles (Kafadar 1971) is equivalent to the theory of dielectrics with the electric field gradient (Landau and Lifshitz 1984; Yang et al. 2004) because electric quadrupoles turn out to be the thermodynamic conjugate parameter of the electric field gradient. The generalized theory of elastic dielectrics with a gradient of the electric field was proposed by Landau and Lifshitz (1984). They used the electric field gradient as an independent constitutive variable. Landau and Lifshitz referred to the resulting theory as the *theory of dielectrics with spatial dispersion*.

Yang et al. (2004) extended the classical theory of dielectrics by allowing the internal energy density to depend on the gradient of the electric field vector, in addition to the strain tensor and electric field vector, that is, $U(\hat{\boldsymbol{e}}, \mathbf{E}, \boldsymbol{\nabla} \otimes \mathbf{E})$. Using the variational principle, they derived the field equations and obtained the following constitutive relations:

$$\hat{\boldsymbol{\sigma}} = \frac{\partial U}{\partial \hat{\boldsymbol{e}}} = \hat{\mathbf{C}}^{(4)} : \hat{\boldsymbol{e}} - \hat{\mathbf{h}}^{(3)} \cdot \mathbf{E},$$

$$\mathbf{P} = -\frac{\partial U}{\partial \mathbf{E}} = \hat{\mathbf{h}}^{(3)} : \hat{e} + \varepsilon_0 \hat{\boldsymbol{\chi}} \cdot \mathbf{E} + \varepsilon_0 \hat{\boldsymbol{\beta}}^{(3)} : (\nabla \otimes \mathbf{E}),$$

$$\hat{\mathbf{Q}} = -\frac{\partial U}{\partial (\nabla \otimes \mathbf{E})} = \varepsilon_0 \hat{\boldsymbol{\beta}}^{(3)} \cdot \mathbf{E} + \varepsilon_0 \hat{\boldsymbol{\alpha}}^{(4)} : (\nabla \otimes \mathbf{E}),$$

$$\boldsymbol{\Pi}_e = \mathbf{P} - \nabla \cdot \hat{\mathbf{Q}}, \quad \mathbf{D} = \varepsilon_0 \mathbf{E} + \boldsymbol{\Pi}_e.$$

Here, **P** and $\hat{\mathbf{Q}}$ are the electric dipoles and quadrupoles; $\hat{\boldsymbol{\chi}}$ denotes the tensor of dielectric susceptibility ($\hat{\boldsymbol{\varepsilon}} = \varepsilon_0(\hat{\mathbf{I}} + \hat{\boldsymbol{\chi}})$); $\hat{\mathbf{h}}^{(3)}$, $\hat{\boldsymbol{\beta}}^{(3)}$, and $\hat{\boldsymbol{\alpha}}^{(4)}$ are the material characteristics.

The theories of elastic dielectrics with electric quadrupoles are also discussed by a number of authors. Prechtl (1980) took into account the electric and magnetic multipoles; Kalpakidis and Massalas (1993) and Massalas et al. (1994) introduced into consideration the thermal effects as well. Hadjigeorgiou et al. (1999a, 1999b) considered the polarization inertia and introduced the polarization gradient and electric quadrupoles into constitutive parameters.

Kalpakides and Agiasofitou (2002) and later Hu and Shen (2009) developed a generalized theory of dielectrics, including both the electric field gradient and the strain gradient. The extended constitutive equations for dielectrics with the incorporation of flexoelectricity can be written as follows (Hu and Shen 2009):

$$\hat{\boldsymbol{\sigma}} = \hat{\underline{\mathbf{C}}}^{(4)} : \hat{e} - \hat{\underline{\mathbf{h}}}^{(3)} \cdot \mathbf{E} - \hat{\underline{\boldsymbol{\gamma}}}^{(4)} : (\nabla \otimes \mathbf{E}) + \hat{\underline{\boldsymbol{\varepsilon}}}^{(5)} \overset{(3)}{\cdot} (\nabla \otimes \hat{e}),$$

$$\hat{\boldsymbol{\tau}}^{(3)} = \hat{e} : \hat{\underline{\boldsymbol{\varepsilon}}}^{(5)} - \hat{\underline{\mathbf{f}}}^{(4)} \cdot \mathbf{E} - (\nabla \otimes \mathbf{E}) : \hat{\underline{\boldsymbol{\eta}}}^{(5)} + \hat{\underline{\mathbf{g}}}^{(6)} \overset{(3)}{\cdot} (\nabla \otimes \hat{e}),$$

$$\mathbf{D} = \hat{e} : \hat{\underline{\mathbf{h}}}^{(3)} + \hat{\underline{\mathbf{a}}} \cdot \mathbf{E} - \hat{\underline{\mathbf{b}}}^{(3)} \cdot (\nabla \otimes \mathbf{E}) + (\nabla \otimes \hat{e}) \overset{(3)}{\cdot} \hat{\underline{\mathbf{f}}}^{(4)},$$

$$\hat{\underline{\mathbf{E}}} = \hat{e} : \hat{\underline{\boldsymbol{\gamma}}}^{(4)} + \mathbf{E} \cdot \hat{\underline{\mathbf{b}}}^{(3)} + \hat{\underline{\boldsymbol{\beta}}}^{(4)} : (\nabla \otimes \mathbf{E}) + \hat{\underline{\boldsymbol{\eta}}}^{(5)} \overset{(3)}{\cdot} (\nabla \otimes \hat{e}).$$

Here, $\hat{\underline{\mathbf{C}}}^{(4)}$, $\hat{\underline{\mathbf{h}}}^{(3)}$, and $\hat{\underline{\mathbf{a}}}$ are the elastic, piezoelectric, and dielectric coefficients; $\hat{\underline{\mathbf{f}}}^{(4)}$ and $\hat{\boldsymbol{\gamma}}^{(4)}$ are the direct and the converse flexoelectric coefficients. Note that tensors $\hat{\underline{\mathbf{f}}}^{(4)}$ and $\hat{\boldsymbol{\gamma}}^{(4)}$ represent "the strain gradient $\nabla \otimes \hat{e}$ and the electric field **E**" and "the electric field gradient $\nabla \otimes \mathbf{E}$ and the elastic strain $\hat{e}$" coupling, respectively. Tensors $\hat{\underline{\mathbf{b}}}^{(3)}$, $\hat{\underline{\boldsymbol{\beta}}}^{(4)}$, and $\hat{\underline{\boldsymbol{\eta}}}^{(5)}$ represent "the electric field $\mathbf{E}$ and the electric field gradient $\nabla \otimes \mathbf{E}$", "the electric field gradient $\nabla \otimes \mathbf{E}$ and the electric field gradient $\nabla \otimes \mathbf{E}$", and "the electric field gradient $\nabla \otimes \mathbf{E}$ and the elastic strain gradient $\nabla \otimes \hat{e}$" coupling, respectively. Tensors $\hat{\underline{\boldsymbol{\varepsilon}}}^{(5)}$ and $\hat{\underline{\mathbf{g}}}^{(6)}$ represent "the strain $\hat{e}$ and the strain gradient $\nabla \otimes \hat{e}$" and "the strain gradient $\nabla \otimes \hat{e}$ and the strain gradient $\nabla \otimes \hat{e}$" coupling, respectively.

Using the variational approach, Bampi and Morro (1986) obtained the Maxwell equation, equation of motion, and constitutive relations for polarizable and magnetizable media with the dipole and quadrupole electric moments and the dipole magnetic moment.

Theory of elastic dielectrics with electric quadrupole have been used in order to investigate the size effect of the capacitance of thin dielectric films and to substantiate the nonlinear dependence of the capacitance on the film thickness (Yang and Yang 2004; Yang et al. 2004). A number of authors (Li et al. 2005; Yang and Yang 2009; et al.) have analyzed the effect of the electric field gradient on the wave processes. They showed that due to accounting for the electric field gradient as a constitutive variable, the short interface waves become dispersive.

The relations of the generalized theory of polarizable media with the gradient of electric field were used to study electromechanical fields in the vicinity of cracks, small cavities, and inclusions in piezoelectric media, as well as at the points of the action of concentrated forces, point charges, and line sources (Wang et al. 2008; Yang 2004; Yang et al. 2004, 2005, 2006; Zeng et al. 2005). Yang (2004b) and Sladek et al. (2018) have studied the effect of the electric field gradient on the electromechanical fields around cracks in piezoelectric solids. Yang et al. (2006) found that the effective dielectric permittivity of a small circular inclusion depends on its diameter. Wang et al. (2008) showed that the displacement and the electric potential are regular functions at the point where the line force and the line charge are located.

Maugin (1980) performed a comparative analysis of the generalized theory of an elastic medium containing the gradients of the electric field (electric quadrupoles) as constitutive parameters and Mindlin's polarization gradient theory of dielectrics.

1.4 Short Discussion

The review of the aforementioned generalized mathematical models of solid dielectric bodies testifies to their effectiveness in describing the surface and size effects. However, the aforementioned theories of polarized media are sometimes incapable of amply and appropriately describing the regularities of a stressed behavior and the processes occurring in solids (Beran and McCoy 1970; Mindlin 1972a; Maugin 1999; Chen 2013). For example, Beran and McCoy (1970) took notice of the problems that arise while using the gradient theory of elasticity to investigate the stress–strain behavior of an infinite isotropic medium subjected to a concentrated force. They showed that this problem will have a physically grounded solution, if the characteristic length that appears in the gradient theory of elasticity is a negative quantity. This undoubtedly contradicts the condition of a positive-definiteness of the deformation energy density. Certain questions arise while solving the wave problems as well. The strain gradient theory of dielectrics predicts a positive curvature of dispersive curves for longitudinal elastic waves, but the laboratory experiments exhibit the opposite trend, namely, the experimentally determined dispersive curves have a negative curvature. Mindlin (1972), Maugin (1999), Jirasek (2004), and Yang (2006) pointed to these and other problems of the generalized theory of elasticity with the first gradient of the strain tensor.

Theories that take account of the dependence of state on the gradients of polarization vector or the electric field vector relate the near-surface and the interface inhomogeneity of physical and mechanical fields exclusively to the presence of an electric field. Having accepted this kind of model description and having neglected the influence of electromagnetic field, the possibilities of describing the near-surface and size effects are lost, which is unacceptable if thin films, fibers, or small particles are investigation objects. Moreover, all the aforementioned nonlocal and gradient models of dielectric media take account of the possibility of the changes in the charge system at interface regions, although they do not take account of the corresponding mass fluxes that accompany the mentioned structural changes of the material in these regions.

In this connection, this book proposes a new continuum-thermodynamic approach to the construction of gradient-type models of mechanics of coupled fields in dielectric bodies that takes account of the changes in the material structure in the vicinity of high field gradients (i.e., in the near-surface regions, as well as in the regions of the action of high-frequency waves, concentrated forces, etc.). This approach is based on the consideration of the mass flux of non-convective and non-diffusive origin associated with the changes in the material structure. Such a process was referred to as local mass displacement. For the first time, such a mass flux was taken into account by Burak (1987) for thermoelastic solids. The generalized mathematical model developed by him was called the local gradient theory of thermoelasticity. In the proposed monograph, the local gradient theory of thermoelastic solids is extended to non-ferromagnetic polarized media.

Thus, this book is focused on:

- The development of an approach to the construction of new continuum-thermodynamic models of thermomechanics of coupled fields in polarized media that takes the local mass displacement into account,
- Using this approach for constructing a number of generalized gradient-type models of mechanics of non-ferromagnetic dielectric media,
- Applying the developed models for the description of near-surface and size effects and the dynamics of formation of a near-surface inhomogeneity of coupled physical and mechanical fields, as well as for the study of electromechanical waves in polarized solids.

An extensive practice of using dielectric materials in ultrasonic, microelectronics, etc., makes these investigations quite important both from a theoretical and a practical viewpoints.

Chapter 2

Thermodynamic Foundations of Local Gradient Electrothermomechanics of Polarized Non-ferromagnetic Solids Taking the Local Mass Displacement into Account

Using the fundamental principles of non-equilibrium thermodynamics, continuum mechanics, and electrodynamics, a new continuum-thermodynamic approach to the construction of a gradient-type theory of thermoelastic polarized medium has been proposed. The approach is based on the accounting for the local displacement of mass, caused by the change in the material structure of a body, and its coupling with the processes of deformation, thermal conductivity, and polarization. To describe the local mass displacement, the new objective physical quantities associated with this process are introduced and an additional balance equation is formulated.

This approach allows for obtaining a complete set of nonlinear equations for the local gradient thermomechanics of non-ferromagnetic polarized medium by taking the local displacements of mass and charges into account. The equations of balance of linear momentum and the conservation of mass are derived by requiring the energy equation to be invariant under the Galilean transformation. The obtained set of governing equations differs

Local Gradient Theory for Dielectrics: Fundamentals and Applications
Olha Hrytsyna and Vasyl Kondrat

ISBN 978-981-4800-62-4 (Hardcover), 978-1-003-00686-2 (eBook)
www.jennystanford.com

from the one used in the classical theory of thermoelastic dielectric materials and includes nonlocal effects.

The governing sets of equations and corresponding initial and boundary conditions are obtained for the linear approximation, which allows for the formulation of the coupled initial boundary-value problems of the local gradient theory of dielectrics. The reciprocity and uniqueness theorems for the non-stationary linear problems of the local gradient electrothermoelasticity are formulated and proved. The coupling of electromechanical processes and the local mass displacement is studied. The Lorentz gauge is generalized to take the local mass displacement into account. The connection of the developed theory with some generalized theories of dielectrics is analyzed. It is shown that the equations of both the local gradient and classical theories of dielectrics coincide if the local mass displacement is neglected.

2.1 Introduction

Thermodynamic processes occurring in solid deformed non-ferromagnetic dielectrics are the object of investigation of this book. The intrinsic resistance of dielectrics is 10^6–10^{15} Ωm, which means that they have very few free electric charges. The changes in the relative positions of positive and negative electric charges result in the polarization of a dielectric. The polarization of a solid body causes the emergence of a polarization current. Apart from the polarization current, the electric currents of convective and diffusive nature (i.e., conductivity current) and a displacement current caused by the change in the electric field may occur in a dielectrics.

In a strong electric field of the order of 10^7–10^8 V/m, there may occur a dielectric breakdown characterized by a high density of conductivity current. Such strong electric fields are not going to be considered herein. Moreover, we will treat weakly magnetic dielectrics as nonmagnetic ones (for which the magnetic permeability is equal to the magnetic permeability of vacuum). As a special case, we are going to examine ideal dielectrics that are completely devoid of free electric charges.

Thus, we consider a thermoelastic polarized non-ferromagnetic solid body that occupies a finite domain (V_*) of Euclidean space and is bounded by a closed smooth boundary (Σ_*). The solid body is subjected to the complex action of external loads of mechanical, thermal, and electromagnetic nature. As a result, mechanical, heat, and electromagnetic processes occur within the solid body, which can also be accompanied by the changes in the material structure. Such changes in material structure can be observed, for instance, in the near-surface regions of newly created surfaces. They are caused by a violation of the atom force balance in these regions. We describe the above structure changes by mass flux of a non-diffusive and non-convective nature. It should be noted that Marchenko et al. (2009) observed such mass fluxes within the near-surface domains of thin films during their formation.

We take into consideration that the mass-center displacement of a small body element, the domain (dV), may be induced, in general, not only by a convective displacement of this element as a rigid entity (i.e., translational displacement of the element's geometric center) but also by the changes in the relative positions of microparticles within this element, that is, the change in its structure (see Fig. 2.1). Such a displacement of non-convective and non-diffusive nature is referred to as a local mass displacement.

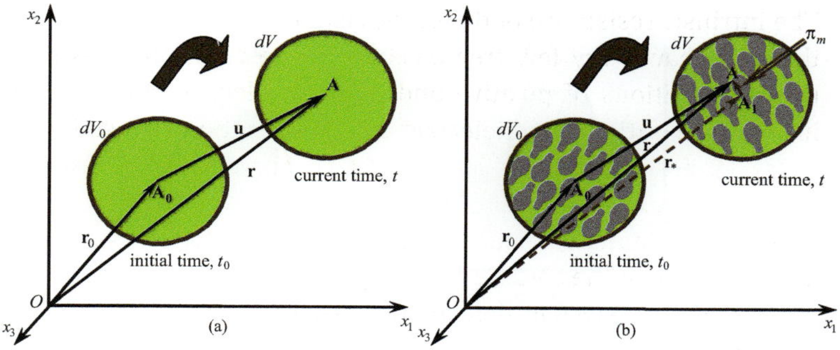

Figure 2.1 Changing the center of mass of a small body element (body particle) due to the body deformation: (a) the classical theory; (b) local gradient theory taking the local mass displacement into account ($\mathbf{r} = \mathbf{r}_* + \boldsymbol{\pi}_m$, where $\boldsymbol{\pi}_m = \boldsymbol{\Pi}_m/\rho$).

Burak (1987) was the first to carry out such a separation of the mass flux (i.e., separating the flux component related to the local mass displacement). He managed to show that such a component of the mass flux (mass dipole moment) being taken into consideration leads us to gradient-type mathematical models of elastic and thermoelastic solids (Burak 1987; Burak et al. 2007). However, it should be noted that taking into consideration only a dipole electric moment within the polarization current, which is known to be caused by the reordering (i.e., by the changing) of the charge structure, does not result in the gradient-type theory of electrothermoelasticity (Toupin 1956; Voigt 1910).

Note also that the processes occurring within the solid body are described herein in the one-continuum approximation.

2.2 Basic Kinematic Relations

Let us consider a deformable body (material continuum) within a curvilinear Eulerian coordinate system $\{x^i\}$.

The individualization of material points of a continuum will be carried out according to their Lagrangian coordinates, namely, the coordinates of material particles at an initial time $t = t_0 : \xi^i \equiv x^i(t_0)$ (Fig. 2.2). We assume that there is a biunique correspondence $x^i = x^i(\xi^1, \xi^2, \xi^3, t)$, $i = \overline{1,3}$, between the Eulerian (laboratory) $\{x^i\}$ and Lagrangian (material) $\{\xi^i\}$ coordinates, at an arbitrary time t.

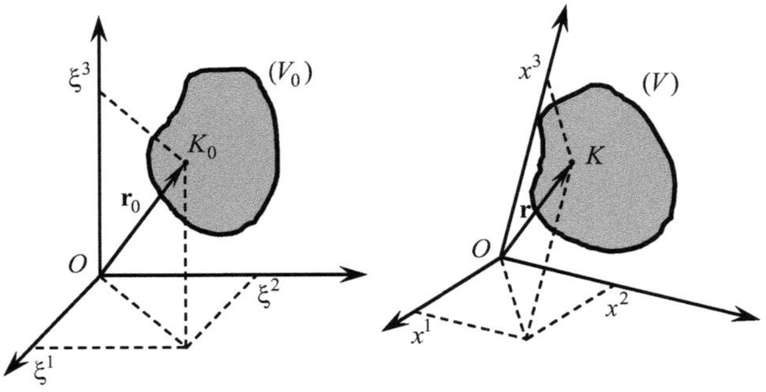

Figure 2.2 Lagrangian and Eulerian coordinates for the reference configuration (V_0) and the current configuration (V).

The motion of points of a material continuum is expressed by the mapping function $\mathbf{r} = \mathbf{r}(\xi^1, \xi^2, \xi^3, t)$. Here, $\mathbf{r}$ is the current position vector of the material point. Note that here bold symbols stand for vector quantities, and italic symbols represent scalar quantities.

Let $\mathbf{r}_0 = \mathbf{r}(\xi^1, \xi^2, \xi^3, t_0)$ be the position vector of a material point at the initial time $t = t_0$. Then, the mechanical displacement $\mathbf{u}$, the velocity $\mathbf{v}$, and the acceleration $\mathbf{w}$ of the material point at the current time are given as follows:

$$\mathbf{u}(\xi^1,\xi^2,\xi^3,t)=\mathbf{r}(\xi^1,\xi^2,\xi^3,t)-\mathbf{r}_0(\xi^1,\xi^2,\xi^3,t_0), \tag{2.1}$$

$$\mathbf{v}(\xi^1,\xi^2,\xi^3,t)=\frac{\partial \mathbf{u}(\xi^1,\xi^2,\xi^3,t)}{\partial t},$$

$$\mathbf{w}(\xi^1,\xi^2,\xi^3,t)=\frac{\partial^2 \mathbf{u}(\xi^1,\xi^2,\xi^3,t)}{\partial t^2}.$$

We also introduce the covariant basis vectors of the Lagrangian coordinate system:

$$\mathbf{i}_i=\frac{\partial \mathbf{r}}{\partial \xi^i},\ \mathbf{i}_i^0=\frac{\partial \mathbf{r}_0}{\partial \xi^i},\ i=\overline{1,3}. \tag{2.2}$$

Taking the formulae (2.1) and (2.2) into account, we get the following relation for vectors $\mathbf{i}_i$ and $\mathbf{i}_i^0$:

$$\mathbf{i}_i^0=\mathbf{i}_i-\frac{\partial \mathbf{u}}{\partial \xi^i}. \tag{2.3}$$

Let $\{\xi^i\}$ and $\{\xi^i + d\xi^i\}$ be two adjacent infinitely close material points of the body (Fig. 2.3). The vectors between these points in the undeformed and deformed states are $d\mathbf{r}_0$ and $d\mathbf{r}$, respectively. At times t and t_0, the squared lengths dl^2 and dl_0^2 between these two points will be determined by the relations $dl^2=|d\mathbf{r}|^2$ and $dl_0^2=|d\mathbf{r}_0|^2$. Here,

$$d\mathbf{r}=\frac{\partial \mathbf{r}}{\partial \xi^i}d\xi^i=d\xi^i\mathbf{i}_i,\ d\mathbf{r}_0=\frac{\partial \mathbf{r}_0}{\partial \xi^i}d\xi^i=d\xi^i\mathbf{i}_i^0. \tag{2.4}$$

Hereinafter, the indices repeated within the same addend denote the summation. Thus, using the formulae (2.4), we have the following expressions to find dl and dl_0:

$$dl^2=g_{ij}d\xi^i d\xi^j,\ dl_0^2=g_{ij}^0 d\xi^i d\xi^j,$$

where

$$g_{ij}=\mathbf{i}_i\cdot\mathbf{i}_j,\ g_{ij}^0=\mathbf{i}_i^0\cdot\mathbf{i}_j^0. \tag{2.5}$$

Here, g_{ij} and g_{ji}^0 are the covariant components of the fundamental metric tensors $\hat{\mathbf{g}}=g_{ij}\mathbf{i}^i\otimes\mathbf{i}^j$ and $\hat{\mathbf{g}}_0=g_{ij}^0\mathbf{i}_0^i\otimes\mathbf{i}_0^j$; $\mathbf{i}_0^i$ and $\mathbf{i}^i$ are

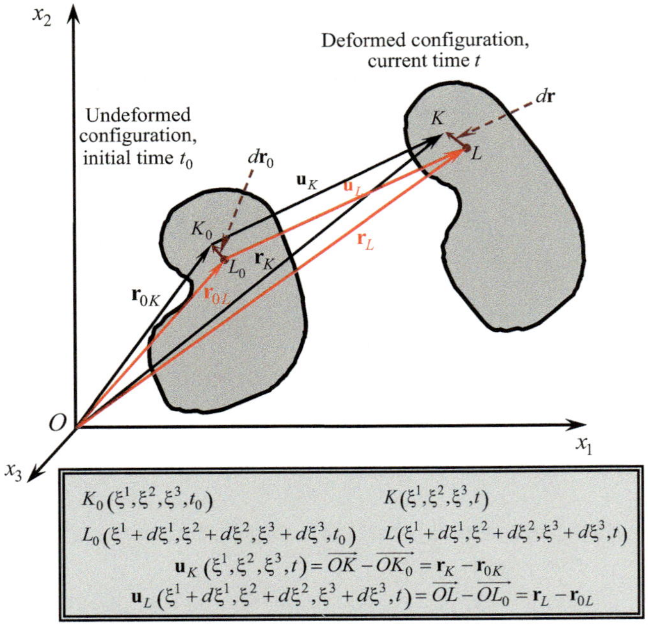

Figure 2.3 Deformation of a continuum.

the vectors of a contravariant basis of the material coordinate system at the initial and current times (i.e., the reciprocal basis vectors to the vectors $\mathbf{i}_i^0$ and $\mathbf{i}_i$).

The change in the metric properties of a body at an arbitrary time t is characterized by the following quadratic form:

$$dl^2 - dl_0^2 = \left(g_{ij} - g_{ij}^0\right) d\xi^i d\xi^j .$$

According to the above relation, the two strain tensors can be defined by the formulae

$$\hat{\boldsymbol{e}}_0 = e_{ij}\, \mathbf{i}^i{}_0 \otimes \mathbf{i}^j{}_0 , \; \hat{\boldsymbol{e}} = e_{ij}\, \mathbf{i}^i \otimes \mathbf{i}^j .$$

Here, the covariance components e_{ij} of these tensors are determined by the covariant components of the metric tensors $\hat{\boldsymbol{g}}$ and $\hat{\boldsymbol{g}}_0$ as follows:

$$e_{ij} = \frac{1}{2}\left(g_{ij} - g_{ij}^0\right). \tag{2.6}$$

Here, $\hat{\boldsymbol{e}}_0$ is the Green strain tensor and $\hat{\boldsymbol{e}}$ is the Almansi strain tensor (Lurie 1990).

Let us represent the components e_{ij} of the strain tensor in terms of a displacement vector. To this end, we substitute the relations (2.5) and (2.3) into (2.6). As a result, to find the covariant components e_{ij}, we get the following relation:

$$e_{ij} = \frac{1}{2}\left(\frac{\partial \mathbf{r}}{\partial \xi^i}\cdot\frac{\partial \mathbf{r}}{\partial \xi^j} - \frac{\partial \mathbf{r}_0}{\partial \xi^i}\cdot\frac{\partial \mathbf{r}_0}{\partial \xi^j}\right).$$

Thus, taking the formula (2.1) into account, we can write:

$$e_{ij} = \frac{1}{2}\left(\frac{\partial \mathbf{u}}{\partial \xi^i}\cdot\mathbf{i}_j^0 + \frac{\partial \mathbf{u}}{\partial \xi^j}\cdot\mathbf{i}_i^0 + \frac{\partial \mathbf{u}}{\partial \xi^i}\cdot\frac{\partial \mathbf{u}}{\partial \xi^j}\right).$$

Hence, in view of formula (2.3), we also get:

$$e_{ij} = \frac{1}{2}\left(\frac{\partial \mathbf{u}}{\partial \xi^i}\cdot\mathbf{i}_j + \frac{\partial \mathbf{u}}{\partial \xi^j}\cdot\mathbf{i}_i - \frac{\partial \mathbf{u}}{\partial \xi^i}\cdot\frac{\partial \mathbf{u}}{\partial \xi^j}\right).$$

By neglecting the quadratic term, we obtain

$$e_{ij} = \frac{1}{2}\left(\frac{\partial \mathbf{u}}{\partial \xi^i}\cdot\mathbf{i}_j + \frac{\partial \mathbf{u}}{\partial \xi^j}\cdot\mathbf{i}_i\right),$$

in a linear approximation for small strains.

Since the correspondence between Eulerian and Lagrangian coordinates is biunique, the expressions for the displacement, velocity, and acceleration fields, as well as for the components of the strain tensor, can be rewritten in the spatial (Eulerian) variables, which are convenient to be taken Cartesian. Then, in the Cartesian coordinate system $\{x^i\}$, we get $\mathbf{u} = \mathbf{u}(x^1, x^2, x^3, t)$ for the displacement field. Since $x^i = x^i(t)$, we obtain

$$\mathbf{v}(x^1,x^2,x^3,t) = \frac{d\mathbf{u}(x^1,x^2,x^3,t)}{dt}, \tag{2.7}$$

$$\mathbf{w}(x^1,x^2,x^3,t) = \frac{d\mathbf{v}(x^1,x^2,x^3,t)}{dt} = \frac{d^2\mathbf{u}(x^1,x^2,x^3,t)}{dt^2},$$

for velocity and acceleration fields, where $\frac{d}{dt} = \frac{\partial}{\partial t} + \mathbf{v}\cdot\boldsymbol{\nabla}$ denotes the material time derivative (the substantial derivative), and $\boldsymbol{\nabla}$ is the Hamilton operator.

In the linear approximation, a convective term in the material time derivative is neglected. Thus, in the Cartesian coordinate system, the strain-displacement relation is as follows:

$$e_{ij} = \frac{1}{2}\left(\frac{\partial u_i}{\partial x^j} + \frac{\partial u_j}{\partial x^i}\right)$$

or, within an invariant form,

$$\hat{\boldsymbol{e}} = \frac{1}{2}\left[\boldsymbol{\nabla} \otimes \mathbf{u} + \left(\boldsymbol{\nabla} \otimes \mathbf{u}\right)^{\mathrm{T}}\right]. \tag{2.8}$$

Here, $\boldsymbol{\nabla} = \frac{\partial}{\partial x^i}\mathbf{e}_i$, and $\mathbf{e}_i = \frac{\partial \mathbf{r}}{\partial x^i}$ is the unit base vector of the Cartesian coordinate system (superscript T stands for transpose).

2.3 Entropy Balance Equation

Let us consider a thermoelastic polarized non-ferromagnetic solid body subjected to external loads of thermal, mechanical, and electromagnetic nature. In response to these loads, the mechanical, heat, and electromagnetic processes occur within a solid. These processes may interact with one another and may lead to the changes in the material structure of a small element of the body.

All the fields that characterize the processes occurring in the solid should obey the fundamental laws of continuum physics, including Maxwell's equations, and the first and second laws of thermodynamics. This section is devoted to the formulation of the entropy balance equation.

We consider the absolute temperature T as the measure of an intensity of thermal motion. To describe the process of heat conductivity, apart from the field of the absolute temperature, we also introduce the entropy S as well as the heat $\mathbf{J}_q$ and entropy $\mathbf{J}_s$ fluxes across the surface of the body. These fluxes are related through the following expression (De Groot and Mazur 1962):

$$\mathbf{J}_s = \frac{1}{T}\mathbf{J}_q. \tag{2.9}$$

Let us separate from the body a fixed volume (V) bounded by closed surface (Σ). The time rate of change of the entropy of a material body element is defined by the convective component of the flux, by the flux of entropy through the bounding surface of the body, by the entropy production η_s due to the irreversible processes occurring within the body, as well as by the action of the distributed thermal sources $\Re$. The general form of the equation of entropy balance for the considered fixed volume (V) is given by:

$$\frac{d}{dt}\int_{(V)} S\,dV = -\oint_{(\Sigma)} (\mathbf{J}_s + S\mathbf{v})\cdot\mathbf{n}\,d\Sigma + \int_{(V)}\left(\eta_s + \rho\frac{\Re}{T}\right)dV\,, \tag{2.10}$$

where ρ is the mass density, and $\mathbf{n}$ is the unit vector normal to the material surface (Σ) that bounds the region (V).

Applying the Ostrogradsky–Gauss divergence theorem (Korn and Korn 1968), Eq. (2.10) in the local form can be written as follows:

$$\frac{\partial S}{\partial t} = -\,\nabla\cdot\mathbf{J}_s - \nabla\cdot(S\mathbf{v}) + \eta_s + \rho\frac{\Re}{T}\,.$$

Hence, taking the relation (2.9) into account, we obtain

$$T\frac{\partial S}{\partial t} = -\,\nabla\cdot\mathbf{J}_q + \frac{1}{T}\mathbf{J}_q\cdot\nabla T - T\,\nabla\cdot(S\mathbf{v}) + T\eta_s + \rho\Re\,. \tag{2.11}$$

2.4 Electromagnetic Field Equations

2.4.1 Maxwell's Equations

The Faraday and Ampère laws, as well as the law of conservation of an electric charge in an integral form, can be written as follows (Landau and Lifshitz 1984):

$$\oint_{(\partial\Sigma)} \mathbf{E}\cdot\mathbf{n}\,dl = -\frac{d}{dt}\int_{(\Sigma)} \mathbf{B}\cdot\mathbf{n}\,d\Sigma\,, \tag{2.12}$$

$$\oint_{(\partial\Sigma)} \mathbf{H}\cdot\mathbf{n}\,dl = \int_{(\Sigma)} \mathbf{J}_{ef}\cdot\mathbf{n}\,d\Sigma\,, \tag{2.13}$$

$$\oint_{(\partial V)} \mathbf{J}_e\cdot\mathbf{n}\,d\Sigma = -\frac{d}{dt}\int_{(V)} \rho_e\,dV\,, \tag{2.14}$$

where **E**, **H**, and **B** are vectors denoting, respectively, the electric and magnetic fields and the magnetic induction; $\mathbf{J}_{ef}$ is a vector denoting the density of the total electric current; $\mathbf{J}_e$ denotes the density of electric current, related to the displacement of free charges (the conduction and convection currents); ρ_e is a scalar quantity denoting the density of a free electric charge; (V) is any fixed volume with closed boundary surface (∂V); and (Σ) is any fixed surface with a closed boundary curve $(\partial\Sigma)$.

Note that Eqs. (2.12) and (2.13) contain line integrals around the boundary curve $(\partial\Sigma)$; here the loop indicates that the curve is closed. Equation (2.14) contains a surface integral over the boundary surface (∂V) (the loop indicates that the surface is closed). Here, dV is the volume element, and $d\Sigma$ and dl denote the elements of surface and loop, respectively.

We represent the vector of density of total electric current as the sum $\mathbf{J}_{ef} = \mathbf{J}_e + \mathbf{J}_{ed} + \mathbf{J}_{es}$, where $\mathbf{J}_{es}$ is the density of current due to the ordering of electric charges (the polarization current), and $\mathbf{J}_{ed}$ is as follows:

$$\mathbf{J}_{ed} = \varepsilon_0 \frac{\partial \mathbf{E}}{\partial t}. \tag{2.15}$$

Here, ε_0 is the electric constant. Note that the vector $\mathbf{J}_{ed}$ is often referred to as a displacement current (Landau and Lifshitz 1984).

The vector of magnetic induction can be represented as:

$$\mathbf{B} = \mu_0 \mathbf{H} + \mathbf{M}, \tag{2.16}$$

where $\mathbf{M}$ is the vector of magnetization, and μ_0 is the magnetic constant.

For a non-ferromagnetic medium, magnetization is equal to zero ($\mathbf{M} = 0$). Hence, for materials without magnetization, the constitutive relation (2.16) becomes:

$$\mathbf{B} = \mu_0 \mathbf{H}. \tag{2.17}$$

Let us introduce a polarization vector $\mathbf{\Pi}_e$, which is related to the vector $\mathbf{J}_{es}$ by the following formula (Bredov et al. 1985):

$$\mathbf{\Pi}_e(\mathbf{r},t) = \int_0^t \mathbf{J}_{es}(\mathbf{r},t')dt'$$

or in the local form:

$$\mathbf{J}_{es} = \frac{\partial \mathbf{\Pi}_e}{\partial t}. \tag{2.18}$$

Note that the vector $\mathbf{\Pi}_e$ can be referred to as a vector of the local displacement of electric charges. It has a dimension of the density of an electric dipole moment, that is, C · m/m³.

In view of the relations (2.15) and (2.18), one can write the density of total electric current as follows:

$$\mathbf{J}_{ef} = \mathbf{J}_e + \varepsilon_0 \frac{\partial \mathbf{E}}{\partial t} + \frac{\partial \mathbf{\Pi}_e}{\partial t}. \tag{2.19}$$

Substituting the expression (2.19) into (2.13) as well as using the Ostrogradsky–Gauss divergence theorem and the Kelvin–Stokes theorem (Marsden and Tromba 2003), we can write the integral equations (2.12)–(2.14) in the local form:

$$\nabla \times \mathbf{E} = -\frac{\partial \mathbf{B}}{\partial t}, \tag{2.20}$$

$$\nabla \times \mathbf{H} = \mathbf{J}_e + \varepsilon_0 \frac{\partial \mathbf{E}}{\partial t} + \frac{\partial \mathbf{\Pi}_e}{\partial t}, \tag{2.21}$$

$$\nabla \cdot \mathbf{J}_e + \frac{\partial \rho_e}{\partial t} = 0. \tag{2.22}$$

Applying the divergence operator to Eqs. (2.20), (2.21) and using Eq. (2.22) and the formula $\nabla \cdot (\nabla \times \mathbf{f}) = 0$ (Korn and Korn 1968) (here, **f** is an arbitrary vector field), we obtain:

$$\frac{\partial}{\partial t}(\nabla \cdot \mathbf{B}) = 0, \quad \frac{\partial}{\partial t}\left[\nabla \cdot \left(\varepsilon_0 \mathbf{E} + \mathbf{\Pi}_e\right) - \rho_e\right] = 0. \tag{2.23}$$

Now we define the vector of electric induction **D** (Landau and Lifshitz 1984) as

$$\mathbf{D} = \varepsilon_0 \mathbf{E} + \mathbf{\Pi}_e. \tag{2.24}$$

Assuming the zero initial conditions for the functions $\nabla \cdot \mathbf{B}$ and $(\nabla \cdot \mathbf{D} - \rho_e)$, from the relations (2.23), we obtain two differential equations

$$\nabla \cdot \mathbf{B} = 0, \tag{2.25}$$

$$\nabla \cdot \mathbf{D} = \rho_e, \tag{2.26}$$

which correspond to the Gauss–Faraday and Gauss–Coulomb laws, respectively. The relations (2.25) and (2.26) together with Eqs. (2.20) and (2.21) form a complete set of Maxwell equations. These equations present a mathematical formulation of basic laws of electrodynamics in a differential form. Equation (2.20) is a differential form of the Faraday law of induction that describes how a time-varying magnetic field induces an electric field. Equation (2.21) is a differential form of the Maxwell–Ampère law that describes the generation of a vortex magnetic field by the electric current and by changing electric fields. Equation (2.25) is a differential form of the Gauss law for magnetism. This law states that there are no "magnetic charges" analogous to electric charges. Equation (2.26) corresponds to the Gauss–Coulomb law, which describes the relationship between a static electric field and the electric charges causing this field.

Let us introduce the density of induced charge $\rho_{e\pi}$, which has a dimension of density of an electric charge, [C/m^3]. We require that for an arbitrary solid of finite size (domain (V_*)), vector $\mathbf{\Pi}_e$ of a local displacement of the electric charge (i.e., polarization vector), and the density of an induced charge $\rho_{e\pi}$ should satisfy the following integral relation (Bredov et al. 1985):

$$\int_{(V_*)} \mathbf{\Pi}_e dV = \int_{(V_*)} \rho_{e\pi} \mathbf{r} dV . \tag{2.27}$$

The integral on the right-hand side of Eq. (2.27) should be independent of the choice of the reference system. Therefore, the following equality should be satisfied (Bredov et al. 1985):

$$\int_{(V_*)} \rho_{e\pi} dV = 0 . \tag{2.28}$$

Let us multiply the left-hand and right-hand sides of the relation (2.27) by an arbitrary constant vector **a** and use the identity $\mathbf{a} \cdot \mathbf{\Pi}_e = (\mathbf{\Pi}_e \cdot \nabla)(\mathbf{a} \cdot \mathbf{r})$. As a result, after some algebra, we obtain

$$\int_{(V_*)} (\mathbf{\Pi}_e \cdot \nabla)(\mathbf{a} \cdot \mathbf{r}) dV =$$

$$-\int_{(V_*)} \nabla \cdot \left[\mathbf{\Pi}_e (\mathbf{a} \cdot \mathbf{r}) \right] dV - \int_{(V_*)} (\mathbf{a} \cdot \mathbf{r})(\nabla \cdot \mathbf{\Pi}_e) dV$$

$$= \int_{(V_*)} \rho_{e\pi} (\mathbf{a} \cdot \mathbf{r}) dV . \tag{2.29}$$

Let us assume that the body comes in contact with vacuum. Since vector $\mathbf{\Pi}_e$ is equal to zero outside the body,

$$\int_{(V_*)} \nabla \cdot \left[\mathbf{\Pi}_e (\mathbf{a} \cdot \mathbf{r}) \right] dV = 0 .$$

Because vector **a** and domain (V) are arbitrary, we get

$$\rho_{e\pi} = -\nabla \cdot \boldsymbol{\Pi}_e \quad (2.30)$$

from the expression (2.29).

By differentiating the formula (2.30) with respect to time and taking relation (2.18) into account, one can obtain a conservation law of an induced electric charge

$$\frac{\partial \rho_{e\pi}}{\partial t} + \nabla \cdot \mathbf{J}_{es} = 0\,. \quad (2.31)$$

2.4.2 Balance Law of Electromagnetic Field Energy

The objective in this subsection is to derive from Maxwell's equations a law of the conservation of energy of an electromagnetic field. To this end, we should calculate the work performed by an electromagnetic field within a certain unit volume per unit time. For the case of distributed electric charges and currents, this work is represented by $\mathbf{J}_e \cdot \mathbf{E}$ (Yariv and Yeh 1984). In view of Eq. (2.21), we obtain

$$\mathbf{J}_e \cdot \mathbf{E} = \left(\nabla \times \mathbf{H} - \varepsilon_0 \frac{\partial \mathbf{E}}{\partial t} - \frac{\partial \boldsymbol{\Pi}_e}{\partial t} \right) \cdot \mathbf{E}\,.$$

Hence, using Eq. (2.20) and the identity

$$\nabla \cdot (\mathbf{E} \times \mathbf{H}) = \mathbf{H} \cdot (\nabla \times \mathbf{E}) - \mathbf{E} \cdot (\nabla \times \mathbf{H}),$$

one can get

$$\mathbf{J}_e \cdot \mathbf{E} = \mathbf{H} \cdot (\nabla \times \mathbf{E}) - \nabla \cdot (\mathbf{E} \times \mathbf{H}) - \left(\varepsilon_0 \frac{\partial \mathbf{E}}{\partial t} + \frac{\partial \boldsymbol{\Pi}_e}{\partial t} \right) \cdot \mathbf{E}$$

$$= -\mathbf{H} \cdot \frac{\partial \mathbf{B}}{\partial t} - \varepsilon_0 \frac{\partial \mathbf{E}}{\partial t} \cdot \mathbf{E} - \nabla \cdot (\mathbf{E} \times \mathbf{H}) - \frac{\partial \boldsymbol{\Pi}_e}{\partial t} \cdot \mathbf{E}.$$

Taking the relation (2.17) into account, this expression can be rewritten as follows:

$$\mathbf{J}_e \cdot \mathbf{E} = -\frac{1}{2} \frac{\partial}{\partial t} (\varepsilon_0 \mathbf{E} \cdot \mathbf{E} + \mu_0 \mathbf{H} \cdot \mathbf{H}) - \nabla \cdot (\mathbf{E} \times \mathbf{H}) - \frac{\partial \boldsymbol{\Pi}_e}{\partial t} \cdot \mathbf{E}\,. \quad (2.32)$$

We define the electromagnetic energy density U_e by the formula

$$U_e = \frac{1}{2} (\varepsilon_0 \mathbf{E}^2 + \mu_0 \mathbf{H}^2)\,. \quad (2.33)$$

Note that $\mathbf{E} \times \mathbf{H} = \mathbf{S}_e$, where $\mathbf{S}_e$ is the Poynting vector. Thus, the relation (2.32) appears as

$$\frac{\partial U_e}{\partial t} + \boldsymbol{\nabla} \cdot \mathbf{S}_e + \left(\mathbf{J}_e + \frac{\partial \boldsymbol{\Pi}_e}{\partial t} \right) \cdot \mathbf{E} = 0. \tag{2.34}$$

We interpret Eq. (2.34) as a conservation law of the electromagnetic field energy. Note that the last term in this equation describes the influence of electromagnetic field on a substance, which together with the field present a single material system. Let us rewrite this term in such a way that it contains the vectors $\mathbf{E}_*$, $\boldsymbol{\Pi}_{e*}$, $\mathbf{J}_{e*}$ of an electric field, polarization, and density of electric current, respectively, in the reference frame of the center of mass moving with the velocity $\mathbf{v}$ relatively to the laboratory reference frame. In this co-moving frame, the vectors of electric field, polarization, and the density of electric current are transformed according to the relations (Bredov et al. 1985)

$$\mathbf{E}_* = \mathbf{E} + \mathbf{v} \times \mathbf{B}, \quad \boldsymbol{\Pi}_{e*} = \boldsymbol{\Pi}_e, \quad \mathbf{J}_{e*} = \mathbf{J}_e - \rho_e \mathbf{v}. \tag{2.35}$$

Finally, using the identity $\mathbf{a} \cdot (\mathbf{b} \times \mathbf{c}) = \mathbf{b} \cdot (\mathbf{c} \times \mathbf{a}) = \mathbf{c} \cdot (\mathbf{a} \times \mathbf{b})$ and the formulae (2.35), we can rewrite the balance law of the electromagnetic field energy (2.34) as follows:

$$\frac{\partial U_e}{\partial t} + \boldsymbol{\nabla} \cdot \mathbf{S}_e + \left(\mathbf{J}_{e*} + \frac{\partial \boldsymbol{\Pi}_e}{\partial t} \right) \cdot \mathbf{E}_*$$

$$+ \mathbf{v} \cdot \left[\rho_e \mathbf{E}_* + \left(\mathbf{J}_{e*} + \frac{\partial \boldsymbol{\Pi}_e}{\partial t} \right) \times \mathbf{B} \right] = 0. \tag{2.36}$$

2.5 Local Mass Displacement: Balance Equation of Induced Mass

To describe the local displacement of mass, we have introduced new quantities associated with this process. Following Burak (1987), we introduce a vector of the local displacement of mass, $\boldsymbol{\Pi}_m$, and a vector of non-diffusive and non-convective mass flux, $\mathbf{J}_{ms}$, caused by the ordering of a structure of a small element of the body (Fig. 2.1). We assume these vectors to be related through the following formula:

$$\boldsymbol{\Pi}_m(\mathbf{r},t) = \int_0^t \mathbf{J}_{ms}(\mathbf{r},t')\, dt',$$

hence,

$$\mathbf{J}_{ms} = \frac{\partial \mathbf{\Pi}_m}{\partial t}. \tag{2.37}$$

Note that the vector of local mass displacement, $\mathbf{\Pi}_m$, has a dimension of the density of a dipole mass moment, [kg · m/m^3].

Besides the vector of local mass displacement, we also introduce the quantity $\rho_{m\pi}$, which has a dimension of mass density. We suppose that for an arbitrary body of finite size (domain (V_*)), the vector $\mathbf{\Pi}_m$ and the density $\rho_{m\pi}$ of the induced mass should satisfy the following integral equation (Burak et al. 2008):

$$\int\limits_{(V_*)} \mathbf{\Pi}_m dV = \int\limits_{(V_*)} \rho_{m\pi} \mathbf{r} dV. \tag{2.38}$$

By analogy with electrodynamics, where the density $\rho_{e\pi}$ of the induced electric charge (Bredov et al. 1985) was introduced by a similar relation (2.27), we refer to $\rho_{m\pi}$ as the density of induced mass.

From the integral relation (2.38) after some calculations in the spirit of those in the previous section 2.4 (see Eqs. (2.28) and (2.30)), we can obtain

$$\int\limits_{(V_*)} \rho_{m\pi} dV = 0, \tag{2.39}$$

$$\rho_{m\pi} = -\mathbf{\nabla} \cdot \mathbf{\Pi}_m. \tag{2.40}$$

By differentiating the formula (2.40) with respect to time and taking the relation (2.37) into account, one can obtain the equation

$$\frac{\partial \rho_{m\pi}}{\partial t} + \mathbf{\nabla} \cdot \mathbf{J}_{ms} = 0, \tag{2.41}$$

which has the form of the conservation law of an induced mass (Burak et al. 2011).

Thus, for quantities introduced for a description of a local mass displacement (i.e., the density of induced mass and mass flux $\mathbf{J}_{ms}$), we obtain a balance-type differential equation (2.41).

Note also that the quantities introduced above (i.e., density of an induced mass, vectors of a mass flux $\mathbf{J}_{ms}$, and local mass displacement $\mathbf{\Pi}_m$) are invariant with respect to the Galilean transformations, spatial and time translations, as well as relative to the spatial rotations, that is, they are objective physical quantities.

2.6 Conservation of Energy for a System "Material Body–Electromagnetic Field"

The conservation energy law should be properly modified when in theoretical description we take into account not only the mechanical, thermal, and polarization processes (being a classical theory of dielectrics) but also the process of a local mass displacement. We do it herein in this section.

Let a heat-conducting, electrically polarizable, and deformable elastic body interact with the electromagnetic field. We assume that for an arbitrary moment of time, the total energy of a system "material body–electromagnetic field" is the sum of internal ρu and kinetic $\rho \mathbf{v}^2/2$ energies and the energy U_e of an electromagnetic field. Thus, the total energy is expressed as follows:

$$\mathcal{E} = \rho u + \frac{1}{2}\rho \mathbf{v}^2 + U_e. \tag{2.42}$$

Here, u is the specific internal energy.

We also assume that the change in the total energy is caused (i) by the action of mass forces $\mathbf{F}$ and distributed thermal sources $\Re$, (ii) by the convective energy transport $\rho(u+\mathbf{v}^2/2)\mathbf{v}$ through the body surface, (iii) by the energy flux $\hat{\boldsymbol{\sigma}}\cdot\mathbf{v}$ due to the power of surface forces, (iv) by the heat flux $\mathbf{J}_q$ and the electromagnetic energy flux $\mathbf{S}_e$, (v) by the energy flux $\mu\mathbf{J}_m$ linked with the mass transport relative to the center of mass of the small body element, and (vi) by the energy flux $\mu_\pi\mathbf{J}_{ms}$ related to the material structure ordering (i.e., local mass displacement). Thus, the balance law for energy of the fixed body volume (V) can be written as:

$$\frac{d}{dt}\int_{(V)} \mathcal{E}\, dV = -\oint_{(\Sigma)}\left[\rho\left(u + \frac{1}{2}\mathbf{v}^2\right)\mathbf{v} - \hat{\boldsymbol{\sigma}}\cdot\mathbf{v}\right.$$

$$\left. + \mathbf{S}_e + \mathbf{J}_q + \mu\mathbf{J}_m + \mu_\pi\mathbf{J}_{ms}\right]\cdot\mathbf{n}\,d\Sigma + \int_{(V)}\left(\rho\mathbf{F}\cdot\mathbf{v} + \rho\Re\right)dV. \tag{2.43}$$

Here, $\hat{\boldsymbol{\sigma}}$ is the Cauchy stress tensor; μ is the chemical potential; μ_π is the energy measure of the influence of the local mass displacement on the internal energy; the convective mass flux $\mathbf{J}_m$ is defined as:

$$\mathbf{J}_m = \rho(\mathbf{v}_* - \mathbf{v}), \tag{2.44}$$

where $\mathbf{v}_*$ is the velocity of convective translations of a small body element and $\mathbf{v}$ is a vector of the velocity of a continuum of the center of mass (see Fig. 2.1). These vectors are related through the formula:

$$\rho\mathbf{v} = \rho\mathbf{v}_* + \mathbf{J}_{ms}. \tag{2.45}$$

Here, in order to describe the process of local mass displacement, we also introduce a new objective physical quantity, namely, the potential μ_π, which has a dimension of [J/kg].

Note that the relations (2.37), (2.44), and (2.45) yield the following useful formula:

$$\mathbf{J}_m = -\frac{\partial \mathbf{\Pi}_m}{\partial t}. \tag{2.46}$$

Making use of the divergence theorem as well as the expressions (2.42) and (2.37), from the integral Eq. (2.43), we obtain the local form of the energy balance law for the local gradient thermoelastic polarized continuum

$$\frac{\partial}{\partial t}\left(\rho u + \frac{1}{2}\rho\mathbf{v}^2 + U_e\right) = -\nabla\cdot\left[\rho\left(u + \frac{1}{2}\mathbf{v}^2\right)\mathbf{v} - \hat{\boldsymbol{\sigma}}\cdot\mathbf{v}\right.$$

$$\left. + \mathbf{S}_e + \mathbf{J}_q + \mu\mathbf{J}_m + \mu_\pi\frac{\partial \mathbf{\Pi}_m}{\partial t}\right] + \rho\mathbf{F}\cdot\mathbf{v} + \rho\Re .$$

By virtue of the equation of entropy balance (2.11), the energy conservation law for electromagnetic field (2.36), as well as Eqs. (2.46) and (2.40), the last relation can be written as follows:

$$\frac{\partial(\rho u)}{\partial t} + \frac{1}{2}\frac{\partial}{\partial t}(\rho\mathbf{v}^2) = -\nabla\cdot\left[\left(u + \frac{1}{2}\mathbf{v}^2\right)\rho\mathbf{v}\right] + \nabla\cdot(\hat{\boldsymbol{\sigma}}\cdot\mathbf{v}) + T\frac{\partial S}{\partial t} + \mathbf{E}_*\cdot\frac{\partial \mathbf{\Pi}_e}{\partial t}$$

$$-\nabla\mu'_\pi\cdot\frac{\partial \mathbf{\Pi}_m}{\partial t} + \mu'_\pi\frac{\partial \rho_{m\pi}}{\partial t} - \frac{1}{T}\mathbf{J}_q\cdot\nabla T + \mathbf{J}_{e*}\cdot\mathbf{E}_* - T\eta_s + T\nabla\cdot(S\mathbf{v})$$

$$+\mathbf{v}\cdot\left[\rho_e\mathbf{E}_* + \left(\mathbf{J}_{e*} + \frac{\partial \mathbf{\Pi}_e}{\partial t}\right)\times\mathbf{B} + \rho\mathbf{F}\right]. \tag{2.47}$$

Here, $\mu'_\pi = \mu_\pi - \mu$.

The entropy s, the density of an induced mass ρ_m, the local mass displacement vector $\boldsymbol{\pi}_m$, and the polarization $\boldsymbol{\pi}_e$ per unit mass are defined as:

$$s = \frac{S}{\rho},\ \rho_m = \frac{\rho_{m\pi}}{\rho},\ \boldsymbol{\pi}_m = \frac{\mathbf{\Pi}_m}{\rho},\ \boldsymbol{\pi}_e = \frac{\mathbf{\Pi}_e}{\rho}. \tag{2.48}$$

Note that ρ_m is a dimensionless quantity and vector $\boldsymbol{\pi}_m$ has the length dimension, [m].

Using the specific quantities (2.48) and material time derivatives, after some algebra, the relation (2.47) can be written as follows:

$$\rho\frac{du}{dt}=\hat{\boldsymbol{\sigma}}_*:(\boldsymbol{\nabla}\otimes\mathbf{v})+\rho T\frac{ds}{dt}+\rho\mathbf{E}_*\cdot\frac{d\boldsymbol{\pi}_e}{dt}+\rho\mu'_\pi\frac{d\rho_m}{dt}-\rho\boldsymbol{\nabla}\mu'_\pi\cdot\frac{d\boldsymbol{\pi}_m}{dt}$$

$$-\left[\frac{\partial\rho}{\partial t}+\boldsymbol{\nabla}\cdot(\rho\mathbf{v})\right]\left[u+\frac{1}{2}\mathbf{v}^2-Ts-\boldsymbol{\pi}_e\cdot\mathbf{E}_*-\rho_m\mu'_\pi+\boldsymbol{\nabla}\mu'_\pi\cdot\boldsymbol{\pi}_m\right]$$

$$+\mathbf{J}_{e*}\cdot\mathbf{E}_*-\mathbf{J}_q\cdot\frac{\boldsymbol{\nabla}T}{T}-T\eta_s+\mathbf{v}\cdot\left(-\rho\frac{d\mathbf{v}}{dt}+\boldsymbol{\nabla}\cdot\hat{\boldsymbol{\sigma}}_*+\mathbf{F}_e+\rho\mathbf{F}_*\right),\qquad(2.49)$$

where

$$\hat{\boldsymbol{\sigma}}_*=\hat{\boldsymbol{\sigma}}-\rho\left[\boldsymbol{\pi}_e\cdot\mathbf{E}_*+\rho_m\mu'_\pi-\boldsymbol{\pi}_m\cdot\boldsymbol{\nabla}\mu'_\pi\right]\hat{\mathbf{I}},\qquad(2.50)$$

$$\mathbf{F}_*=\mathbf{F}+\underbrace{\rho_m\boldsymbol{\nabla}\mu'_\pi}-\underbrace{(\boldsymbol{\nabla}\otimes\boldsymbol{\nabla}\mu'_\pi)\cdot\boldsymbol{\pi}_m},\qquad(2.51)$$

$$\mathbf{F}_e=\overbrace{\rho_e\mathbf{E}_*}+\left[\mathbf{J}_{e*}+\frac{\partial(\rho\boldsymbol{\pi}_e)}{\partial t}\right]\times\mathbf{B}+\overbrace{\rho(\boldsymbol{\nabla}\otimes\mathbf{E}_*)\cdot\boldsymbol{\pi}_e}.\qquad(2.52)$$

Here, $\mathbf{F}_e$ is the ponderomotive force, and $\hat{\mathbf{I}}$ is the unit tensor.

Relations (2.50) and (2.51) define the modified stress tensor $\hat{\boldsymbol{\sigma}}_*$ and additional mass force $\mathbf{F}_*$ for an elastic dielectric body.

It is obvious from Eqs. (2.51) and (2.52) that the additional mass force $\mathbf{F}_*$ and ponderomotive force $\mathbf{F}_e$ are of similar structure (in the case where we substitute formula $\mathbf{E}_*=-\boldsymbol{\nabla}\varphi_e$ into Eq. (2.51), where φ_e is the electric potential, and make a replacement in Eqs. (2.51) and (2.52) $\boldsymbol{\nabla}\varphi_e\leftrightarrow\boldsymbol{\nabla}\mu'_\pi$, $\rho_e\leftrightarrow\rho\rho_m$, $\boldsymbol{\pi}_e\leftrightarrow\boldsymbol{\pi}_m$).

Note that in this section, we used the following identities (Korn and Korn 1968):

$$\boldsymbol{\nabla}(ab)=(\boldsymbol{\nabla}a)b+a\boldsymbol{\nabla}b,$$

$$\boldsymbol{\nabla}(\mathbf{a}\cdot\mathbf{b})=(\boldsymbol{\nabla}\otimes\mathbf{a})\cdot\mathbf{b}+\mathbf{a}\cdot(\boldsymbol{\nabla}\otimes\mathbf{b}),$$

$$\boldsymbol{\nabla}\cdot(a\hat{\mathbf{A}})=(\boldsymbol{\nabla}a)\cdot\hat{\mathbf{A}}+a\boldsymbol{\nabla}\cdot\hat{\mathbf{A}},$$

$$\boldsymbol{\nabla}\cdot(\hat{\mathbf{A}}\cdot\mathbf{a})=(\boldsymbol{\nabla}\cdot\hat{\mathbf{A}})\cdot\mathbf{a}+\hat{\mathbf{A}}^{\mathrm{T}}:(\boldsymbol{\nabla}\otimes\mathbf{a}).$$

2.7 Conservation Laws of Mass and Momentum

Equation (2.49) should satisfy the principle of a frame "indifference" in a rigid translation. Therefore, using the method presented by Green and Rivlin (1964), we assume the energy balance equation (2.49) to be valid under superimposed rigid body translation, when $\mathbf{v} \to \mathbf{v} + \mathbf{a}$. Here, **a** is an arbitrary constant velocity vector. After such a replacement, Eq. (2.49) appears as follows:

$$\rho \frac{du}{dt} = \hat{\boldsymbol{\sigma}}_* : (\boldsymbol{\nabla} \otimes \mathbf{v}) + \rho T \frac{ds}{dt} + \rho \mathbf{E}_* \cdot \frac{d\boldsymbol{\pi}_e}{dt} + \rho \mu'_\pi \frac{d\rho_m}{dt} - \rho \boldsymbol{\nabla} \mu'_\pi \cdot \frac{d\boldsymbol{\pi}_m}{dt}$$

$$-\left[\frac{\partial \rho}{\partial t} + \boldsymbol{\nabla} \cdot (\rho \mathbf{v})\right]\left[u + \frac{1}{2}\mathbf{v}^2 + \frac{1}{2}\mathbf{a}^2 + \mathbf{v} \cdot \mathbf{a} - Ts\right.$$

$$\left. - \mathbf{E}_* \cdot \boldsymbol{\pi}_e - \rho_m \mu'_\pi + \boldsymbol{\nabla} \mu'_\pi \cdot \boldsymbol{\pi}_m \right]$$

$$+ \mathbf{J}_{e*} \cdot \mathbf{E}_* - \mathbf{J}_q \cdot \frac{\boldsymbol{\nabla} T}{T} - T\eta_s$$

$$+ (\mathbf{v} + \mathbf{a}) \cdot \left(-\rho \frac{d\mathbf{v}}{dt} + \boldsymbol{\nabla} \cdot \hat{\boldsymbol{\sigma}}_* + \mathbf{F}_e + \rho \mathbf{F}_*\right). \tag{2.53}$$

Subtracting Eq. (2.49) from its incremented version (2.53), we get

$$\frac{1}{2}\mathbf{a}^2 \left[\frac{\partial \rho}{\partial t} + \boldsymbol{\nabla} \cdot (\rho \mathbf{v})\right]$$

$$= \mathbf{a} \cdot \left[-\rho \frac{d\mathbf{v}}{dt} + \boldsymbol{\nabla} \cdot \hat{\boldsymbol{\sigma}}_* + \mathbf{F}_e + \rho \mathbf{F}_* - \mathbf{v} \cdot \left(\frac{\partial \rho}{\partial t} + \boldsymbol{\nabla} \cdot (\rho \mathbf{v})\right)\right].$$

Since **a** is an arbitrary vector, the coefficients of the quadratic and the linear summands (quantities in square brackets) should be equal to zero. As a result, we get the momentum balance equation

$$\rho \frac{d\mathbf{v}}{dt} = \boldsymbol{\nabla} \cdot \hat{\boldsymbol{\sigma}}_* + \mathbf{F}_e + \rho \mathbf{F}_* \tag{2.54}$$

and the mass balance equation

$$\frac{\partial \rho}{\partial t} + \boldsymbol{\nabla} \cdot (\rho \mathbf{v}) = 0. \tag{2.55}$$

Note that the balance of momentum is formulated for a modified stress tensor $\hat{\boldsymbol{\sigma}}_*$ (see formula (2.50)).

We see that the local displacement of mass being taken into consideration leads to the appearance of an additional nonlinear mass force in the equation of motion (2.54), namely:

$$\mathbf{F}_*' = \rho_m \nabla \mu_\pi' - \boldsymbol{\pi}_m \cdot \nabla \otimes \nabla \mu_\pi'$$

and to the redefinition of the stress tensor $\hat{\boldsymbol{\sigma}}_*$ because this process also induces additional stresses within the body

$$\hat{\boldsymbol{\sigma}}_*' = -\rho\left(\rho_m \mu_\pi' - \boldsymbol{\pi}_m \cdot \nabla \mu_\pi'\right)\hat{\mathbf{I}} \ .$$

In view of Eqs. (2.54) and (2.55), the balance equation (2.49) of internal energy is simplified and may be written as follows:

$$\rho \frac{du}{dt} = \hat{\boldsymbol{\sigma}}_* : (\nabla \otimes \mathbf{v}) + \rho T \frac{ds}{dt} + \rho \mathbf{E}_* \cdot \frac{d\boldsymbol{\pi}_e}{dt} + \rho \mu_\pi' \frac{d\rho_m}{dt} - \rho \nabla \mu_\pi' \cdot \frac{d\boldsymbol{\pi}_m}{dt}$$

$$+\mathbf{J}_{e*} \cdot \mathbf{E}_* - \mathbf{J}_q \cdot \frac{\nabla T}{T} - T\eta_s \ . \tag{2.56}$$

2.8 Symmetry of the Stress Tensor

Equation (2.56) should be satisfied for the case of the body motion that differs from the given motion only by a superposed uniform rigid body angular velocity $\boldsymbol{\Omega}$. In the expression (2.56), therefore, we substitute $\mathbf{v}$ with $\mathbf{v} + \boldsymbol{\Omega} \times \mathbf{r}$ and take

$$\nabla \otimes (\mathbf{v} + \boldsymbol{\Omega} \times \mathbf{r}) = \nabla \otimes \mathbf{v} - \hat{\in}^{(3)} \cdot \boldsymbol{\Omega}$$

into account (Nowacki 1983). Here, $\hat{\in}^{(3)} = \left\{\in_{ijk}\right\}$ is the Levi–Civita symbol defined by the formula:

$$\in_{ijk} = \begin{cases} 1, & \text{if} \quad (i\,j\,k) = 123, 231, 312, \\ -1, & \text{if} \quad (i\,j\,k) = 321, 213, 132, \\ 0, & \qquad \text{otherwise.} \end{cases} \tag{2.57}$$

As a result, we can obtain the following:

$$\rho \frac{du}{dt} = \hat{\boldsymbol{\sigma}}_* : (\nabla \otimes \mathbf{v} - \hat{\in}^{(3)} \cdot \boldsymbol{\Omega})$$

$$+\rho T \frac{ds}{dt} + \rho \mathbf{E}_* \cdot \frac{d\boldsymbol{\pi}_e}{dt} + \rho \mu_\pi' \frac{d\rho_m}{dt} - \rho \nabla \mu_\pi' \cdot \frac{d\boldsymbol{\pi}_m}{dt}$$

$$+\mathbf{J}_{e*} \cdot \mathbf{E}_* - \mathbf{J}_q \cdot \frac{\nabla T}{T} - T\eta_s \ . \tag{2.58}$$

Subtracting the expressions (2.56) and (2.58), we get

$$\hat{\boldsymbol{\sigma}}_* : \hat{\boldsymbol{\in}}^{(3)} \cdot \boldsymbol{\Omega} = 0.$$

This equality is valid for all arbitrary constant values of angular velocity $\boldsymbol{\Omega}$ if $\in_{ijk} \sigma_{*jk} = 0$. As a result, we obtain $\sigma_{*jk} = \sigma_{*kj}$, where σ_{*jk} are the components of the tensor $\hat{\boldsymbol{\sigma}}_*$. Thus, the generalized stress tensor $\hat{\boldsymbol{\sigma}}_*$ is a symmetric tensor of the second order.

Now, let us use the identity

$$\boldsymbol{\nabla} \otimes \mathbf{u} = \frac{1}{2}\left[\boldsymbol{\nabla} \otimes \mathbf{u} + (\boldsymbol{\nabla} \otimes \mathbf{u})^{\mathrm{T}}\right] + \frac{1}{2}\left[\boldsymbol{\nabla} \otimes \mathbf{u} - (\boldsymbol{\nabla} \otimes \mathbf{u})^{\mathrm{T}}\right].$$

Hence, taking formula (2.8) into account, we obtain

$$\boldsymbol{\nabla} \otimes \mathbf{u} = \hat{\boldsymbol{e}} + \hat{\boldsymbol{\omega}}, \tag{2.59}$$

where by the formula

$$\hat{\boldsymbol{\omega}} = \frac{1}{2}\left[\boldsymbol{\nabla} \otimes \mathbf{u} - (\boldsymbol{\nabla} \otimes \mathbf{u})^{\mathrm{T}}\right] \tag{2.60}$$

we introduce an antisymmetric tensor of rotation $\hat{\boldsymbol{\omega}}$. The consequence of the formula (2.59) is

$$\boldsymbol{\nabla} \otimes \mathbf{v} = \frac{d\hat{\boldsymbol{e}}}{dt} + \frac{d\hat{\boldsymbol{\omega}}}{dt}. \tag{2.61}$$

Since the tensor $\hat{\boldsymbol{\sigma}}_*$ is symmetric and a convolution of a symmetric and antisymmetric tensors is equal to zero, $\hat{\boldsymbol{\sigma}}_* : \dfrac{d\hat{\boldsymbol{\omega}}}{dt} = 0$. Therefore, using formula (2.61), we can write the balance equation of the internal energy (2.56) as follows:

$$\rho\frac{du}{dt} = \hat{\boldsymbol{\sigma}}_* : \frac{d\hat{\boldsymbol{e}}}{dt} + \rho T\frac{ds}{dt} + \rho\mathbf{E}_* \cdot \frac{d\boldsymbol{\pi}_e}{dt} + \rho\mu'_\pi\frac{d\rho_m}{dt} - \rho\boldsymbol{\nabla}\mu'_\pi \cdot \frac{d\boldsymbol{\pi}_m}{dt}$$
$$+\mathbf{J}_{e*} \cdot \mathbf{E}_* - \mathbf{J}_q \cdot \frac{\boldsymbol{\nabla} T}{T} - T\eta_s. \tag{2.62}$$

2.9 Gibbs Equation and Entropy Production

Using the Legendre transformation

$$f = u - Ts - \mathbf{E}_* \cdot \boldsymbol{\pi}_e + \boldsymbol{\nabla}\mu'_\pi \cdot \boldsymbol{\pi}_m, \tag{2.63}$$

we define the new thermodynamic potential, namely, the generalized Helmholtz free energy per unit mass. After the introduction of a thermodynamic function (2.63), Eq. (2.62) is transformed into

$$\rho\frac{df}{dt}=\hat{\boldsymbol{\sigma}}_*:\frac{d\hat{e}}{dt}-\rho s\frac{dT}{dt}-\rho\boldsymbol{\pi}_e\cdot\frac{d\mathbf{E}_*}{dt}+\rho\mu_\pi'\frac{d\rho_m}{dt}+\rho\boldsymbol{\pi}_m\cdot\frac{d\boldsymbol{\nabla}\mu_\pi'}{dt}$$
$$+\mathbf{J}_{e*}\cdot\mathbf{E}_*-\mathbf{J}_q\cdot\frac{\boldsymbol{\nabla}T}{T}-T\eta_s. \quad (2.64)$$

While inspecting Eq. (2.64), we assume that the free energy density depends not only on the strain tensor $\hat{\boldsymbol{e}}$, temperature T, and electric field vector $\mathbf{E}_*$, as it follows from the classical theory, but also on the parameters ρ_m and $\boldsymbol{\nabla}\mu_\pi'$ related to the process of local mass displacement

$$f=f(\hat{e},T,\mathbf{E}_*,\rho_m,\boldsymbol{\nabla}\mu_\pi').$$

The scalar T, ρ_m, vector $\mathbf{E}_*,\boldsymbol{\nabla}\mu_\pi'$, and tensor $\hat{e}$ quantities are independent parameters. Therefore, using Eq. (2.64), we get a generalized Gibbs equation

$$df=\frac{1}{\rho}\hat{\boldsymbol{\sigma}}_*:d\hat{e}-sdT-\boldsymbol{\pi}_e\cdot d\mathbf{E}_*+\mu_\pi'd\rho_m+\boldsymbol{\pi}_m\cdot d\boldsymbol{\nabla}\mu_\pi' \quad (2.65)$$

and the following relation for entropy production

$$\eta_s=\frac{1}{T}\left(-\mathbf{J}_q\cdot\frac{\boldsymbol{\nabla}T}{T}+\mathbf{J}_{e*}\cdot\mathbf{E}_*\right). \quad (2.66)$$

It should be noted that according to the second law of thermodynamics, the production of entropy satisfies the following inequality: $\eta_s \geq 0$. Note also that in the formula (2.66) for entropy production, there are no terms caused by polarization and local mass displacement because we describe these processes as reversible. In ideal dielectrics, there are no free electrical charges, and, therefore, in the expression (2.66) for such materials, it should be assumed that $\mathbf{J}_{e*} = 0$.

The generalized Gibbs equation (2.65) and the expression for entropy production (2.66) obtained here are the basis for the formulation of constitutive relations.

2.10 Constitutive Equations

2.10.1 General Form

Considering the Helmholtz free energy f to be a function of scalar T, ρ_m, vector $\mathbf{E}_*, \nabla\mu'_\pi$, and tensor $\hat{e}$ arguments, its full differential looks as follows:

$$df = \frac{\partial f}{\partial \hat{e}} d\hat{e} + \frac{\partial f}{\partial T} dT + \frac{\partial f}{\partial \mathbf{E}_*} \cdot d\mathbf{E}_* + \frac{\partial f}{\partial \rho_m} d\rho_m + \frac{\partial f}{\partial \nabla\mu'_\pi} \cdot d\nabla\mu'_\pi. \quad (2.67)$$

Subtracting the generalized Gibbs equation (2.65) and the formula (2.67), we get

$$\left(\frac{\partial f}{\partial \hat{e}} - \frac{1}{\rho}\hat{\boldsymbol{\sigma}}_*\right) d\hat{e} + \left(\frac{\partial f}{\partial T} + s\right) dT + \left(\frac{\partial f}{\partial \mathbf{E}_*} + \boldsymbol{\pi}_e\right) \cdot d\mathbf{E}_*$$

$$+ \left(\frac{\partial f}{\partial \rho_m} - \mu'_\pi\right) d\rho_m + \left(\frac{\partial f}{\partial \nabla\mu'_\pi} - \boldsymbol{\pi}_m\right) \cdot d\nabla\mu'_\pi = 0.$$

The obtained relation must hold for arbitrary T, ρ_m, $\mathbf{E}_*, \nabla\mu'_\pi$, and $\hat{e}$. It follows that

$$\hat{\boldsymbol{\sigma}}_* = \rho \left.\frac{\partial f}{\partial \hat{e}}\right|_{T, \mathbf{E}_*, \rho_m, \nabla\mu'_\pi}, \quad s = -\left.\frac{\partial f}{\partial T}\right|_{\hat{e}, \mathbf{E}_*, \rho_m, \nabla\mu'_\pi}, \quad \boldsymbol{\pi}_e = -\left.\frac{\partial f}{\partial \mathbf{E}_*}\right|_{\hat{e}, T, \rho_m, \nabla\mu'_\pi},$$

$$\mu'_\pi = \left.\frac{\partial f}{\partial \rho_m}\right|_{\hat{e}, T, \mathbf{E}_*, \nabla\mu'_\pi}, \quad \boldsymbol{\pi}_m = \left.\frac{\partial f}{\partial (\nabla\mu'_\pi)}\right|_{\hat{e}, T, \mathbf{E}_*, \rho_m}. \quad (2.68)$$

Here, the subscripts to the right from the vertical lines indicate the variables held fixed during differentiation.

We shall treat ($\hat{\boldsymbol{\sigma}}_*$, $\hat{e}$), (s, T), ($\boldsymbol{\pi}_e$, $\mathbf{E}_*$), (μ'_π, ρ_m), and ($\boldsymbol{\pi}_m, \nabla\mu'_\pi$) as a set of conjugate variables. Note that the space of constitutive parameters in the classical theory of electrothermoelastic dielectrics contains three pairs of conjugate constitutive parameters, namely, the stress and the strain tensors, the entropy and the temperature, as well as the polarization and the electric field vectors. Within the framework of the local gradient theory of dielectrics, the set of conjugate variables is complemented by two pairs of variables (μ'_π, ρ_m) and ($\boldsymbol{\pi}_m$, $\nabla\mu'_\pi$), related to the local mass displacement. The presence of both the gradient of a modified chemical potential $\nabla\mu'_\pi$

and the divergence of the vector of local mass displacement $\rho_m = \nabla\cdot(\rho\boldsymbol{\pi}_m)/\rho$ in the space of constitutive parameters indicates that the constructed theory is spatially nonlocal.

2.10.2 Linear Constitutive Equations for Anisotropic Media

Let us represent the constitutive equations (2.68) in an explicit form. To this end, we should concretize the function of the free energy f. Let us decompose the free energy density in the Taylor series about the small perturbations of the constitutive parameters with respect to the original state of an infinite anisotropic homogeneous medium with $\hat{e}=0$, $\hat{\boldsymbol{\sigma}}_*=0$, $\mathbf{E}_*=0$, $\boldsymbol{\pi}_e=0$, $\boldsymbol{\pi}_m=0$, $\nabla\mu'_\pi=0$, $T=T_0$, $s=s_0$, $\rho_m=0$, and $\mu'_\pi=\mu'_{\pi0}$. Here, T_0 is the reference temperature and $\mu'_{\pi0}$ is the modified chemical potential μ'_π of an infinite medium. Let

$$\frac{T-T_0}{T_0}\ll 1,\ e_{ij}\ll 1,\ E_{*i}\ll 1,\ \rho_m\ll 1,\ \nabla\mu'_\pi\ll 1.$$

For small perturbations, we retain quadratic terms in this decomposition, which enables us to get linear constitutive equations. For the linear approximation, $\mathbf{E}_*=\mathbf{E}$. Therefore, the density of free energy f for an anisotropic media may be expressed in the form

$$\begin{aligned}
f &= f_0 - s_0(T-T_0) + \mu'_{\pi0}\rho_m - \frac{C_V}{2T_0}(T-T_0)^2 + \frac{1}{2}d_\rho\rho_m^2 \\
&+\frac{1}{2\rho_0}\hat{\mathbf{C}}^{(4)}\overset{(4)}{\cdot}(\hat{e}\otimes\hat{e}) - \frac{1}{2}\hat{\boldsymbol{\chi}}^m:(\nabla\mu'_\pi\otimes\nabla\mu'_\pi) - \frac{1}{2}\hat{\boldsymbol{\chi}}^E:(\mathbf{E}\otimes\mathbf{E}) \\
&-\frac{1}{\rho_0}\hat{\boldsymbol{\beta}}:\hat{e}(T-T_0) - \frac{1}{\rho_0}\hat{\boldsymbol{\alpha}}^\rho:\hat{e}\,\rho_m - \frac{1}{\rho_0}\hat{\mathbf{f}}^{(3)}\overset{(3)}{\cdot}(\hat{e}\otimes\mathbf{E}) \\
&-\boldsymbol{\beta}^E\cdot\mathbf{E}(T-T_0) - \boldsymbol{\beta}^\mu\cdot\nabla\mu'_\pi(T-T_0) - \boldsymbol{\gamma}^E\cdot\mathbf{E}\,\rho_m \\
&-\frac{1}{\rho_0}\hat{\mathbf{g}}^{(3)}\overset{(3)}{\cdot}(\hat{e}\otimes\nabla\mu'_\pi) - \beta_{T\rho}\rho_m(T-T_0) \\
&+\hat{\boldsymbol{\chi}}^{Em}:(\mathbf{E}\otimes\nabla\mu'_\pi) - \boldsymbol{\gamma}^\rho\cdot\nabla\mu'_\pi\,\rho_m.
\end{aligned}\tag{2.69}$$

Here and in what follows, the superscript in parentheses indicates the order of the tensor quantity (we omit these superscripts for the tensors of the zeroth, first, and second orders, retaining them for the tensor quantities of the third and higher orders); the dot denotes the convolution of the corresponding indices; while the number in parentheses above the dot indicates the number of the index convolution. We also introduced the following notations:

$$s_0 = -\left.\frac{\partial f}{\partial T}\right|_{(0,T_0,0,0,0)}, \quad \mu'_{\pi 0} = \left.\frac{\partial f}{\partial \rho_m}\right|_{(0,T_0,0,0,0)},$$

$$C_V = -T_0 \left.\frac{\partial^2 f}{\partial T^2}\right|_{(0,T_0,0,0,0)}, \quad \hat{\mathbf{C}}^{(4)} = \rho_0 \left.\frac{\partial^2 f}{\partial \hat{e}^2}\right|_{(0,T_0,0,0,0)},$$

$$d_\rho = -\left.\frac{\partial^2 f}{\partial \rho_m^2}\right|_{(0,T_0,0,0,0)}, \quad \hat{\boldsymbol{\beta}} = -\rho_0 \left.\frac{\partial^2 f}{\partial \hat{e}\,\partial T}\right|_{(0,T_0,0,0,0)},$$

$$\hat{\boldsymbol{\alpha}}^{\rho} = -\rho_0 \left.\frac{\partial^2 f}{\partial \hat{e}\,\partial \rho_m}\right|_{(0,T_0,0,0,0)}, \quad \beta_{T\rho} = -\left.\frac{\partial^2 f}{\partial \rho_m\,\partial T}\right|_{(0,T_0,0,0,0)},$$

$$\hat{\mathbf{f}}^{(3)} = -\rho_0 \left.\frac{\partial^2 f}{\partial \hat{e}\,\partial \mathbf{E}}\right|_{(0,T_0,0,0,0)}, \quad \hat{\mathbf{g}}^{(3)} = -\rho_0 \left.\frac{\partial^2 f}{\partial \hat{e}\,\partial(\boldsymbol{\nabla}\mu'_\pi)}\right|_{(0,T_0,0,0,0)},$$

$$\boldsymbol{\beta}^{E} = -\left.\frac{\partial^2 f}{\partial \mathbf{E}\,\partial T}\right|_{(0,T_0,0,0,0)}, \quad \boldsymbol{\beta}^{\mu} = -\left.\frac{\partial^2 f}{\partial T\,\partial(\boldsymbol{\nabla}\mu'_\pi)}\right|_{(0,T_0,0,0,0)},$$

$$\hat{\boldsymbol{\chi}}^{E} = -\left.\frac{\partial^2 f}{\partial \mathbf{E}^2}\right|_{(0,T_0,0,0,0)}, \quad \hat{\boldsymbol{\chi}}^{m} = -\left.\frac{\partial^2 f}{\partial(\boldsymbol{\nabla}\mu'_\pi)^2}\right|_{(0,T_0,0,0,0)},$$

$$\hat{\boldsymbol{\chi}}^{Em} = \left.\frac{\partial^2 f}{\partial \mathbf{E}\,\partial(\boldsymbol{\nabla}\mu'_\pi)}\right|_{(0,T_0,0,0,0)},$$

$$\boldsymbol{\gamma}^{E} = -\left.\frac{\partial^2 f}{\partial \mathbf{E}\,\partial \rho_m}\right|_{(0,T_0,0,0,0)}, \quad \boldsymbol{\gamma}^{\rho} = -\left.\frac{\partial^2 f}{\partial \rho_m\,\partial(\boldsymbol{\nabla}\mu'_\pi)}\right|_{(0,T_0,0,0,0)}.$$

Using Eqs. (2.68) and expression (2.69), we obtain the following constitutive equations for anisotropic media:

$$\hat{\boldsymbol{\sigma}}_* = \hat{\mathbf{C}}^{(4)} : \hat{e} - \hat{\boldsymbol{\beta}}\theta - \hat{\boldsymbol{\alpha}}^{\rho}\rho_m - \mathbf{E}\cdot\hat{\mathbf{f}}^{(3)} - \boldsymbol{\nabla}\mu'_\pi \cdot \hat{\mathbf{g}}^{(3)}, \tag{2.70}$$

$$s = s_0 + \frac{C_V}{T_0}\theta + \beta_{T\rho}\rho_m + \boldsymbol{\beta}^{\mu} \cdot \nabla\mu'_{\pi} + \boldsymbol{\beta}^{E} \cdot \mathbf{E} + \frac{1}{\rho_0}\hat{\boldsymbol{\beta}} : \hat{e}, \qquad (2.71)$$

$$\mu'_{\pi} = \mu'_{\pi 0} + d_{\rho}\rho_m - \beta_{T\rho}\theta - \boldsymbol{\gamma}^{E} \cdot \mathbf{E} - \boldsymbol{\gamma}^{\rho} \cdot \nabla\mu'_{\pi} - \frac{1}{\rho_0}\hat{\boldsymbol{\alpha}}^{\rho} : \hat{e}, \qquad (2.72)$$

$$\boldsymbol{\pi}_e = \hat{\boldsymbol{\chi}}^{E} \cdot \mathbf{E} - \nabla\mu'_{\pi} \cdot \hat{\boldsymbol{\chi}}^{Em} + \boldsymbol{\beta}^{E}\theta + \boldsymbol{\gamma}^{E}\rho_m + \frac{1}{\rho_0}\hat{\mathbf{f}}^{(3)} : \hat{e}, \qquad (2.73)$$

$$\boldsymbol{\pi}_m = -\hat{\boldsymbol{\chi}}^{m} \cdot \nabla\mu'_{\pi} + \hat{\boldsymbol{\chi}}^{Em} \cdot \mathbf{E} - \boldsymbol{\gamma}^{\rho}\rho_m - \boldsymbol{\beta}^{\mu}\theta - \frac{1}{\rho_0}\hat{\mathbf{g}}^{(3)} : \hat{e}, \qquad (2.74)$$

Where $\theta = T - T_0$; $\hat{\mathbf{C}}^{(4)}$ is the fourth-order tensor of the elastic constants; $\hat{\boldsymbol{\beta}}$ is the tensor of thermal expansion coefficients; $\hat{\boldsymbol{\alpha}}^{\rho}$ denotes the tensor of volumetric expansion caused by the local mass displacement; $\mathbf{f}^{(3)}$ and $\hat{\mathbf{g}}^{(3)}$ are the tensors of piezoelectric and piezomass constants, respectively; C_V stands for the specific heat capacity at a constant deformation and specific density of induced mass; $\beta_{T\rho}$ and d_{ρ} are the isothermal–isochoric coefficients of the dependency of entropy and potential μ'_{π} on specific density of the induced mass; $\boldsymbol{\beta}^{E}$ and $\boldsymbol{\beta}^{\mu}$ are the pyroelectric and pyromass coefficients; $\hat{\boldsymbol{\chi}}^{E}$ is the dielectric susceptibility tensor; $\hat{\boldsymbol{\chi}}^{m}$ and $\hat{\boldsymbol{\chi}}^{Em}$ denote the tensors characterizing the dependence of vectors of the local mass displacement and the polarization on $\nabla\mu'_{\pi}$; $\boldsymbol{\gamma}^{\rho}$ and $\boldsymbol{\gamma}^{E}$ are the coefficients characterizing the dependence of the potential μ'_{π} on its gradient $\nabla\mu'_{\pi}$ and the vector of electric field.

The constitutive equations (2.70)–(2.74) connect the mechanical, thermal, and electromagnetic properties of elastic polarized solids taking the process of local mass displacement into consideration.

Note that the relation (2.70) presents the generalized Hooke's law for thermoelastic polarized solids with local mass displacement.

Due to the introduction of the local mass displacement into the mathematical model of the dielectrics, the constitutive equations (2.70)–(2.74) contain new material constants $\boldsymbol{\beta}_{T\rho}$, d_{ρ}, $\boldsymbol{\beta}^{\mu}$, $\boldsymbol{\gamma}^{\rho}$, $\boldsymbol{\gamma}^{E}$, $\hat{\boldsymbol{\alpha}}^{\rho}$, $\hat{\mathbf{g}}^{(3)}$, $\hat{\boldsymbol{\chi}}^{m}$, and $\hat{\boldsymbol{\chi}}^{Em}$ characterizing the physical properties of the medium.

Let us consider the state of an anisotropic non-ferromagnetic dielectric devoid of the action of the mechanical, thermal, and electromagnetic fields, so that there are no deformations and perturbations of temperature and electric field in it. From the relations (2.70)–(2.74), it follows that even in the absence of an external load, a perturbation of electromechanical fields may arise in the dielectric body. These perturbations are caused by the local displacement of mass (i.e., changes in the material structure). A change in the material structure, for example, is observed in the near-surface regions of the body. This change causes a polarization of the body surface $\boldsymbol{\pi}_e' = \boldsymbol{\gamma}^E \rho_m - \hat{\boldsymbol{\chi}}^{Em} \cdot \nabla\mu_\pi'$ as well as the appearance of surface stresses $\hat{\boldsymbol{\sigma}}_*' = -\nabla\mu_\pi' \cdot \hat{\mathbf{g}}^{(3)} - \hat{\boldsymbol{\alpha}}^{\rho} \rho_m$ (see formulae (2.70) and (2.73)).

The fourth-order tensor of elastic moduli $\hat{\mathbf{C}}^{(4)}$ contains, in general, 81 coefficients, and the third-order tensors of the piezoelectric $\hat{\mathbf{f}}^{(3)}$ and piezomass $\hat{\mathbf{g}}^{(3)}$ coefficients contain 27 components. Every tensor of the second order $\hat{\boldsymbol{\alpha}}^{\rho}$, $\hat{\boldsymbol{\beta}}$, $\hat{\boldsymbol{\chi}}^{E}$, $\hat{\boldsymbol{\chi}}^{m}$, $\hat{\boldsymbol{\chi}}^{Em}$ has nine components and each of the vectors $\boldsymbol{\beta}^{\mu}$, $\boldsymbol{\beta}^{E}$, $\boldsymbol{\gamma}^{E}$, $\boldsymbol{\gamma}^{\rho}$ has three components. It is easy to show that the number of tensor components $\hat{\mathbf{C}}^{(4)}$, $\hat{\mathbf{f}}^{(3)}$, $\hat{\mathbf{g}}^{(3)}$, $\hat{\boldsymbol{\alpha}}^{\rho}$, and $\hat{\boldsymbol{\beta}}$ decreases if we take the symmetry of stress and strain tensors into account. Indeed, from the condition of symmetry of a stress tensor $\hat{\boldsymbol{\sigma}}_*$, we obtain the equalities: $C_{ijkl} = C_{jikl}$, $C_{kij} = C_{kji}$, $g_{kij} = g_{kji}$, $\beta_{ij} = \beta_{ji}$, and $\alpha_{ij}^{\rho} = \alpha_{ji}^{\rho}$. The symmetry properties of the strain tensor ($e_{ij} = e_{ji}$) enforce the following additional symmetry properties: $C_{ijkl} = C_{jikl}$. Moreover, since the Gibbs equation is written for the total differential of function f, the following conditions should be satisfied:

$$\frac{\partial^2 f}{\partial e_{ij}\partial e_{kl}} = \frac{\partial^2 f}{\partial e_{kl}\partial e_{ij}}, \quad \frac{\partial^2 f}{\partial E_i \partial E_j} = \frac{\partial^2 f}{\partial E_j \partial E_i},$$

$$\frac{\partial^2 f}{\partial(\nabla_i \mu_\pi')\partial(\nabla_j \mu_\pi')} = \frac{\partial^2 f}{\partial(\nabla_j \mu_\pi')\partial(\nabla_i \mu_\pi')},$$

$$\frac{\partial^2 f}{\partial E_i \partial(\nabla_j \mu_\pi')} = \frac{\partial^2 f}{\partial(\nabla_j \mu_\pi')\partial E_i}.$$

Here, $\nabla_i \mu_\pi'$ and E_i, $i = \overline{1,3}$ are components of the vector of the spatial gradient of a modified chemical potential μ_π' and of the electric field vector $\mathbf{E}$.

Using the constitutive equations (2.68), we get

$$\frac{\partial \sigma_{*ij}}{\partial e_{kl}} = \frac{\partial \sigma_{*kl}}{\partial e_{ij}}, \quad \frac{\partial \pi_{ei}}{\partial E_j} = \frac{\partial \pi_{ej}}{\partial E_i},$$

$$\frac{\partial \pi_{mi}}{\partial (\nabla_j \mu'_\pi)} = \frac{\partial \pi_{mj}}{\partial (\nabla_i \mu'_\pi)}, \quad -\frac{\partial \pi_{ei}}{\partial (\nabla_j \mu'_\pi)} = \frac{\partial \pi_{mj}}{\partial E_i}.$$

From these formulae, it follows that the tensors $\hat{\mathbf{C}}^{(4)}$, $\hat{\boldsymbol{\chi}}^E$, $\hat{\boldsymbol{\chi}}^m$, and $\hat{\boldsymbol{\chi}}^{Em}$ obey the following symmetry properties: $C_{ijkl} = C_{klij}$, $\chi_{ij}^E = \chi_{ji}^E$, $\chi_{ij}^m = \chi_{ji}^m$, and $\chi_{ij}^{Em} = \chi_{ji}^{Em}$. Thus, the number of independent components of the elastic moduli tensor $\hat{\mathbf{C}}^{(4)}$ decreased to 21. Each of the third-order tensors, $\hat{\mathbf{f}}^{(3)}$ and $\hat{\mathbf{g}}^{(3)}$, has 18 independent components (they are symmetric about the permutation of the second and third subscripts) and each of the tensors $\hat{\boldsymbol{\alpha}}^\rho$, $\hat{\boldsymbol{\beta}}$, $\hat{\boldsymbol{\chi}}^E$, $\hat{\boldsymbol{\chi}}^m$, and $\hat{\boldsymbol{\chi}}^{Em}$ has six independent components.

Note that the number of material constants is much reduced for isotropic bodies. In the next subsection, an explicit form of linear constitutive relations for such materials will be presented.

2.10.3 Linear Constitutive Equations for Isotropic Media

Let the material of the body in the reference state be homogeneous and centrosymmetric (isotropic). For such materials, all components of the third-order tensors vanish. The components of the second-order tensors $\hat{\boldsymbol{\alpha}}^\rho$, $\hat{\boldsymbol{\beta}}$, $\hat{\boldsymbol{\chi}}^E$, $\hat{\boldsymbol{\chi}}^m$, and $\hat{\boldsymbol{\chi}}^{Em}$ are given by the formulae: $\alpha_{ij} = \gamma_\rho \delta_{ij}$, $\beta_{ij} = \gamma_T \delta_{ij}$, $\chi_{ij}^E = \chi_E \delta_{ij}$, $\chi_{ij}^{Em} = \chi_{Em} \delta_{ij}$, and $\chi_{ij}^m = \chi_m \delta_{ij}$. For the fourth-order elastic moduli tensor $\mathbf{C}^{(4)}$, we have (Nowacki 1970): $C_{ijkl} = \underline{\lambda} \delta_{ij} \delta_{kl} + \underline{\mu}(\delta_{jk}\delta_{il} + \delta_{ik}\delta_{jl})$, $\underline{\mu} \equiv (C_{1111} - C_{1122})/2$, $\underline{\lambda} \equiv C_{1122}$, where $\underline{\lambda}$ and $\underline{\mu}$ are the Lamé moduli, and δ_{ij} denotes the Kronecker delta.

Thus, the density of free energy (2.69) for a linear isotropic media is as follows:

$$f = f_0 - s_0 \theta + \mu'_{\pi 0} \rho_m - \frac{C_V}{2T_0} \theta^2 + \frac{1}{2\rho_0}\left(K - \frac{2}{3}G\right) I_1^2$$

$$+\frac{G}{\rho_0}I_2+\frac{d_\rho}{2}\rho_m^2-\frac{K\alpha_T}{\rho_0}I_1\,\theta-\frac{K\alpha_\rho}{\rho_0}I_1\,\rho_m-\beta_{T\rho}\,\rho_m\theta$$

$$-\frac{\chi_m}{2}\nabla\mu'_\pi\cdot\nabla\mu'_\pi-\frac{\chi_E}{2}\mathbf{E}\cdot\mathbf{E}+\chi_{Em}\mathbf{E}\cdot\nabla\mu'_\pi. \qquad (2.75)$$

Here, $I_1=\hat{\boldsymbol{e}}:\hat{\mathbf{I}}\equiv e$ and $I_2=\hat{\boldsymbol{e}}:\hat{\boldsymbol{e}}$ are the first and second invariants of strain tensor, respectively; $K=\underline{\lambda}+2\underline{\mu}/3$ is the modulus of volume elasticity at a constant temperature and specific density of the induced mass; $G=\underline{\mu}$ is the shear modulus; $K\alpha_T = \gamma_T$, where α_T is the temperature coefficient of volume dilatation at uniform specific density of induced mass; $K\alpha_\rho = \gamma_\rho$, where α_ρ is the coefficient of volume dilatation caused by the local mass displacement at uniform temperature; χ_E is the dielectric susceptibility; χ_m and χ_{Em} are the coefficients that characterize the local mass displacement and body polarization due to the gradient of potential μ'_π, respectively.

Taking the expression (2.75) for the free energy into account and using the formulae (2.68), for an isotropic material, we obtain the following linear constitutive equations:

$$\hat{\boldsymbol{\sigma}}_*=2G\hat{e}+\left[\left(K-\frac{2}{3}G\right)e-K(\alpha_T\theta+\alpha_\rho\rho_m)\right]\hat{\mathbf{I}}, \qquad (2.76)$$

$$s=s_0+\frac{C_V}{T_0}\theta+\frac{1}{\rho_0}K\alpha_T\,e+\beta_{T\rho}\,\rho_m, \qquad (2.77)$$

$$\mu'_\pi=\mu'_{\pi0}+d_\rho\rho_m-\beta_{T\rho}\theta-\frac{1}{\rho_0}K\alpha_\rho\,e, \qquad (2.78)$$

$$\boldsymbol{\pi}_e=\chi_E\mathbf{E}-\chi_{Em}\nabla\mu'_\pi, \qquad (2.79)$$

$$\boldsymbol{\pi}_m=-\chi_m\nabla\mu'_\pi+\chi_{Em}\mathbf{E}. \qquad (2.80)$$

The analysis of the formulae (2.76) and (2.77) shows that the stress tensor and entropy are related not only to the strain $\hat{\boldsymbol{e}}$ and perturbation of temperature θ, as it follows from the classical theory of dielectrics (Toupin 1956; Voigt 1910), but also to the specific density of induced mass. Note also that according to the obtained constitutive equations, the polarization and the local mass displacement are coupled: The body polarization is caused not only by the electric field but also by the gradient of the modified chemical potential μ'_π. In those areas where the value $\nabla\mu'_\pi$ is nonzero, the body will be polarized even in the absence of an external electric field. Thus, Eq. (2.79) reveals the fact that piezoelectricity appears even in isotropic materials if we consider the process of local mass displacement.

It is worth noting that there is no direct coupling between the electric field $\mathbf{E}$ and the modified stress tensor $\hat{\boldsymbol{\sigma}}_*$. The couplings are only between the electric field $\mathbf{E}$ and the vector $\boldsymbol{\pi}_m$ of local mass displacement and the stress tensor $\hat{\boldsymbol{\sigma}}_*$ and the density of induced mass $\rho_m = \nabla \cdot (\rho\boldsymbol{\pi}_m)/\rho$.

The above constitutive relations are gradient-type and, in comparison to the classical theory of dielectrics, contain new material characteristics: α_ρ (a dimensionless scalar quantity), γ_ρ [N/m^2], d_ρ [J/kg], $\beta_{T\rho}$ [J/kg·grad], χ_m [s^2], and χ_{Em} [m^2/V].

It is known that in the state of thermodynamic equilibrium, the free energy $F = U - Ts$ of a body acquires a minimal value (Gyarmati 1970). Thermodynamics requires the energy to be positive defined. In our case, the free energy is a function of five arguments, namely, $\hat{\mathbf{e}}, T, \boldsymbol{\pi}_e, \boldsymbol{\pi}_m$, and ρ_m. In the absence of perturbations of deformation, temperature, polarization $\boldsymbol{\pi}_e$, and vector of local mass displacement $\boldsymbol{\pi}_m$, we obtain

$$\mathsf{F}(t) = \int_{(V)} \left[F(\mathbf{r},t) - F_0\right] d\mathbf{r} = d_\rho \int_{(V)} \rho_m^2(\mathbf{r},t)\, d\mathbf{r} \geq 0$$

using the formula (2.39). It follows from the positive-definiteness of the free energy that coefficient d_ρ is positive defined. Similarly, we can get that coefficient χ_m should be positive defined ($\chi_m > 0$) as well.

From the relation (2.78), we can determine the specific density of an induced mass as a function of the first invariant of the strain tensor e, and perturbations of both the potential $\tilde{\mu}'_\pi = \mu'_\pi - \mu'_{\pi 0}$ and the temperature θ, that is,

$$\rho_m = \frac{1}{d_\rho}\tilde{\mu}'_\pi + \frac{\beta_{T\rho}}{d_\rho}\theta + \frac{K\alpha_\rho}{\rho_0 d_\rho}e\,. \tag{2.81}$$

With the substitution of relation (2.81), Eqs. (2.76) and (2.77) can be written as follows:

$$\hat{\boldsymbol{\sigma}}_* = 2G\hat{\mathbf{e}} + \left[\left(\bar{K} - \frac{2}{3}G\right)e - K\bar{\alpha}_T\theta - \frac{K\alpha_\rho}{d_\rho}\tilde{\mu}'_\pi\right]\hat{\mathbf{I}}\,, \tag{2.82}$$

$$s = s_0 + \frac{\bar{C}_V}{T_0}\theta + \frac{K\bar{\alpha}_T}{\rho_0}e + \frac{\beta_{T\rho}}{d_\rho}\tilde{\mu}'_\pi . \quad (2.83)$$

Here,

$$\bar{K} = K - \frac{K^2\alpha_\rho^2}{\rho_0 d_\rho}, \ \bar{\alpha}_T = \alpha_T + \beta_{T\rho}\frac{\alpha_\rho}{d_\rho}, \ \bar{C}_V = C_V + T_0\frac{\beta_{T\rho}^2}{d_\rho}. \quad (2.84)$$

The linear constitutive relations (2.82) and (2.83) determine the stress tensor $\hat{\boldsymbol{\sigma}}_*$ and the entropy s as functions of deformation $\hat{\boldsymbol{e}}$, as well as the perturbations of temperature θ and modified chemical potential $\tilde{\mu}'_\pi$.

2.10.4 Kinetic Relations

Along with the thermodynamic parameters of state, we should also introduce the thermodynamic parameters of the process, namely, thermodynamic fluxes $\mathbf{j}_k$ and thermodynamic forces $\mathbf{X}_k$. The above fluxes and forces characterize an intensity of thermodynamic processes and the causes of their occurrence. To define these quantities, we represent the entropy production (2.66) as follows:

$$\eta_s = \frac{1}{T}(\mathbf{j}_1 \cdot \mathbf{X}_1 + \mathbf{j}_2 \cdot \mathbf{X}_2). \quad (2.85)$$

Here,

$$\mathbf{j}_1 = \mathbf{J}_q, \quad \mathbf{X}_1 = -\frac{\nabla T}{T}, \quad \mathbf{j}_2 = \mathbf{J}_{e*}, \quad \mathbf{X}_2 = \mathbf{E}_* \quad (2.86)$$

are the thermodynamic fluxes and forces that are mutually conjugated process parameters.

Since the thermodynamic fluxes are emanated by thermodynamic forces, we define the fluxes as functions of thermodynamic forces, that is,

$$\mathbf{j}_i = \mathbf{j}_i(\mathbf{X}_1, \mathbf{X}_2; A), \quad i = 1, 2. \quad (2.87)$$

The parameter A in Eq. (2.87) indicates the dependence of the fluxes on the characteristics of the local state (i.e., the dependence on the constitutive parameters). Note that the functions $\mathbf{j}_i = \mathbf{j}_i(\mathbf{X}_1, \mathbf{X}_2; A)$, $i = 1, 2$, should satisfy the following relations (De Groot and Mazur 1962):

$$\mathbf{j}_i(0, 0; A) = 0, \quad i = 1,2, \quad \mathbf{j}_1 \cdot \mathbf{X}_1 + \mathbf{j}_2 \cdot \mathbf{X}_2 \geq 0.$$

In general, the relations (2.87) are nonlinear. In a linear approximation, for anisotropic materials, from the relations (2.85) and (2.86), we obtain the following expressions for fluxes:

$$\mathbf{j}_1 = \hat{\mathbf{L}}_{11} \cdot \mathbf{X}_1 + \hat{\mathbf{L}}_{12} \cdot \mathbf{X}_2, \ \mathbf{j}_2 = \hat{\mathbf{L}}_{21} \cdot \mathbf{X}_1 + \hat{\mathbf{L}}_{22} \cdot \mathbf{X}_2, \tag{2.88}$$

where $\hat{\mathbf{L}}_{ij}$ $(i, j = 1, 2)$ are the kinetic coefficients (the second-order tensor quantities, which in general, can depend on constitutive parameters).

For linear isotropic materials, the relations (2.88) can be expressed as:

$$\mathbf{j}_1 = L_{11}\mathbf{X}_1 + L_{12}\mathbf{X}_2, \quad \mathbf{j}_2 = L_{21}\mathbf{X}_1 + L_{22}\mathbf{X}_2. \tag{2.89}$$

Note that the coefficients L_{11}, L_{12}, L_{22}, and L_{21} are considered here to be constants.

In view of the Onsager theorem (Gyarmati 1970), we obtain:

$$\frac{\partial \mathbf{j}_1}{\partial \mathbf{X}_2} = \frac{\partial \mathbf{j}_2}{\partial \mathbf{X}_1}.$$

From Onsager reciprocity relations, it follows that:

$$L_{12} = L_{21}. \tag{2.90}$$

For a linear approximation, using the notation (2.86), one can write the kinetic equation (2.89) as follows:

$$\mathbf{J}_q = -\lambda \nabla T + \pi_t \mathbf{J}_e, \quad \mathbf{J}_e = \sigma_e \mathbf{E} - \eta \nabla T, \tag{2.91}$$

where $\lambda = \lambda'(1 - L_{12}^2 / L_{22}L_{11})$ is the modified coefficient of thermal conductivity; $\lambda' = L_{11}/T_0$ and $\sigma_e = L_{22}$ are thermal and electric conductivity, respectively; and the coefficients $\pi_t = L_{12}/L_{22}$ and $\eta = L_{12}/T_0$ characterize thermoelectric phenomena.

Note that in Eqs. (2.91), we take into account that $\mathbf{E}_* = \mathbf{E}$ and $\mathbf{J}_{e*} = \mathbf{J}_e$ for the linear approximation.

In view of the relation (2.90), we get the following formula: $\pi_t = \eta T_0/\sigma_e$. The second law of thermodynamics implies $\eta_s \geq 0$ (De Groot and Mazur 1962), which means that the coefficients σ_e, λ, η, and π_t should be positive defined, that is, $\sigma_e > 0$, $\lambda > 0$, $\eta > 0$, and $\pi_t > 0$.

The first relation of the set (2.91) can be written as follows:

$$\mathbf{J}_q = -(\lambda + \pi_t \eta) \nabla T + \pi_t \sigma_e \mathbf{E}.$$

This relation generalizes the well-known Fourier law of heat conduction (Kovalenko 1969) for the case of nonideal dielectric media. For ideal dielectrics, which are characterized by the absence of free charges, the relations (2.91) acquire the form:

$$\mathbf{J}_q = -\lambda \nabla T, \quad \mathbf{J}_e = 0. \tag{2.92}$$

2.11 Linear Set of Governing Equations in Terms of Displacement

2.11.1 Governing Equations

The balance of momentum (2.54), the Maxwell equations (2.20), (2.21), (2.25), and (2.26), the conservation laws of mass (2.55), entropy (2.11), electric charge (2.22), induced mass (2.41), and induced charge (2.31) together with the expression for entropy production (2.66), the constitutive relations (2.17), (2.24), (2.68), (2.87), the strain-displacement relations (2.8), the formulae (2.7), (2.18), (2.37), (2.48), (2.51), and (2.52), as well as the corresponding representation for the free energy density form a complete set of equations of the local gradient electrothermoelasticity of polarized non-ferromagnetic solid bodies. In general, this set is nonlinear. The number of equations in this set can be reduced by substituting the constitutive and geometric relations, as well as the expression for the entropy production into balance laws and Maxwell equations.

Now, we obtain a final form of the governing equations for linear isotropic medium.

Note that the substitution of the formulae (2.48) into relations (2.40) yields the following nonlinear equation:

$$\rho_m = -\frac{1}{\rho} \nabla \rho \cdot \pi_m - \nabla \cdot \boldsymbol{\pi}_m .$$

Hence, in a linear approximation, we obtain:

$$\rho_m = -\nabla \cdot \boldsymbol{\pi}_m . \tag{2.93}$$

We take the vector of displacement **u**, the electric field vector **E**, the vector of magnetic induction **B**, the temperature perturbation

$\theta = T - T_0$, and the density of induced mass ρ_m as the key functions. Substituting the formulae (2.37), (2.48), (2.93), the geometric relations (2.7), (2.8), the expressions for fluxes (2.91) and the constitutive equations (2.17), (2.24), (2.76)–(2.80) into the balance equations (2.11), (2.41), (2.54) and into the Maxwell equations (2.20), (2.21), (2.25), (2.26), we obtain the final set of linear equations to determine the functions **u**, θ, ρ_m, **E**, and **B**:

$$\rho_0 \frac{\partial^2 \mathbf{u}}{\partial t^2} = \left(K + \frac{1}{3} G \right) \nabla(\nabla \cdot \mathbf{u}) + G\Delta \mathbf{u} - K\alpha_T \nabla\theta - K\alpha_\rho \nabla\rho_m + \rho_0 \mathbf{F}, \tag{2.94}$$

$$\rho_0 C_V \frac{\partial \theta}{\partial t} + T_0 K \alpha_T \frac{\partial(\nabla \cdot \mathbf{u})}{\partial t} + \rho_0 T_0 \beta_{T\rho} \frac{\partial \rho_m}{\partial t} = (\lambda + \pi_t \eta)\Delta\theta - \sigma_e \pi_t \nabla \cdot \mathbf{E} + \rho_0 \Re, \tag{2.95}$$

$$\Delta\rho_m - \frac{1}{d_\rho \chi_m}\rho_m = \frac{1}{d_\rho}\left[K\frac{\alpha_\rho}{\rho_0}\Delta(\nabla \cdot \mathbf{u}) + \beta_{T\rho}\Delta\theta + \frac{\chi_{Em}}{\chi_m}\nabla \cdot \mathbf{E} \right], \tag{2.96}$$

$$\nabla \times \mathbf{E} = -\frac{\partial \mathbf{B}}{\partial t}, \quad \nabla \cdot \mathbf{B} = 0, \tag{2.97}$$

$$\nabla \times \mathbf{B} = \varepsilon\mu_0 \frac{\partial \mathbf{E}}{\partial t} + \mu_0(\sigma_e \mathbf{E} - \eta\nabla\theta) + \varepsilon\mu_0\kappa_E \left[K\frac{\alpha_\rho}{\rho_0}\frac{\partial \nabla(\nabla \cdot \mathbf{u})}{\partial t} + \beta_{T\rho}\frac{\partial(\nabla\theta)}{\partial t} - d_\rho \frac{\partial(\nabla\rho_m)}{\partial t} \right], \tag{2.98}$$

$$\nabla \cdot \mathbf{E} + \kappa_E \left[K\frac{\alpha_\rho}{\rho_0}\Delta(\nabla \cdot \mathbf{u}) + \beta_{T\rho}\Delta\theta - d_\rho \Delta\rho_m \right] = \frac{\rho_e}{\varepsilon}. \tag{2.99}$$

Here, ρ_0 is the mass density at the initial time, and coefficients ε and κ_E are defined as follows:

$$\varepsilon = \varepsilon_0 + \rho_0 \chi_E, \quad \kappa_E = \frac{\rho_0 \chi_{Em}}{\varepsilon}. \tag{2.100}$$

Note that the ponderomotive force is absent in the momentum equation (2.94), as is the Joule heat in the heat conduction equation (2.95). This is because in the chosen reference state, the ponderomotive force and the Joule heat are nonlinear functions of the perturbation of fields. We can see that compared to the classical theory of electrothermoelastic dielectrics, as a result of accounting for the process of the local mass displacement, an additional equation (2.96) appears in the set of governing equations. Another consequence of the accounting for the local mass displacement is a modification of the equations of motion (2.94) and heat conduction (2.95), as well as the electrodynamics equations (2.98), (2.99), which now also contain the addends related to the local mass displacement. The consideration of the impact of the gradient of a modified chemical potential in the motion equation (2.94) may be quantitatively interpreted as the emergence of an additional mass force within the body, proportional to $\nabla\rho_m$, while in the equation of heat conduction, it may be interpreted as the emergence of a source of heat in the body of the power $-\rho_0 T_0 \beta_{T\rho} \dfrac{\partial \rho_m}{\partial t}$.

It is worth noting that the mechanical (2.94), thermal (2.95), and electromagnetic equations (2.97), (2.98) are dynamic, while the equation (2.96) is static. Thus, the electromagnetic and thermomechanical fields is not dynamically coupled to the local mass displacement. Indeed, Eq. (2.96) is stationary because we consider here the local mass displacement as a reversible process. The solution to this equation depends on the sign of the coefficient $(d_\rho \chi_m)^{-1}$. Since the coefficients χ_m and d_ρ are positive valued, $(d_\rho \chi_m)^{-1} > 0$. We denote

$$\lambda_\mu = \left|\sqrt{d_\rho \chi_m}\right|^{-1}. \tag{2.101}$$

Here, λ_μ is the material length scale parameter accounting for the nonlocal effects in continuum. The appearance of such a constant is related to the consideration of the local mass displacement in the material model. The parameter $l_* = 1/\lambda_\mu$ is a material constant in the dimension of length and is a characteristic length for near-surface phenomena. The emergence of such a constant is typical of the gradient-type theories, while an intrinsic length scale is absent from classical theories.

Let us now write a governing set of equations of local gradient electrothermoelasticity, taking the perturbation of the modified chemical potential $\tilde{\mu}'_\pi = \mu'_\pi - \mu'_{\pi 0}$ as a key function instead of the density of the induced mass

$$\rho_0 \frac{\partial^2 \mathbf{u}}{\partial t^2} = \left(\bar{K} + \frac{1}{3} G \right) \boldsymbol{\nabla}(\boldsymbol{\nabla} \cdot \mathbf{u}) + G \Delta \mathbf{u} - K \bar{\alpha}_T \boldsymbol{\nabla} \theta - K \frac{\alpha_\rho}{d_\rho} \boldsymbol{\nabla} \tilde{\mu}'_\pi + \rho_0 \mathbf{F}, \tag{2.102}$$

$$\begin{aligned} \rho_0 \bar{C}_V \frac{\partial \theta}{\partial t} + K T_0 \bar{\alpha}_T \frac{\partial (\boldsymbol{\nabla} \cdot \mathbf{u})}{\partial t} + \rho_0 T_0 \frac{\beta_{T\rho}}{d_\rho} \frac{\partial \tilde{\mu}'_\pi}{\partial t} \\ = (\lambda + \pi_t \eta) \Delta \theta - \sigma_e \pi_t \, \boldsymbol{\nabla} \cdot \mathbf{E} + \rho_0 \Re, \end{aligned} \tag{2.103}$$

$$\Delta \tilde{\mu}'_\pi - \lambda_\mu^2 \tilde{\mu}'_\pi = \lambda_\mu^2 \left(K \frac{\alpha_\rho}{\rho_0} \boldsymbol{\nabla} \cdot \mathbf{u} + \beta_{T\rho} \theta \right) + \frac{\chi_{Em}}{\chi_m} \boldsymbol{\nabla} \cdot \mathbf{E}, \tag{2.104}$$

$$\boldsymbol{\nabla} \times \mathbf{E} = -\frac{\partial \mathbf{B}}{\partial t}, \quad \boldsymbol{\nabla} \cdot \mathbf{B} = 0, \tag{2.105}$$

$$\boldsymbol{\nabla} \times \mathbf{B} = \mu_0 \sigma_e \mathbf{E} - \mu_0 \eta \boldsymbol{\nabla} \theta + \varepsilon \mu_0 \frac{\partial}{\partial t} (\mathbf{E} - \kappa_E \boldsymbol{\nabla} \tilde{\mu}'_\pi), \tag{2.106}$$

$$\boldsymbol{\nabla} \cdot \mathbf{E} - \kappa_E \Delta \tilde{\mu}'_\pi = \frac{\rho_e}{\varepsilon}. \tag{2.107}$$

The obtained set of governing equations (2.102)–(2.107) can be easily divided into two subsets, which can be then solved consecutively. Indeed, we can eliminate the electric field from the second and third equations of this set. Based on Eq. (2.107), one can write:

$$\boldsymbol{\nabla} \cdot \mathbf{E} = \kappa_E \Delta \tilde{\mu}'_\pi + \frac{\rho_e}{\varepsilon}. \tag{2.108}$$

Now, using the relation (2.108), Eqs. (2.103) and (2.104) can be presented as follows:

$$\begin{aligned} \rho_0 \bar{C}_V \frac{\partial \theta}{\partial t} + K T_0 \bar{\alpha}_T \frac{\partial (\boldsymbol{\nabla} \cdot \mathbf{u})}{\partial t} + \rho_0 T_0 \frac{\beta_{T\rho}}{d_\rho} \frac{\partial \tilde{\mu}'_\pi}{\partial t} \\ = (\lambda + \pi_t \eta) \Delta \theta - \sigma_e \pi_t \kappa_E \left(\Delta \tilde{\mu}'_\pi + \frac{\rho_e}{\rho_0 \chi_{Em}} \right) + \rho_0 \Re, \end{aligned} \tag{2.109}$$

$$\Delta \tilde{\mu}'_\pi - \lambda_{\mu E}^2 \tilde{\mu}'_\pi = \lambda_{\mu E}^2 \left(K \frac{\alpha_\rho}{\rho_0} \boldsymbol{\nabla} \cdot \mathbf{u} + \beta_{T\rho} \theta \right) + \lambda_{\mu E}^2 \chi_{Em} d_\rho \frac{\rho_e}{\varepsilon}. \tag{2.110}$$

Here,

$$\lambda_{\mu E}^2 = \frac{\lambda_\mu^2}{1 - \kappa_E \chi_{Em} / \chi_m}. \tag{2.111}$$

Thus, the formulated problem can be solved consecutively. In order to determine the functions **u**, θ, and $\tilde{\mu}'_\pi$, we use the related set of equations (2.102), (2.109), and (2.110). Once the functions **u**, θ, and $\tilde{\mu}'_\pi$ are found, they can be substituted into Eqs. (2.105)–(2.107) to determine the vectors of electromagnetic field **E** and **B**.

In a similar manner, using Eq. (2.99), one can obtain:

$$\rho_0 C_V \frac{\partial \theta}{\partial t} + T_0 K \alpha_T \frac{\partial(\nabla \cdot \mathbf{u})}{\partial t} + \rho_0 T_0 \beta_{T\rho} \frac{\partial \rho_m}{\partial t} = (\lambda + \pi_t \eta + \pi_t \sigma_e \kappa_E \beta_{T\rho}) \Delta\theta$$

$$-\sigma_e \pi_t \kappa_E \left[d_\rho \Delta \rho_m - K \frac{\alpha_\rho}{\rho_0} \Delta(\nabla \cdot \mathbf{u}) \right] - \sigma_e \pi_t \frac{\rho_e}{\varepsilon} + \rho_0 \Re, \qquad (2.112)$$

$$\Delta \rho_m - \lambda_{\mu E}^2 \rho_m = \lambda_\mu^2 \chi_m \left[K \frac{\alpha_\rho}{\rho_0} \Delta(\nabla \cdot \mathbf{u}) + \beta_{T\rho} \Delta\theta \right] + \lambda_{\mu E}^2 \chi_{Em} \frac{\rho_e}{\varepsilon}$$

$$(2.113)$$

from Eqs. (2.95) and (2.96).

The functions **u**, θ, and ρ_m, are found from Eqs. (2.94), (2.112), and (2.113). Then, the vectors of the electromagnetic field are derived from Eqs. (2.97)–(2.99), the right-hand sides of which contain the known functions **u**, θ, and ρ_m.

2.11.2 Governing Equations for Ideal Dielectrics

In the case of ideal dielectrics, the aforementioned governing sets of equations significantly simplify.

For such materials, the set of Eqs. (2.102), (2.105)–(2.107), (2.109), and (2.110) can be reduced to

$$\rho_0 \frac{\partial^2 \mathbf{u}}{\partial t^2} = \left(\bar{K} + \frac{1}{3} G \right) \nabla(\nabla \cdot \mathbf{u}) + G \Delta \mathbf{u} - K \bar{\alpha}_T \nabla\theta - K \frac{\alpha_\rho}{d_\rho} \nabla \tilde{\mu}'_\pi + \rho_0 \mathbf{F},$$

$$(2.114)$$

$$\rho_0 \bar{C}_V \frac{\partial \theta}{\partial t} + K T_0 \bar{\alpha}_T \frac{\partial(\nabla \cdot \mathbf{u})}{\partial t} + \rho_0 T_0 \frac{\beta_{T\rho}}{d_\rho} \frac{\partial \tilde{\mu}'_\pi}{\partial t} = \lambda \Delta\theta + \rho_0 \Re, \qquad (2.115)$$

$$\Delta \tilde{\mu}'_\pi - \lambda_{\mu E}^2 \tilde{\mu}'_\pi = \lambda_{\mu E}^2 \left(K \frac{\alpha_\rho}{\rho_0} \nabla \cdot \mathbf{u} + \beta_{T\rho} \theta \right), \qquad (2.116)$$

$$\nabla\times\mathbf{E}=-\frac{\partial\mathbf{B}}{\partial t},\quad \nabla\cdot\mathbf{B}=0, \tag{2.117}$$

$$\nabla\times\mathbf{B}=\varepsilon\mu_0\frac{\partial}{\partial t}(\mathbf{E}-\kappa_E\nabla\tilde{\mu}'_\pi), \tag{2.118}$$

$$\nabla\cdot\mathbf{E}-\kappa_E\Delta\tilde{\mu}'_\pi=0. \tag{2.119}$$

Correspondently, the set of differential relations (2.94), (2.97)–(2.99), (2.112), and (2.113) can be given as follows:

$$\rho_0\frac{\partial^2\mathbf{u}}{\partial t^2}=\left(K+\frac{1}{3}G\right)\nabla(\nabla\cdot\mathbf{u})+G\Delta\mathbf{u}-K\alpha_T\nabla\theta-K\alpha_\rho\nabla\rho_m+\rho_0\mathbf{F}, \tag{2.120}$$

$$\rho_0C_V\frac{\partial\theta}{\partial t}+T_0K\alpha_T\frac{\partial(\nabla\cdot\mathbf{u})}{\partial t}+\rho_0T_0\beta_{T\rho}\frac{\partial\rho_m}{\partial t}=\lambda\Delta\theta+\rho_0\Re, \tag{2.121}$$

$$\Delta\rho_m-\lambda_{\mu E}^2\rho_m=\lambda_\mu^2\chi_m\left[K\frac{\alpha_\rho}{\rho_0}\Delta(\nabla\cdot\mathbf{u})+\beta_{T\rho}\Delta\theta\right], \tag{2.122}$$

$$\nabla\times\mathbf{E}=-\frac{\partial\mathbf{B}}{\partial t},\quad \nabla\cdot\mathbf{B}=0, \tag{2.123}$$

$$\nabla\times\mathbf{B}=\varepsilon\mu_0\left[\frac{\partial\mathbf{E}}{\partial t}+\kappa_E\left(K\frac{\alpha_\rho}{\rho_0}\frac{\partial(\nabla(\nabla\cdot\mathbf{u}))}{\partial t}+\beta_{T\rho}\frac{\partial(\nabla\theta)}{\partial t}-d_\rho\frac{\partial(\nabla\rho_m)}{\partial t}\right)\right], \tag{2.124}$$

$$\nabla\cdot\mathbf{E}+\kappa_E\left[K\frac{\alpha_\rho}{\rho_0}\Delta(\nabla\cdot\mathbf{u})+\beta_{T\rho}\Delta\theta-d_\rho\Delta\rho_m\right]=0. \tag{2.125}$$

Let us show that from the field equations (2.94)–(2.99) and (2.102)–(2.107), one can eliminate the functions ρ_m and $\tilde{\mu}'_\pi$, respectively. Indeed, for the ideal dielectrics, Eq. (2.107) yields

$$\Delta\tilde{\mu}'_\pi=\frac{1}{\kappa_E}\nabla\cdot\mathbf{E}.$$

Substituting this relation into Eq. (2.104), we get:

$$\tilde{\mu}'_\pi=-\frac{K\alpha_\rho}{\rho_0}\nabla\cdot\mathbf{u}-\beta_{T\rho}\theta+\frac{1}{\kappa_E\lambda_{\mu E}^2}\nabla\cdot\mathbf{E}. \tag{2.126}$$

The expression (2.126) defines the potential $\tilde{\mu}'_\pi$ through the functions **u**, θ, and **E**. Substituting the formula (2.126) into

Eqs. (2.102), (2.103), (2.106), and (2.107), we obtain the following set of equations in order to find the functions **u**, θ, **B**, and **E**:

$$\rho_0 \frac{\partial^2 \mathbf{u}}{\partial t^2} = \left(K + \frac{1}{3} G \right) \boldsymbol{\nabla}(\boldsymbol{\nabla} \cdot \mathbf{u}) + G\Delta \mathbf{u} - K\alpha_T \boldsymbol{\nabla}\theta$$

$$-K\alpha_\rho \left(\frac{\chi_m}{\kappa_E} - \chi_{Em} \right) \boldsymbol{\nabla}(\boldsymbol{\nabla} \cdot \mathbf{E}) + \rho_0 \mathbf{F}, \tag{2.127}$$

$$\lambda\Delta\theta + \rho_0 \Re = \rho_0 C_V \frac{\partial \theta}{\partial t} + T_0 K \alpha_T \frac{\partial(\boldsymbol{\nabla} \cdot \mathbf{u})}{\partial t} + \rho_0 T_0 \beta_{T\rho} \left(\frac{\chi_m}{\kappa_E} - \chi_{Em} \right) \frac{\partial(\boldsymbol{\nabla} \cdot \mathbf{E})}{\partial t}, \tag{2.128}$$

$$\boldsymbol{\nabla} \times \mathbf{E} = -\frac{\partial \mathbf{B}}{\partial t}, \quad \boldsymbol{\nabla} \cdot \mathbf{B} = 0, \tag{2.129}$$

$$\boldsymbol{\nabla} \times \mathbf{B} = \varepsilon\mu_0 \left\{ \frac{\partial \mathbf{E}}{\partial t} - \frac{1}{\lambda_{\mu E}^2} \frac{\partial \boldsymbol{\nabla}(\boldsymbol{\nabla} \cdot \mathbf{E})}{\partial t} + \kappa_E \left[K \frac{\alpha_\rho}{\rho_0} \frac{\partial \boldsymbol{\nabla}(\boldsymbol{\nabla} \cdot \mathbf{u})}{\partial t} + \beta_{T\rho} \frac{\partial(\boldsymbol{\nabla}\theta)}{\partial t} \right] \right\}, \tag{2.130}$$

$$\boldsymbol{\nabla} \cdot \mathbf{E} - \frac{1}{\lambda_{\mu E}^2} \Delta(\boldsymbol{\nabla} \cdot \mathbf{E}) + \kappa_E \left[K \frac{\alpha_\rho}{\rho_0} \Delta(\boldsymbol{\nabla} \cdot \mathbf{u}) + \beta_{T\rho} \Delta\theta \right] = 0. \tag{2.131}$$

Thus, to determine the functions **u**, θ, **E**, and **B**, we have coupled set of Eqs. (2.127)–(2.131). The functions $\tilde{\mu}'_\pi$, ρ_m, and $\boldsymbol{\pi}_m$, which characterize the local mass displacement, can be found from the expression (2.126) and the relations:

$$\boldsymbol{\pi}_m = \chi_{Em} \mathbf{E} - \frac{\chi_m}{\kappa_E \lambda_{\mu E}^2} \boldsymbol{\nabla}(\boldsymbol{\nabla} \cdot \mathbf{E}) + \chi_m K \frac{\alpha_\rho}{\rho_0} \boldsymbol{\nabla}(\boldsymbol{\nabla} \cdot \mathbf{u}) + \chi_m \beta_{T\rho} \boldsymbol{\nabla}\theta,$$

$$\rho_m = \left(\frac{\chi_m}{\kappa_E} - \chi_{Em} \right) \boldsymbol{\nabla} \cdot \mathbf{E}, \tag{2.132}$$

which follow from formulae (2.80), (2.93), (2.107), and (2.126).

We can see that elimination of the modified chemical potential from the field equations increases the order of differential equations (2.130) and (2.131). Compared to the corresponding equations of classical theory, Eqs. (2.130) and (2.131) contain the third-order mixed time-spatial and spatial derivatives of the key functions. Now, the equation of motion (2.127) contains the addends with electric field vector **E**. It means that the developed theory allows us to describe the flexoelectric effect in high-symmetry crystalline dielectrics, including isotropic materials.

2.11.3 Governing Equations for Stationary State

To determine the stationary state of thermoelastic ideal dielectrics, we have the following set of equations:

$$\left(\bar{K}+\frac{1}{3}G\right)\nabla(\nabla\cdot\mathbf{u})+G\Delta\mathbf{u}-K\frac{\alpha_\rho}{d_\rho}\nabla\tilde{\mu}'_\pi-K\bar{\alpha}_T\nabla\theta+\rho_0\mathbf{F}=0\,, \quad (2.133)$$

$$\lambda\Delta\theta+\rho_0\Re=0\,, \quad (2.134)$$

$$\Delta\tilde{\mu}'_\pi-\lambda_{\mu E}^2\tilde{\mu}'_\pi=\lambda_{\mu E}^2\left(K\frac{\alpha_\rho}{\rho_0}\nabla\cdot\mathbf{u}+\beta_{T\rho}\theta\right), \quad (2.135)$$

$$\nabla\times\mathbf{E}=0\,,\ \nabla\cdot\mathbf{E}=\kappa_E\Delta\tilde{\mu}'_\pi\,. \quad (2.136)$$

From Eqs. (2.120)–(2.125), we get:

$$\left(K+\frac{1}{3}G\right)\nabla(\nabla\cdot\mathbf{u})+G\Delta\mathbf{u}-K\alpha_T\nabla\theta-K\alpha_\rho\nabla\rho_m+\rho_0\mathbf{F}=0\,, \quad (2.137)$$

$$\lambda\Delta\theta+\rho_0\Re=0\,, \quad (2.138)$$

$$\Delta\rho_m-\lambda_{\mu E}^2\rho_m=\lambda_\mu^2\chi_m\left[K\frac{\alpha_\rho}{\rho_0}\Delta(\nabla\cdot\mathbf{u})+\beta_{T\rho}\Delta\theta\right], \quad (2.139)$$

$$\nabla\times\mathbf{E}=0,\quad \nabla\cdot\mathbf{E}+\kappa_E\left[K\frac{\alpha_\rho}{\rho_0}\Delta(\nabla\cdot\mathbf{u})+\beta_{T\rho}\Delta\theta-d_\rho\Delta\rho_m\right]=0\,. \quad (2.140)$$

We see that the heat equation is not related to the rest of the equations of the sets (2.133)–(2.136) and (2.137)–(2.140). Hence, the solution method for the corresponding problems of local gradient electrothermoelasticity becomes much simpler. First, from Eq. (2.134), we can find the temperature field. Afterward, from the coupled set of equations (2.133), (2.135) (or (2.137), (2.139)), we can determine the displacement vector **u** and the modified chemical potential $\tilde{\mu}'_\pi$ (or the induced mass density ρ_m). Finally, to find the electric field vector, we have the equation of electrostatics (2.136) (or (2.140)).

2.11.4 Isothermal Approximation

For isothermal approximation, the governing set of equations (2.120)–(2.125) can be simplified to

$$\rho_0 \frac{\partial^2 \mathbf{u}}{\partial t^2} = \left(K + \frac{G}{3}\right)\nabla(\nabla\cdot\mathbf{u}) + G\Delta\mathbf{u} - K\alpha_\rho \nabla\rho_m + \rho_0\mathbf{F}\,, \quad (2.141)$$

$$\Delta\rho_m - \lambda_{\mu E}^2\,\rho_m = K\frac{\alpha_\rho}{\rho_0 d_\rho}\Delta(\nabla\cdot\mathbf{u})\ , \quad (2.142)$$

$$\nabla\times\mathbf{E} = -\frac{\partial\mathbf{B}}{\partial t},\ \nabla\cdot\mathbf{B} = 0, \quad (2.143)$$

$$\nabla\times\mathbf{B} = \varepsilon\mu_0\left[\frac{\partial\mathbf{E}}{\partial t} + \kappa_E\left(K\frac{\alpha_\rho}{\rho_0}\frac{\partial(\nabla(\nabla\cdot\mathbf{u}))}{\partial t} - d_\rho\frac{\partial(\nabla\rho_m)}{\partial t}\right)\right], \quad (2.144)$$

$$\nabla\cdot\mathbf{E} + \kappa_E\left[K\frac{\alpha_\rho}{\rho_0}\Delta(\nabla\cdot\mathbf{u}) - d_\rho\Delta\rho_m\right] = 0\ . \quad (2.145)$$

Now, instead of the set (2.114)–(2.119), we have the following equations:

$$\rho_0\frac{\partial^2\mathbf{u}}{\partial t^2} = \left(\bar{K} + \frac{1}{3}G\right)\nabla(\nabla\cdot\mathbf{u}) + G\Delta\mathbf{u} - K\frac{\alpha_\rho}{d_\rho}\nabla\tilde{\mu}'_\pi + \rho_0\mathbf{F}\,, \quad (2.146)$$

$$\Delta\tilde{\mu}'_\pi - \lambda_{\mu E}^2\,\tilde{\mu}'_\pi = \lambda_{\mu E}^2\frac{K\alpha_\rho}{\rho_0}\nabla\cdot\mathbf{u}\,, \quad (2.147)$$

$$\nabla\times\mathbf{E} = -\frac{\partial\mathbf{B}}{\partial t}\,, \quad \nabla\cdot\mathbf{B} = 0\,, \quad (2.148)$$

$$\nabla\times\mathbf{B} = \varepsilon\mu_0\frac{\partial}{\partial t}(\mathbf{E} - \kappa_E\nabla\tilde{\mu}'_\pi)\,, \quad \nabla\cdot\mathbf{E} - \kappa_E\Delta\tilde{\mu}'_\pi = 0\,. \quad (2.149)$$

For stationary problems based on Eqs. (2.141)–(2.145), we get:

$$\left(K + \frac{G}{3}\right)\nabla(\nabla\cdot\mathbf{u}) + G\Delta\mathbf{u} - K\alpha_\rho\nabla\rho_m + \rho_0\mathbf{F} = 0\ , \quad (2.150)$$

$$\Delta\rho_m - \lambda_{\mu E}^2\,\rho_m = K\frac{\alpha_\rho}{\rho_0 d_\rho}\Delta(\nabla\cdot\mathbf{u})\,, \quad (2.151)$$

$$\nabla\times\mathbf{E} = 0, \quad \nabla\cdot\mathbf{E} + \kappa_E\left[K\frac{\alpha_\rho}{\rho_0}\Delta(\nabla\cdot\mathbf{u}) - d_\rho\Delta\rho_m\right] = 0\ . \quad (2.152)$$

Based on the set of Eqs. (2.146)–(2.149), we can write:

$$\left(\bar{K}+\frac{1}{3}G\right)\nabla(\nabla\cdot\mathbf{u})+G\Delta\mathbf{u}-K\frac{\alpha_\rho}{d_\rho}\nabla\tilde{\mu}'_\pi+\rho_0\mathbf{F}=0\,, \qquad (2.153)$$

$$\Delta\tilde{\mu}'_\pi-\lambda_{\mu E}^2\tilde{\mu}'_\pi=\lambda_{\mu E}^2 K\frac{\alpha_\rho}{\rho_0}\nabla\cdot\mathbf{u}\,, \qquad (2.154)$$

$$\nabla\times\mathbf{E}=0\,,\quad \nabla\cdot\mathbf{E}=\kappa_E\Delta\tilde{\mu}'_\pi\,. \qquad (2.155)$$

Applying the operator $\mathbf{L}=\Delta-\lambda_{\mu E}^2$ to Eq. (2.146) and taking Eq. (2.147) into account, we obtain the fourth-order partial differential equation of motion in terms of displacement components, which governs the deformation process within the framework of this gradient theory

$$l_*'^2\left[\left(\bar{K}+\frac{1}{3}G\right)\Delta\nabla(\nabla\cdot\mathbf{u})+G\Delta\Delta\mathbf{u}+\rho_0\Delta\mathbf{F}-\rho_0\frac{\partial^2\Delta\mathbf{u}}{\partial t^2}\right]$$
$$-\left[\left(K+\frac{1}{3}G\right)\nabla(\nabla\cdot\mathbf{u})+G\Delta\mathbf{u}+\rho_0\mathbf{F}-\rho_0\frac{\partial^2\mathbf{u}}{\partial t^2}\right]=0.$$

It is easy to verify that when the characteristic length $l_*'=\lambda_{\mu E}^{-1}\,l_*$ vanishes ($l_*'\to 0$), the obtained equation reduces to the classical equation of motion.

2.12 Linear Set of Governing Equations in Terms of Stress Tensor

2.12.1 Beltrami–Michell Equations. Governing Equations

If we take the stress tensor as a key function instead of the displacement vector, then the set of equations of the local gradient theory should be supplemented by the condition of strain compatibility by Saint-Venant (Nowacki 1970).

Let us derive the Beltrami–Michell equations corresponding to the linear theory of non-ferromagnetic dielectric bodies with a local mass displacement. To this end, we use the equilibrium equation, constitutive relations, and the Saint-Venant strain compatibility equations

$$\nabla \otimes (\nabla \otimes \hat{\boldsymbol{e}})^{\mathrm{T}} = 0 . \qquad (2.156)$$

In this section, we restrict ourselves to the stationary approximation.

Using the constitutive equation (2.76), we define the strain tensor as a function of the stress tensor, temperature perturbation, and the specific density of the induced mass

$$\hat{\boldsymbol{e}} = 2G'\hat{\boldsymbol{\sigma}}_* + \left[K'\sigma_* + \frac{1}{3}(\alpha_T\theta + \alpha_\rho\rho_m)\right]\hat{\mathbf{I}} . \qquad (2.157)$$

Here,

$$\sigma_* = \sigma_{*kk} = \hat{\boldsymbol{\sigma}}_* : \hat{\mathbf{I}} ,\ G' = \frac{1}{4G} ,\ K' = \frac{2G - 3K}{18KG} .$$

For the volumetric deformation $e = e_{kk}$, from the relation (2.157), one can obtain:

$$e = \frac{1}{3K}\sigma_* + \alpha_T\theta + \alpha_\rho\rho_m . \qquad (2.158)$$

Introducing the expression (2.158) into constitutive equations (2.77) and (2.78), we get

$$s = s_0 + \frac{\breve{C}_V}{T_0}\theta + \frac{\alpha_T}{3\rho_0}\sigma_* + \breve{\beta}_{T\rho}\rho_m , \qquad (2.159)$$

$$\mu'_\pi = \mu'_{\pi 0} + \breve{d}_\rho\rho_m - \frac{\alpha_\rho}{3\rho_0}\sigma_* - \breve{\beta}_{T\rho}\theta . \qquad (2.160)$$

The formulae (2.159) and (2.160) allow us to define the entropy s and potential μ'_π as functions of temperature field, stress tensor, and the density of induced mass. Here,

$$\breve{\beta}_{T\rho} = \beta_{T\rho} + K\frac{\alpha_T\alpha_\rho}{\rho_0} ,\ \breve{d}_\rho = d_\rho - K\frac{\alpha_\rho^2}{\rho_0} ,\ \breve{C}_V = C_V + KT_0\frac{\alpha_T^2}{\rho_0} .$$

Let us substitute the relation (2.157) into the Saint-Venant strain compatibility equation (2.156). As a result, using the linearized equilibrium equation

$$\nabla \cdot \hat{\boldsymbol{\sigma}}_* + \rho_0\mathbf{F}_* = 0 , \qquad (2.161)$$

after some algebra, we obtain the generalized Beltrami–Michell equation:

$$\Delta\hat{\boldsymbol{\sigma}}_* + \frac{2G'+K'}{2G'}\boldsymbol{\nabla}\otimes\boldsymbol{\nabla}(\hat{\boldsymbol{\sigma}}_*:\hat{\mathbf{I}}) + \frac{\alpha_T}{6G'}\left(\boldsymbol{\nabla}\otimes\boldsymbol{\nabla}\theta + \frac{G'}{G'+K'}\Delta\theta\hat{\mathbf{I}}\right)$$

$$+\frac{\alpha_\rho}{6G'}\left(\boldsymbol{\nabla}\otimes\boldsymbol{\nabla}\rho_m + \frac{G'}{G'+K'}\Delta\rho_m\hat{\mathbf{I}}\right)$$

$$+\rho_0\left[\boldsymbol{\nabla}\otimes\mathbf{F}_* + (\boldsymbol{\nabla}\otimes\mathbf{F}_*)^{\mathrm{T}} - \frac{G'}{2(G'+K')}(\boldsymbol{\nabla}\cdot\mathbf{F}_*)\hat{\mathbf{I}}\right] = 0. \qquad (2.162)$$

It should be noted that Eq. (2.162) can be used only within the framework of the linear theory of isotropic dielectrics.

The obtained equation (2.162), in comparison to the similar equation of the classical theory, additionally contains the terms proportional to the spatial derivatives of the specific density of induced mass. The relation (2.162) is a generalization of the Beltrami–Michell equation for a gradient-type model of deformable non-ferromagnetic dielectric media with local mass displacement.

Thus, the stationary problems, which are formulated in terms of the stress tensor, are reduced to solving three equations of equilibrium (2.161), the six Beltrami–Michell equations (2.162), as well as the following equations:

$$(\lambda + \pi_t\eta)\Delta\theta - \sigma_e\pi_t\boldsymbol{\nabla}\cdot\mathbf{E} + \rho_0\Re = 0, \qquad (2.163)$$

$$\Delta\rho_m - \breve{\lambda}_\mu^2\rho_m = \frac{1}{\breve{d}_\rho}\left(\frac{\alpha_\rho}{3\rho_0}\Delta\sigma_* + \breve{\beta}_{T\rho}\Delta\theta + \frac{\chi_{Em}}{\chi_m}\boldsymbol{\nabla}\cdot\mathbf{E}\right), \qquad (2.164)$$

$$\boldsymbol{\nabla}\times\mathbf{E} = 0, \qquad (2.165)$$

$$\boldsymbol{\nabla}\cdot\mathbf{E} + \kappa_E\left(\frac{\alpha_\rho}{3\rho_0}\Delta\sigma_* + \breve{\beta}_{T\rho}\Delta\theta - \breve{d}_\rho\Delta\rho_m\right) = \frac{\rho_e}{\varepsilon}. \qquad (2.166)$$

The last ones are obtained by means of substitution of the relations (2.24), (2.79), (2.80), (2.91), (2.159), and (2.160) into the Maxwell equations, formula (2.93), and the linearized equation of entropy balance. In Eq. (2.164), the following notation is used:

$$\breve{\lambda}_\mu^2 = \frac{1}{\chi_m\breve{d}_\rho}. \qquad (2.167)$$

2.12.2 Ideal Dielectrics

For ideal dielectrics, the coupled set of equations (2.161)–(2.166) can be solved step by step. First, from the heat equation

$$\lambda\Delta\theta + \rho_0\Re = 0, \tag{2.168}$$

we can find the temperature field.

Taking Eqs. (2.168) and (2.166) into account, we can eliminate the electric field and temperature from Eq. (2.164). As a result, we obtain the following equation to find the specific density of induced mass:

$$\left[1 + \frac{\alpha_\rho^2}{9\rho_0 \breve{d}_\rho (K'+G')}\right]\Delta\rho_m - \frac{\breve{\lambda}_\mu^2 \chi_m}{\chi_m - \kappa_E \chi_{Em}}\rho_m$$

$$= -\frac{\rho_0}{\lambda \breve{d}_\rho}\Re\left[\breve{\beta}_{T\rho} - \frac{\alpha_\rho \alpha_T}{9\rho_0 (K'+G')}\right] - \frac{\alpha_\rho G'}{3\breve{d}_\rho (K'+G')}\nabla \cdot \mathbf{F}_* \,. \tag{2.169}$$

From the Beltrami–Michell equation (2.162), we can write

$$\Delta\sigma_* = \frac{1}{3(K'+G')}\left(\rho_0 \frac{\alpha_T}{\lambda}\Re - \alpha_\rho \Delta\rho_m\right) - \rho_0 \frac{G'}{K'+G'}\nabla \cdot \mathbf{F}_* \,. \tag{2.170}$$

Using Eqs. (2.168)–(2.170), we may also eliminate from Eq. (2.166) the summands proportional to $\Delta\sigma_*$, $\Delta\theta$, and $\Delta\rho_m$. Thus, basing on relations (2.165) and (2.166), we obtain the following equations to find the electric field vector:

$$\nabla \times \mathbf{E} = 0, \quad \nabla \cdot \mathbf{E} = \frac{\kappa_E}{\chi_m - \kappa_E \chi_{Em}}\rho_m. \tag{2.171}$$

The right-hand side of the second equation of this set contains the function ρ_m, which can be found from Eq. (2.169). The equilibrium equation (2.161) and the Beltrami–Michell equation (2.162) allow us to find the component of stress tensor. Note that although Eqs. (2.161), (2.162), (2.168), (2.169), and (2.171) are not coupled, the fields of stress, temperature, and specific density of induced mass may be coupled through the boundary conditions. The last ones ensure the uniqueness of solution to the formulated problems of mathematical physics.

2.13 Potential Methods: Mechanical and Electromagnetic Interaction in Isotropic Dielectrics

2.13.1 Electromagnetic Potentials and Displacement–Potential Relations

Let us attempt to uncouple the foregoing field equations. To this end, we decompose the vectors of displacement **u** and mass force **F** as the sum of its potential φ_u, $\mathbf{\Phi}$ and solenoidal $\boldsymbol{\psi}_u$, $\boldsymbol{\Psi}$ components, as well as the electric field **E** and the magnetic induction **B** in terms of the scalar (electrical) potential φ_e and the vector potential **A**, namely (Nowacki 1983):

$$\mathbf{u} = \boldsymbol{\nabla}\,\varphi_u + \boldsymbol{\nabla}\times\boldsymbol{\psi}_u,\ \boldsymbol{\nabla}\cdot\boldsymbol{\psi}_u = 0\,, \tag{2.172}$$

$$\mathbf{E} = -\boldsymbol{\nabla}\varphi_e - \frac{\partial \mathbf{A}}{\partial t}\,,\ \mathbf{B} = \boldsymbol{\nabla}\times\mathbf{A}\,, \tag{2.173}$$

$$\mathbf{F} = \boldsymbol{\nabla}\,\mathbf{\Phi} + \boldsymbol{\nabla}\times\boldsymbol{\Psi},\ \boldsymbol{\nabla}\cdot\boldsymbol{\Psi} = 0\,. \tag{2.174}$$

Here, φ_u, φ_e, and $\mathbf{\Phi}$ are the scalar fields, and $\boldsymbol{\psi}_u$, **A**, and $\boldsymbol{\Psi}$ are the vector fields.

For simplicity, we use the governing set of equations (2.114)–(2.119) for ideal dielectrics. Substituting the decompositions (2.172) and (2.174) into Eq. (2.114) yields the following equations:

$$\left(\bar{K} + \frac{4}{3}G\right)\Delta\varphi_u - \rho_0\frac{\partial^2\varphi_u}{\partial t^2} + \rho_0\mathbf{\Phi} = K\frac{\alpha_\rho}{d_\rho}\tilde{\mu}'_\pi + K\bar{\alpha}_T\theta\,, \tag{2.175}$$

$$G\Delta\boldsymbol{\psi}_u - \rho_0\frac{\partial^2\boldsymbol{\psi}_u}{\partial t^2} + \rho_0\boldsymbol{\Psi} = 0\,. \tag{2.176}$$

In view of the expression (2.172), the heat equation becomes:

$$\rho_0\bar{C}_V\frac{\partial\theta}{\partial t} = \lambda\Delta\theta - KT_0\bar{\alpha}_T\frac{\partial\Delta\varphi_u}{\partial t} - \rho_0T_0\frac{\beta_{T\rho}}{d_\rho}\frac{\partial\tilde{\mu}'_\pi}{\partial t} + \rho_0\mathfrak{R}\,. \tag{2.177}$$

Substituting formulae (2.172) and (2.173) into Eqs. (2.116) and (2.118), we get

$$\Delta\tilde{\mu}'_\pi - \lambda_{\mu E}^2\,\tilde{\mu}'_\pi = \lambda_{\mu E}^2\,K\frac{\alpha_\rho}{\rho_0}\Delta\varphi_u + \lambda_{\mu E}^2\,\beta_{T\rho}\theta\,, \tag{2.178}$$

$$\Delta \mathbf{A} - \varepsilon\mu_0 \left(\frac{\partial^2 \mathbf{A}}{\partial t^2} - \kappa_E \frac{\partial \nabla \tilde{\mu}'_\pi}{\partial t} \right) = \nabla \left(\nabla \cdot \mathbf{A} + \varepsilon\mu_0 \frac{\partial \varphi_e}{\partial t} \right). \quad (2.179)$$

2.13.2 Lorentz Gauge Condition

Let us assume that the Lorenz gauge condition is satisfied (Nowacki 1983):

$$\nabla \cdot \mathbf{A} + \varepsilon\mu_0 \frac{\partial \varphi_e}{\partial t} = 0. \quad (2.180)$$

With regard for the relation (2.180), Eq. (2.179) takes the form:

$$\Delta \mathbf{A} - \varepsilon\mu_0 \frac{\partial^2 \mathbf{A}}{\partial t^2} = \varepsilon\mu_0 \kappa_E \frac{\partial \nabla \tilde{\mu}'_\pi}{\partial t}. \quad (2.181)$$

Substituting the representation (2.173) for the electric field vector in Eq. (2.119) and taking the Lorenz gauge condition (2.180) into account, we finally obtain

$$\Delta \varphi_e - \varepsilon\mu_0 \frac{\partial^2 \varphi_e}{\partial t^2} = -\kappa_E \Delta \tilde{\mu}'_\pi. \quad (2.182)$$

Thus, taking the Lorentz gauge condition (2.180) and formulae (2.172)–(2.174) into account, the governing equations (2.114)–(2.119) of the local gradient theory of dielectrics are transformed into Eqs. (2.175)–(2.178), (2.181), and (2.182). The shear mechanical motion is described by the vector field $\boldsymbol{\psi}_u$. To determine the vector $\boldsymbol{\psi}_u$, we obtain Eq. (2.176), which is identical to the equation for transverse waves in the classical elasticity. Note that Eq. (2.176), representing this motion, and Eqs. (2.175), (2.177), (2.178), (2.181), (2.182) are uncoupled. This fact demonstrates that the shape change is associated neither with the change in the volume (the compression and tension) nor with the local mass displacement or the heat conduction and the electromagnetic processes. The mechanical process of the change in the body volume is connected with the scalar field φ_u. Within the framework of the local gradient theory, the mechanical process of the change in the volume can be produced not only by scalar field Φ of the mass force but also by the local mass displacement and by the temperature field. Thereby, the local mass displacement can affect the velocity of longitudinal waves.

The local mass displacement is the cause of the perturbation of the electromagnetic field (see Eqs. (2.181) and (2.182)). In this case, we obtain a homogeneous equation for the perturbation of the magnetic induction. Indeed, taking the curl $[\mathrm{rot}(\ldots)=\nabla\times(\ldots)]$ of Eq. (2.181), and using the formulae $\mathbf{B}=\nabla\times\mathbf{A}$ and $\nabla\times(\nabla\tilde{\mu}'_\pi)=0$, we get

$$\Delta\mathbf{B}-\varepsilon\mu_0\frac{\partial^2\mathbf{B}}{\partial t^2}=0.$$

This indicates that the magnetic component of the electromagnetic field is connected neither with the compression–tension processes, nor with the shear processes or the heat conduction, or the local mass displacement. Meanwhile, to determine the electric field $\mathbf{E}=-\nabla\varphi_e-\dfrac{\partial\mathbf{A}}{\partial t}$ from Eqs. (2.181) and (2.182), we obtain the following inhomogeneous equation:

$$\Delta\mathbf{E}-\varepsilon\mu_0\frac{\partial^2\mathbf{E}}{\partial t^2}=\kappa_E\left[\Delta(\nabla\tilde{\mu}'_\pi)-\varepsilon\mu_0\frac{\partial^2(\nabla\tilde{\mu}'_\pi)}{\partial t^2}\right]. \tag{2.183}$$

The perturbation of electric field is connected neither with mechanical and thermal processes nor with the local mass displacement. Note that for the vector quantity

$$\mathbf{E}_\mu=\mathbf{E}-\kappa_E\nabla\tilde{\mu}'_\pi \tag{2.184}$$

from (2.183), we obtain the following homogeneous wave equation:

$$\Delta\mathbf{E}_\mu-\varepsilon\mu_0\frac{\partial^2\mathbf{E}_\mu}{\partial t^2}=0\ .$$

This means that the perturbation of the vector of the modified electric field $\mathbf{E}_\mu$ propagates in the body region with the velocity of an electromagnetic wave. The mechanical and thermal processes as well as the local mass displacement do not affect the parameters of this wave.

Using Eq. (2.175), we can define the modified chemical potential

$$\tilde{\mu}'_\pi=\frac{\rho_0 d_\rho}{K\alpha_\rho}\left(c_1^2\Delta\varphi_u-\frac{\partial^2\varphi_u}{\partial t^2}+\Phi\right)-d_\rho\frac{\bar{\alpha}_T}{\alpha_\rho}\theta\,, \tag{2.185}$$

where

$$c_1 = \sqrt{\left(K + \frac{4}{3}G - \frac{K^2\alpha_\rho^2}{\rho_0 d_\rho}\right) \Big/ \rho_0} \, .$$

The function $\tilde{\mu}'_\pi$ can be eliminated from Eqs. (2.175) and (2.177). Substituting the expression (2.185) into these equations, we obtain the following set of relations to determine the functions φ_u and θ

$$\left(1 - \frac{1}{\lambda_{\mu E}^2}\Delta\right)\left(c_1^2 \Delta\varphi_u - \frac{\partial^2 \varphi_u}{\partial t^2} + \Phi - K\frac{\bar{\alpha}_T}{\rho_0}\theta\right)$$
$$= -\frac{K\alpha_\rho}{\rho_0 d_\rho}\left(K\frac{\alpha_\rho}{\rho_0}\Delta\varphi_u + \beta_{T\rho}\theta\right), \qquad (2.186)$$

$$\rho_0\left(C_V - T_0\beta_{T\rho}\frac{\alpha_T}{\alpha_\rho}\right)\frac{\partial\theta}{\partial t} = \lambda\Delta\theta - KT_0\left(\bar{\alpha}_T + \beta_{T\rho}\frac{c_1^2\rho_0^2}{K^2\alpha_\rho}\right)\frac{\partial\Delta\varphi_u}{\partial t}$$
$$+ T_0\beta_{T\rho}\frac{\rho_0^2}{K\alpha_\rho}\frac{\partial}{\partial t}\left(\frac{\partial^2\varphi_u}{\partial t^2} - \Phi\right) + \rho_0\Re \, . \qquad (2.187)$$

In the case of an isothermal approximation (θ = 0), the set of governing equations (2.176), (2.181), (2.182), (2.185), and (2.186) can be solved consecutively. First, from Eq. (2.186), one can define the scalar field φ_u. As the next step, from Eq. (2.185), we can find the function $\tilde{\mu}'_\pi$ and based on Eqs. (2.181) and (2.182), we determine the vector field **A** and the scalar field φ_e.

For static problems, Eqs. (2.178), (2.186), and (2.187) transform to the following set of equations to determine the temperature θ, the scalar field φ_u, and the modified chemical potential $\tilde{\mu}'_\pi$:

$$\Delta\theta + \frac{\rho_0}{\lambda}\Re = 0, \qquad (2.188)$$

$$(\Delta - \breve{\lambda}^2)\Delta\varphi_u = -\frac{1}{\left(\bar{K} + \frac{4}{3}G\right)}\Bigg[\rho_0(\Delta\Phi - \lambda_{\mu E}^2\Phi)$$
$$+ \lambda_{\mu E}^2 K\alpha_T\theta + \rho_0\frac{K\bar{\alpha}_T}{\lambda}\Re\Bigg], \qquad (2.189)$$

$$\Delta\tilde{\mu}'_\pi - \breve{\lambda}^2\,\tilde{\mu}'_\pi = -\lambda^2_{\mu E}\mathfrak{M}\frac{\rho_0 d_\rho}{K\alpha_\rho}\Phi + \lambda^2_{\mu E}\left(\beta_{T\rho} + \bar{\alpha}_T\mathfrak{M}\frac{d_\rho}{\alpha_\rho}\right)\theta\,. \quad (2.190)$$

Here,

$$\breve{\lambda}^2 = \lambda^2_{\mu E}(1+\mathfrak{M})\;,\; \mathfrak{M} = \left(K + \frac{4}{3}G - \frac{K^2\alpha_\rho^2}{\rho_0 d_\rho}\right)^{-1}\frac{K^2\alpha_\rho^2}{\rho_0 d_\rho}\,.$$

Hence, in the case of an isothermal approximation and in the absence of mass forces, to define the potentials φ_u and $\tilde{\mu}'_\pi$, we obtained the following set of uncoupled differential equations of the fourth and second orders, respectively:

$$\Delta\Delta\varphi_u - \breve{\lambda}^2\Delta\varphi_u = 0, \quad (2.191)$$

$$\Delta\tilde{\mu}'_\pi - \breve{\lambda}^2\,\tilde{\mu}'_\pi = 0\,. \quad (2.192)$$

From these equations, it follows that in this case the function $\tilde{\mu}'_\pi$ satisfies the Helmholtz equation. The scalar field φ_u can be found as the sum of the solutions to the Laplace equation and the Helmholtz equation. Note that within the framework of classic (local) theory, the scalar field φ_u satisfies the Laplace equation. It is known that the fundamental solution to the Laplace equation is singular at the origin (Nowacki 1970). For example, in the case of a three-dimensional problem, the solution to the Laplace equation is proportional to a function $1/R$, where $R = \sqrt{x_1^2 + x_2^2 + x_3^2}$. The solution to a two-dimensional problem is proportional to $\ln r$, where $r = \sqrt{x_1^2 + x_2^2}$. The fundamental solutions to Eq. (2.191) for three- and two-dimensional problems are $(1 - e^{-\breve{\lambda}R})/R$ and $\ln r + K_0(\breve{\lambda}r)$, respectively. Here, $K_0(\breve{\lambda}r)$ is the modified Bessel function of the order 0 of the second kind and $K_0(r) \to -\ln r$ if $r \to 0$. It is easy to see that the fundamental solutions to Eq. (2.191) are bounded at the origin.

Table 2.1 The scalar displacement-potential within the local gradient and classical theories

	The local gradient theory	The classical theory (Nowacki 1970)
Differential equation for the displacement potential φ_u	$(1-l_*^2\Delta)\Delta\varphi_u = \dfrac{\rho_0}{K+\frac{4}{3}G}\left[l_*^2(1+\mathfrak{M})\Delta\Phi-\Phi\right]$	$\Delta\varphi_u = -\dfrac{\rho_0}{K+\frac{4}{3}G}\Phi$
Differential equation for Green's function	$(1-\Lambda^2\Delta)\Delta G(\mathbf{x},\boldsymbol{\xi}) = 4\pi\Lambda^2\delta(\mathbf{x}-\boldsymbol{\xi})$	$\Delta G(\mathbf{x},\boldsymbol{\xi}) = 4\pi\delta(\mathbf{x}-\boldsymbol{\xi})$
Green's function for boundless medium (three-dimensional problem)	$G(\mathbf{x},\boldsymbol{\xi}) = \dfrac{1-e^{-\Lambda R(\mathbf{x},\boldsymbol{\xi})}}{R(\mathbf{x},\boldsymbol{\xi})}$, $R=\left[(x_1-\xi_1)^2+(x_2-\xi_2)^2 +(x_3-\xi_3)^2\right]^{\frac{1}{2}}$	$G(\mathbf{x},\boldsymbol{\xi}) = \dfrac{1}{R(\mathbf{x},\boldsymbol{\xi})}$, $R=\left[(x_1-\xi_1)^2+(x_2-\xi_2)^2 +(x_3-\xi_3)^2\right]^{\frac{1}{2}}$
Green's function for boundless medium, (two-dimensional problem)	$G(\mathbf{x},\boldsymbol{\xi}) = \ln r(\mathbf{x},\boldsymbol{\xi}) + K_0\left(\Lambda r(\mathbf{x},\boldsymbol{\xi})\right)$, $\left(K_0(r)\to -\ln r,\ \text{if } r\to 0\right)$ $r=\sqrt{(x_1-\xi_1)^2+(x_2-\xi_2)^2}$	$G(\mathbf{x},\boldsymbol{\xi}) = \ln r(\mathbf{x},\boldsymbol{\xi})$, $r=\sqrt{(x_1-\xi_1)^2+(x_2-\xi_2)^2}$

2.13.3 Generalized Lorentz Gauge Condition

In view of the relation (2.184), we introduce a function $\varphi_{e\mu}$ so that

$$\varphi_{e\mu} = \varphi_e + \kappa_E \tilde{\mu}_\pi' \tag{2.193}$$

and call it the modified electric potential. As a gauge condition, we now take the relation

$$\nabla\cdot\mathbf{A} + \varepsilon\mu_0\frac{\partial\varphi_{e\mu}}{\partial t} = 0\,, \tag{2.194}$$

which generalizes the Lorentz gauge condition (2.180). Then, from Eqs. (2.179) and (2.182), to determine the vector potential $\mathbf{A}$ and the modified electric potential $\varphi_{e\mu}$, we obtain homogeneous similar equations

$$\Delta\mathbf{A} - \varepsilon\mu_0 \frac{\partial^2 \mathbf{A}}{\partial t^2} = 0\,, \tag{2.195}$$

$$\Delta\varphi_{e\mu} - \varepsilon\mu_0 \frac{\partial^2 \varphi_{e\mu}}{\partial t^2} = 0\,. \tag{2.196}$$

Equations (2.175)–(2.178), which are used to calculate the potentials $\boldsymbol{\psi}_u$, φ_u, $\tilde{\mu}'_\pi$ and temperature disturbance θ, remain unchanged. Thus, the processes that are described by the vector fields $\boldsymbol{\psi}_u$, $\mathbf{A}$ and scalar fields φ_{em} are related neither with one another nor with the local mass displacement and nor with the processes of compression and tension.

2.13.4 Coupling Factors between Electromechanical Fields and Local Mass Displacement

The analysis of Eq. (2.178) shows that the electric field does not affect the character of the distribution of the potential $\tilde{\mu}'_\pi$ and the density of the induced mass ρ_m. However, the local mass displacement may affect the quantitative value of the characteristic length of the near-surface inhomogeneity

$$l'_* = \frac{1}{\lambda_{\mu E}} = \frac{\sqrt{1 - \kappa_E\, \chi_{Em} / \chi_m}}{\lambda_\mu} \approx l_* \left(1 - \frac{\rho_0 \chi_{Em}^2}{2\varepsilon\chi_m}\right).$$

The source of an electric field is the change in potential μ'_π, that is, the local mass displacement (see Eq. (2.183)). Quantitatively, the effect of the local mass displacement on the electromagnetic processes is determined by the parameter χ_{Em}. This parameter characterizes the interlinking of the processes of polarization and the local mass displacement (see Eqs. (2.79) and (2.80)). Equations (2.193), (2.195), and (2.196) show that the non-dimensional coupling factor between the local displacement of mass and the electric field may be taken as follows:

$$\varepsilon_E = \kappa_E \frac{\mu_\pi^*}{\varphi_e^*} = \frac{\rho_0 \chi_{Em} \mu_\pi^*}{\varepsilon \varphi_e^*}, \tag{2.197}$$

where μ_π^* and φ_e^* are the characteristic values of the modified chemical potential μ_π' and scalar electric potential φ_e.

Let us set up the coupling factor between the local mass displacement and the process of deformation. Assuming an isothermal approximation, we consider the set of Eqs. (2.175), (2.176), and (2.178).

We introduce dimensionless quantities

$$\mathbf{R} = \frac{\mathbf{r}}{L^*}, \quad \tau = \frac{t}{t^*}, \quad \underline{\varphi_u} = \frac{\varphi_u}{\varphi_u^*}, \quad \underline{\boldsymbol{\psi}_u} = \frac{\boldsymbol{\psi}_u}{\psi_u^*}, \quad \underline{\mu_\pi'} = \frac{\tilde{\mu}_\pi'}{\mu_\pi^*},$$

$$\underline{\Phi} = \frac{\Phi}{\Phi^*}, \quad \underline{\boldsymbol{\Psi}} = \frac{\boldsymbol{\Psi}}{\psi^*}. \tag{2.198}$$

Here, L^*, t^*, φ_u^*, ψ_u^*, μ_π^*, Φ^*, and ψ^* are the corresponding characteristic values of the considered quantities. The Laplace operator and the time derivative in a dimensionless form look as below:

$$\underline{\Delta} = (L^*)^2 \Delta, \quad \frac{\partial}{\partial \tau} = t^* \frac{\partial}{\partial t}.$$

In a dimensionless form, the set of Eqs. (2.175), (2.176), and (2.178) can be written as

$$\underline{\Delta} \underline{\varphi_u} - \left(\frac{v_*}{c_1} \right)^2 \frac{\partial^2 \underline{\varphi_u}}{\partial \tau^2} + \underline{\Phi} = \frac{\mathfrak{M}}{\mathfrak{I}} \underline{\mu_\pi'}, \tag{2.199}$$

$$\underline{\Delta} \underline{\boldsymbol{\psi}_u} - \left(\frac{v_*}{c_2} \right)^2 \frac{\partial^2 \underline{\boldsymbol{\psi}_u}}{\partial \tau^2} + \underline{\boldsymbol{\Psi}} = 0, \tag{2.200}$$

$$\underline{\Delta} \underline{\mu_\pi'} - \Lambda^2 \underline{\mu_\pi'} = \Lambda^2 \mathfrak{I} \underline{\Delta} \underline{\varphi_u}. \tag{2.201}$$

Here,

$$\mathfrak{I} = \frac{K \alpha_\rho \varphi_u^*}{\rho_0 L^{*2} \mu_\pi^*}, \quad \Lambda = L^* \lambda_{\mu E},$$

$c_2 = \sqrt{G/\rho_0}$ is the velocity of propagation of a transverse wave, $v_* = L^*/t^*$ is the characteristic velocity, and the characteristic values Φ^* and ψ^* are chosen as follows: $\Phi^* = c_1^2 \varphi_u^* / L^{*2}$, $\Psi^* = c_2^2 \psi_u^* / L^{*2}$.

If μ_π^* is defined by the formula $\mu_\pi^* = K\alpha_\rho / \rho_0$, and choosing $\varphi_u^* = L^{*2}$ and $\psi_u^* = L^{*2}$, we get $\Im = 1$, $\Phi^* = c_1^2$, and $\Psi^* = c_2^2$. Thus, Eqs. (2.199) and (2.201) can be written as follows:

$$\underline{\Delta}\underline{\varphi}_u - \left(\frac{v_*}{c_1}\right)^2 \frac{\partial^2 \underline{\varphi}_u}{\partial \tau^2} + \underline{\Phi} = \mathfrak{M}\underline{\mu}_\pi',$$

$$\underline{\Delta}\underline{\mu}_\pi' - \Lambda^2 \underline{\mu}_\pi' = \Lambda^2 \underline{\Delta}\underline{\varphi}_u.$$

Therefore, within the framework of the constructed theory, the process of local mass displacement influences the propagation of waves of compression and tension but does not affect the shear wave (see Eqs. (2.199) and (2.200)).

Let us apply the operator $\mathbf{L}_2 \equiv \underline{\Delta} - \Lambda^2$ to Eq. (2.199) and the operator $\mathbf{L}_1 \equiv \underline{\Delta} - \frac{v_*^2}{c_1^2}\frac{\partial^2}{\partial \tau^2}$ to Eq. (2.201). As a result, we get

$$\mathbf{L}_1\mathbf{L}_2(\underline{\varphi}_u) - \varepsilon_\mu \underline{\Delta}\underline{\varphi}_u = -\mathbf{L}_2(\underline{\Phi}), \tag{2.202}$$

$$\mathbf{L}_1\mathbf{L}_2(\underline{\mu}_\pi') - \varepsilon_\mu \underline{\Delta}\underline{\mu}_\pi' = 0. \tag{2.203}$$

Hence, the coupling between the mechanical fields and the local mass displacement is effected through the dimensionless factor $\varepsilon_\mu = \mathfrak{M}\Lambda^2$.

Note that Eq. (2.202), describing the motion of longitudinal waves within the framework of local gradient theory, can be written as follows:

$$\underbrace{\left[\underline{\Delta}\underline{\varphi}_u - \frac{v_*^2}{c_L^2}\frac{\partial^2 \underline{\varphi}_u}{\partial \tau^2} + \underline{\Phi}\right]}_{\substack{\text{equation}\\ \text{for longitudinal wave,}\\ \text{according to "classical" theory}}} \underbrace{- \frac{1}{\Lambda^2}\underline{\Delta}\left[\mathbf{L}_1\left(\underline{\varphi}_u\right) + \underline{\Phi}\right] + \mathfrak{M}\underline{\Delta}\underline{\varphi}_u}_{\substack{\text{effect}\\ \text{of local mass displacement}\\ \text{on longitudinal wave}}} = 0, \tag{2.204}$$

where

$$c_L = \sqrt{\left(K + \frac{4}{3}G\right)\Big/\rho_0}.$$

2.14 Initial Boundary-Value Problems of Local Gradient Electrothermoelasticity

In order to accomplish the formulation of the problems of the local gradient electrothermoelasticity, the boundary (or jump) conditions and associated initial conditions are to be imposed along with the governing set of differential equations. These conditions ensure the uniqueness of the solution to the formulated problem.

On the material surface between two different media (bodies), the field's interface conditions should be satisfied. These conditions should agree with the corresponding balance laws, electrodynamics field equations, and continuity condition. Assume that the contact between two bodies is ideal. Hence, on the interface (Σ^{int}), the following quantities undergo a jump:

- The displacement vector and surface traction (kinematic and mechanical conditions):

$$[\![\mathbf{u}]\!]=0,\quad [\![\hat{\boldsymbol{\sigma}}_* + \hat{\mathbf{T}}]\!]\cdot\mathbf{n}=0. \tag{2.205}$$

- The temperature and normal component of the heat flux (thermal conditions):

$$[\![T]\!]=0,\quad [\![\mathbf{J}_q]\!]\cdot\mathbf{n}=0. \tag{2.206}$$

- The potential μ'_π and normal component of vector of the local mass displacement (condition for local mass displacement):

$$[\![\mu'_\pi]\!]=0,\quad [\![\boldsymbol{\pi}_m]\!]\cdot\mathbf{n}=0. \tag{2.207}$$

- The vectors of electromagnetic field (electromagnetic conditions):

$$[\![\mathbf{E}]\!]\times\mathbf{n}=0,\quad [\![\mathbf{H}]\!]\times\mathbf{n}=\mathbf{I}_s+\rho_{es}\mathbf{v}_s, \tag{2.208}$$

$$[\![\mathbf{D}]\!]\cdot\mathbf{n}=-\rho_{es},\quad [\![\mathbf{B}]\!]\cdot\mathbf{n}=0. \tag{2.209}$$

Here, the brackets indicate the jump of the enclosed quantity across (Σ^{int}), that is, $[\![\mathbf{G}]\!]=\mathbf{G}^{(1)}-\mathbf{G}^{(2)}$ and $[\![G]\!]=G^{(1)}-G^{(2)}$, where $\mathbf{G}$ and G are the vector and scalar quantities and $\hat{\mathbf{T}}$ is the Maxwell stress tensor given by

$$\hat{\mathbf{T}}=\mathbf{E}\otimes\mathbf{D}+\mathbf{H}\otimes\mathbf{B}-\frac{1}{2}(\mathbf{E}\cdot\mathbf{D}+\mathbf{H}\cdot\mathbf{B})\hat{\mathbf{I}}\ \ (i=1,2),$$

ρ_{es} and $\mathbf{I}_s$ are the surface densities of electric charges and current; $\mathbf{v}_s$ is a tangential component of velocity to the body surface; and $\mathbf{n}$ is the normal to the surface (Σ^{int}) directed from medium (1) to medium (2).

Note that the second relation of Eqs. (2.208) and the first relation of Eqs. (2.209) are used to determine the surface charges and the surface current. In the absence of surface charges and current, the mentioned relations state that the corresponding vectors or their components are continuous across the boundary surface (Σ^{int}).

Assume that the polarized solid is subjected to the following boundary conditions on the external surface (Σ^{ext}):

- The mechanical boundary conditions (either the displacement or the traction (force per unit area) is prescribed):

$$\mathbf{u} = \mathbf{u}_a \text{ or } (\hat{\boldsymbol{\sigma}}_* + \hat{\mathbf{T}})\cdot\mathbf{n} = \boldsymbol{\sigma}_a \quad \forall\mathbf{r}\in(\Sigma^{\text{ext}}). \tag{2.210}$$

- The thermal boundary conditions (either the temperature or the normal heat flux, or the condition of convective heat exchange is prescribed):

$$T = T_a \text{ or } \mathbf{J}_q\cdot\mathbf{n} = J_{qa}, \text{ or } \mathbf{J}_q\cdot\mathbf{n} = H_*(T - T_c) \quad \forall\mathbf{r}\in(\Sigma^{\text{ext}}). \tag{2.211}$$

- The boundary condition for local mass displacement:

$$\mu'_\pi = \mu'_{\pi a} \quad \forall\mathbf{r}\in(\Sigma^{\text{ext}}). \tag{2.212}$$

- The electromagnetic boundary conditions are written as prescriptive for tangential components of vectors of electric and magnetic fields:

$$\mathbf{E}\times\mathbf{n} = \mathbf{E}_a, \ \mathbf{H}\times\mathbf{n} = \mathbf{H}_a \quad \forall\mathbf{r}\in(\Sigma^{\text{ext}}). \tag{2.213}$$

In the relations (2.210)–(2.213): $\mathbf{n}$ is the outward unit vector normal to the smooth boundary (Σ^{ext}); $\mathbf{u}_a$, σ_a, $\mathbf{E}_a$, $\mathbf{H}_a$, J_{qa}, T_a, and $\mu'_{\pi a}$ are the values of the displacement vector, of traction, of the electric and magnetic fields, of the normal component of heat flux, of temperature, and of the modified chemical potential μ'_π given on the surface (Σ^{ext}); H_* is a heat transfer coefficient from the surface; and T_c is the environmental temperature.

Sometimes different boundary conditions are imposed on different parts of the body surface. For example, the displacements $\mathbf{u}_a$ can be imposed on the portion of the surface of the body Σ_u^{ext}, while the tractions σ_a are imposed on the portion of the surface of the body $(\Sigma_\sigma^{\text{ext}})$. Then the mechanical boundary conditions can be written as follows:

$$\mathbf{u} = \mathbf{u}_a \quad \forall \mathbf{r} \in (\Sigma_u^{\text{ext}}), \tag{2.214}$$

$$(\hat{\boldsymbol{\sigma}}_* + \hat{\mathbf{T}}) \cdot \mathbf{n} = \boldsymbol{\sigma}_a \quad \forall \mathbf{r} \in (\Sigma_\sigma^{\text{ext}}). \tag{2.215}$$

Here, $(\Sigma_u^{\text{ext}}) \cup (\Sigma_\sigma^{\text{ext}}) = (\Sigma^{\text{ext}})$ and $(\Sigma_u^{\text{ext}}) \cap (\Sigma_\sigma^{\text{ext}}) = \varnothing$.

Assume the thermal boundary conditions to be such that the temperature T_a is specified over a part of the surface (Σ_T^{ext}), the normal heat flux is specified over a part (Σ_J^{ext}), and the Newton condition is specified over a part of the surface (Σ_H^{ext}) with the total surface $(\Sigma_T^{\text{ext}}) \cup (\Sigma_J^{\text{ext}}) \cup (\Sigma_H^{\text{ext}}) = (\Sigma^{\text{ext}})$. Thus,

$$T = T_a \quad \forall \mathbf{r} \in (\Sigma_T^{\text{ext}}); \quad \mathbf{J}_q \cdot \mathbf{n} = J_{qa} \quad \forall \mathbf{r} \in (\Sigma_J^{\text{ext}});$$

$$\mathbf{J}_q \cdot \mathbf{n} = H_*(T - T_c) \quad \forall \mathbf{r} \in (\Sigma_H^{\text{ext}}), \tag{2.216}$$

where $(\Sigma_T^{\text{ext}}) \cap (\Sigma_J^{\text{ext}}) = \varnothing$, $(\Sigma_T^{\text{ext}}) \cap (\Sigma_H^{\text{ext}}) = \varnothing$,

and $(\Sigma_H^{\text{ext}}) \cap (\Sigma_J^{\text{ext}}) = \varnothing$.

In some cases, certain conditions for the body surface can be formulated as the boundary conditions, while the others as jump conditions. Indeed, let the body be in contact with vacuum or an environment with similar properties. In this case, mechanical conditions may be formulated as kinematic boundary conditions (if displacements are known on the body surface) or natural (traction) boundary conditions corresponding to a traction-free surface. Thermal boundary conditions should correspond to the prescription of the surface temperature (if we can control it) or to the flux from the surface. Now, the equality between the potential μ'_π and zero is a condition for the local mass displacement. Since the perturbation of electrothermomechanical processes within the body will cause a radiation of the electromagnetic field into the vacuum, the electromagnetic conditions on the body surface should be formulated as contact conditions. Therefore, the Maxwell equations in vacuum (domain (V_v)) should be added to the governing set of equations,

$$\nabla\times\mathbf{E}_\mathrm{v}=-\frac{\partial\mathbf{B}_\mathrm{v}}{\partial t},\quad \nabla\times\mathbf{H}_\mathrm{v}=\frac{\partial\mathbf{D}_\mathrm{v}}{\partial t},$$

$$\nabla\cdot\mathbf{B}_\mathrm{v}=0,\quad \nabla\cdot\mathbf{D}_\mathrm{v}=0, \tag{2.217}$$

$$\mathbf{D}_\mathrm{v}=\varepsilon_0\mathbf{E}_\mathrm{v},\quad \mathbf{B}_\mathrm{v}=\mu_0\mathbf{H}_\mathrm{v}, \tag{2.218}$$

where $\mathbf{E}_\mathrm{v}$, $\mathbf{H}_\mathrm{v}$, $\mathbf{D}_\mathrm{v}$, and $\mathbf{B}_\mathrm{v}$ are the electric and magnetic fields and inductions in vacuum.

The jump conditions on (Σ^{ext}) take the following form:

$$(\mathbf{E}-\mathbf{E}_\mathrm{v})\times\mathbf{n}=0,\quad (\mathbf{H}-\mathbf{H}_\mathrm{v})\times\mathbf{n}=\mathbf{I}_s+\rho_{es}\mathbf{v}_s, \tag{2.219}$$

$$(\mathbf{D}-\mathbf{D}_\mathrm{v})\cdot\mathbf{n}=-\rho_{es},\quad (\mathbf{B}-\mathbf{B}_\mathrm{v})\cdot\mathbf{n}=0. \tag{2.220}$$

If the governing equations (2.102)–(2.107) are chosen as the key ones, then the first relations in boundary conditions (2.205) and (2.210), as well as the conditions (2.208), (2.209), and (2.213), do not change, while the second relations from the conditions (2.205) and (2.210) as well as the conditions (2.206), (2.207), (2.211), and (2.212) transform as follows:

$$\Bigg\langle 2G^{(1)}\left[\nabla\otimes\mathbf{u}^{(1)}+(\nabla\otimes\mathbf{u}^{(1)})^T\right]+\left\{\left[K^{(1)}-\frac{2}{3}G^{(1)}-\frac{\left(K^{(1)}\alpha_\rho^{(1)}\right)^2}{\rho_0^{(1)}d_\rho^{(1)}}\right]\nabla\cdot\mathbf{u}^{(1)}\right.$$
$$\left.-K^{(1)}\left(\bar{\alpha}_T^{(1)}\theta^{(1)}+\frac{\alpha_\rho^{(1)}}{d_\rho^{(1)}}\tilde{\mu}_\pi'^{(1)}\right)\right\}\hat{\mathbf{I}}-2G^{(2)}\left[\nabla\otimes\mathbf{u}^{(2)}+(\nabla\otimes\mathbf{u}^{(2)})^T\right]$$
$$-\left[K^{(2)}-\frac{2}{3}G^{(2)}-\frac{\left(K^{(2)}\alpha_\rho^{(2)}\right)^2}{\rho_0^{(2)}d_\rho^{(2)}}\right]\nabla\cdot\mathbf{u}^{(2)}$$
$$-\left[-K^{(2)}\left(\bar{\alpha}_T^{(2)}\theta^{(2)}+\frac{\alpha_\rho^{(2)}}{d_\rho^{(2)}}\tilde{\mu}_\pi'^{(2)}\right)\right]\hat{\mathbf{I}}\Bigg\rangle\cdot\mathbf{n}=0, \tag{2.221}$$

$$\theta^{(1)}-\theta^{(2)}=0, \tag{2.222}$$

$$\left\{\left[\left(\lambda^{(1)}+\pi_t^{(1)}\eta^{(1)}\right)\nabla\theta^{(1)}-\pi_t^{(1)}\sigma_e^{(1)}\mathbf{E}^{(1)}\right]\right.$$
$$\left.-\left[\left(\lambda^{(2)}+\pi_t^{(2)}\eta^{(2)}\right)\nabla\theta^{(2)}-\pi_t^{(2)}\sigma_e^{(2)}\mathbf{E}^{(2)}\right]\right\}\cdot\mathbf{n}=0, \tag{2.223}$$

$$\tilde{\mu}_{\pi}^{\prime(1)} - \tilde{\mu}_{\pi}^{\prime(2)} = \mu_{\pi 0}^{\prime(2)} - \mu_{\pi 0}^{\prime(1)}, \tag{2.224}$$

$$\left[-\chi_m^{(1)} \nabla \tilde{\mu}_{\pi}^{\prime(1)} + \chi_{Em}^{(1)}\mathbf{E}^{(1)} + \chi_m^{(2)} \nabla \tilde{\mu}_{\pi}^{\prime(2)} - \chi_{Em}^{(2)}\mathbf{E}^{(2)}\right]\cdot\mathbf{n} = 0, \tag{2.225}$$

$$\Big\langle 2G\left[\nabla \otimes \mathbf{u} + (\nabla \otimes \mathbf{u})^T\right]$$

$$+\left\{\left(\bar{K} - \frac{2}{3}G\right)(\nabla \cdot \mathbf{u}) - K\left(\bar{\alpha}_T\theta + \frac{\alpha_\rho}{d_\rho}\tilde{\mu}_{\pi}^{\prime}\right)\right\}\mathbf{I}\Big\rangle\cdot\mathbf{n} = \boldsymbol{\sigma}_a \tag{2.226}$$

$$\theta = T_a - T_0, \tag{2.227}$$

$$\left[-(\lambda + \pi_t\eta)\nabla\theta + \pi_t\sigma_e\mathbf{E}\right]\cdot\mathbf{n} = J_{qa}, \tag{2.228}$$

$$\left[-(\lambda + \pi_t\eta)\nabla\theta + \pi_t\sigma_e\mathbf{E}\right]\cdot\mathbf{n} = H_*\left[\theta - (T_c - T_0)\right], \tag{2.229}$$

$$\tilde{\mu}_{\pi}^{\prime} = \mu_{\pi a}^{\prime} - \mu_{\pi 0}^{\prime}. \tag{2.230}$$

If the governing equations (2.94)–(2.99) are chosen as a key set, then the perturbation of the modified chemical potential $\tilde{\mu}_{\pi}^{\prime}$ should be replaced in conditions (2.221), (2.224)–(2.226), and (2.230) by the expression:

$$\tilde{\mu}_{\pi}^{\prime} = d_\rho\rho_m - \frac{1}{\rho_0}K\alpha_\rho\nabla\cdot\mathbf{u} - \beta_{T\rho}\theta.$$

In order to solve non-stationary problems, it is necessary to impose the relevant initial conditions. For a linear governing set of equations (2.94)–(2.99), these conditions look as follows:

$$\mathbf{u} = \mathbf{u}_0,\ \frac{\partial\mathbf{u}}{\partial t} = \mathbf{v}_0,\ \theta = 0,\ \rho_m = \rho_{m0},\ \mathbf{E} = \mathbf{E}_0,\ \mathbf{B} = \mathbf{B}_0 \text{ at } t = t_0. \tag{2.231}$$

If we take the set of equations (2.102)–(2.107) as the key set, then the initial conditions for a displacement vector, displacement rate, temperature, electric field and magnetic field induction will be of the form (2.231), while the condition for the specific density of induced mass should be replaced by the initial condition for the modified chemical potential $\tilde{\mu}_{\pi}^{\prime}$:

$$\tilde{\mu}_{\pi}^{\prime} = 0 \text{ at } t = t_0. \tag{2.232}$$

Note that in conditions (2.232), it is assumed that the initial time corresponds to the reference equilibrium state of the thermodynamic system. Note also that the boundary and initial conditions should be compatible, in particular, $\mathbf{u}_0\big|_{\mathbf{r}\in\Sigma} = \mathbf{u}_a\big|_{t=t_0}$ and $\mathbf{v}_0\big|_{\mathbf{r}\in\Sigma} = \dfrac{\partial \mathbf{u}_a}{\partial t}\bigg|_{t=t_0}$.

Regarding the problems formulated in terms of scalar φ_u, φ_e and vector $\boldsymbol{\psi}_u$, $\mathbf{A}$ fields of the displacement and electromagnetic field, the above conditions should be rewritten using the relations (2.172) and (2.173).

The coupled initial boundary-value problem lies in determining the displacement vector $\mathbf{u}(\mathbf{r}, t)$, temperature field $\theta(\mathbf{r}, t)$, modified chemical potential $\tilde{\mu}'_\pi(\mathbf{r}, t)$ (or density of induced mass $\rho_m(\mathbf{r}, t)$), electric field $\mathbf{E}(\mathbf{r}, t)$, and magnetic induction $\mathbf{B}(\mathbf{r}, t)$ of $C^{(2)}$ in the medium, governed by Eqs. (2.102)–(2.107) (or (2.94)–(2.99)) and subjected to suitable boundary, jump, and initial conditions. To find the electric field $\mathbf{E}_v(\mathbf{r}, t)$ and magnetic induction $\mathbf{B}_v(\mathbf{r}, t)$ in vacuum, we have Eqs. (2.217) and (2.218).

2.15 Uniqueness Theorem[1]

As shown in section 2.11, the set of differential equations of local gradient electrothermoelasticity can be divided into two uncoupled subsets. Therefore, the formulated problem can be solved consecutively. To determine the functions $\mathbf{u}$, θ, and $\tilde{\mu}'_\pi$, we use the set of equations (2.114)–(2.116). The vectors of the electromagnetic field are derived from Eqs. (2.117)–(2.119), where the functions $\mathbf{u}$, θ, and $\tilde{\mu}'_\pi$ are known. In view of this, we study the conditions of uniqueness of the solution to the corresponding problems in mathematical physics in two stages: separately for the equations of motion (2.102), heat conduction (2.109), and modified chemical potential (2.110), and separately for the equations of electrodynamics (2.117)–(2.119).

Theorem 1. For the domain (V) bounded by a smooth surface (Σ), and positive G, $K - \dfrac{2}{3}G - \dfrac{K^2\alpha_\rho^2}{\rho_0 d_\rho}$, C_V, d_ρ, χ_m, and H_*, there is only one set of functions $\mathbf{u}(\mathbf{r}, t)$, $\theta(\mathbf{r}, t)$, and $\tilde{\mu}'_\pi(\mathbf{r}, t)$ such that

- $\forall \mathbf{r} \in (V) \cup (\Sigma)$, $\{\mathbf{u}, \theta, \tilde{\mu}'_\pi\} \in C^{(2)}$;

[1]First published in *Journal of Mechanics of Materials and Structures* in Vol. 14 (2019), published by Mathematical Sciences Publishers.

- $\forall \mathbf{r} \in (V)$, it satisfies the set of differential equations (2.114)–(2.116);
- $\forall \mathbf{r} \in (V) \cup (\Sigma)$, it satisfies the stress–displacement relation (2.8), the constitutive equations (2.76)–(2.78), (2.80), and the kinetic relation (2.92);
- It satisfies the boundary and initial conditions:

$$\forall \mathbf{r} \in (\Sigma): \; \hat{\boldsymbol{\sigma}}_* \cdot \mathbf{n} = \boldsymbol{\sigma}^a, \quad \mathbf{J}_q \cdot \mathbf{n} - H_*(\theta - \theta_c) = 0, \quad \tilde{\mu}'_\pi = \mu'_{\pi a},$$

$$\text{at } t = t_0: \; \mathbf{u} = \mathbf{u}_0, \quad \frac{\partial \mathbf{u}}{\partial t} = \mathbf{v}_0, \quad \theta = 0, \quad \tilde{\mu}'_\pi = 0.$$

Proof. For the linear problems considered here, we replace the mass density ρ with its initial value ρ_0. We also neglect the convective time derivative by assuming the material time derivative to be equal to the Eulerian time derivative. Then, using Eqs. (2.62) and (2.66), for thermoelastic media, we can write:

$$\rho_0 \frac{\partial u}{\partial t} = \hat{\boldsymbol{\sigma}}_* : \frac{\partial \hat{e}}{\partial t} + \rho_0 T \frac{\partial s}{\partial t} + \rho_0 \mu'_\pi \frac{\partial \rho_m}{\partial t} - \rho_0 \boldsymbol{\nabla} \mu'_\pi \cdot \frac{\partial \boldsymbol{\pi}_m}{\partial t}.$$

Here, u is the specific internal energy.

Proceeding to the perturbations of the temperature field $\theta = T - T_0$ and the modified chemical potential $\tilde{\mu}'_\pi = \mu'_\pi - \mu'_{\pi 0}$, we get

$$\rho_0 \frac{\partial u}{\partial t} = \rho_0 T_0 \frac{\partial s}{\partial t} + \rho_0 \mu'_{\pi 0} \frac{\partial \rho_m}{\partial t} + \hat{\boldsymbol{\sigma}}_* : \frac{\partial \hat{e}}{\partial t} + \rho_0 \theta \frac{\partial s}{\partial t} + \rho_0 \tilde{\mu}'_\pi \frac{\partial \rho_m}{\partial t} - \rho_0 \boldsymbol{\nabla} \tilde{\mu}'_\pi \cdot \frac{\partial \boldsymbol{\pi}_m}{\partial t}.$$

Substitute the constitutive relations (2.76)–(2.78), (2.80) into nonlinear summands of the obtained relation. After some transformations, we obtain

$$\rho_0 \frac{\partial u}{\partial t} = \rho_0 T_0 \frac{\partial s}{\partial t} + \rho_0 \mu'_{\pi 0} \frac{\partial \rho_m}{\partial t} + \frac{1}{2} \frac{\partial}{\partial t} \left[\left(\bar{K} - \frac{2}{3} G \right) I_1^2 + 2 G I_2 + \rho_0 \frac{\bar{C}_V}{T_0} \theta^2 + \frac{\rho_0}{d_\rho} (\tilde{\mu}'_\pi)^2 + \rho_0 \chi_m \boldsymbol{\nabla} \tilde{\mu}'_\pi \cdot \boldsymbol{\nabla} \tilde{\mu}'_\pi + 2 \rho_0 \frac{\beta_{T\rho}}{d_\rho} \theta \tilde{\mu}'_\pi \right]. \quad (2.233)$$

The proof of the theorem will be based on the energy balance equation (2.43), which for the model of the thermoelastic solid is of the form:

$$\frac{\partial}{\partial t}\int_{(V)}\left(\rho_0 u+\frac{1}{2}\rho_0\mathbf{v}^2\right)dV=\int_{(V)}\left(\rho_0\mathbf{F}\cdot\mathbf{v}+\rho_0\Re\right)dV$$

$$-\oint_{(\Sigma)}\left(\mathbf{J}_q-\hat{\boldsymbol{\sigma}}\cdot\mathbf{v}+\mu_\pi'\frac{\partial\boldsymbol{\Pi}_m}{\partial t}\right)\cdot\mathbf{n}\,d\Sigma.$$

Hence, making use of the expression (2.233), formulae (2.93) and (2.92), the entropy balance equation

$$\rho_0\frac{\partial s}{\partial t}=-\boldsymbol{\nabla}\cdot\mathbf{J}_s+\eta_s+\rho_0\frac{\Re}{T}$$

and the divergence theorem, one can write

$$\frac{\partial\mathcal{E}_*}{\partial t}=\int_{(V)}\left(\rho_0\mathbf{F}\cdot\mathbf{v}+\rho_0\Re\frac{\theta}{T}-T_0\eta_s\right)dV$$

$$+\oint_{(\Sigma)}\left[\boldsymbol{\sigma}_n\cdot\mathbf{v}+\frac{\lambda\theta}{T}(\boldsymbol{\nabla}\theta\cdot\mathbf{n})-\rho_0\tilde{\mu}_\pi'\frac{\partial(\boldsymbol{\pi}_m\cdot\mathbf{n})}{\partial t}\right]d\Sigma\,, \qquad (2.234)$$

where

$$\mathcal{E}_*=\frac{1}{2}\int_{(V)}\left[\rho_0\mathbf{v}^2+\left(\bar{K}-\frac{2}{3}G\right)I_1^2+2GI_2+\rho_0\frac{C_V}{T_0}\theta^2\right.$$

$$\left.+\rho_0\chi_m(\boldsymbol{\nabla}\tilde{\mu}_\pi')^2+\frac{\rho_0}{d_\rho}(\tilde{\mu}_\pi'+\beta_{T\rho}\theta)^2\right]dV\,. \qquad (2.235)$$

For convenience, we write the energy balance equation (2.234) as follows:

$$\frac{\partial\mathcal{E}_*}{\partial t}=\int_{(V)}\left(\rho_0\mathbf{F}\cdot\mathbf{v}+\rho_0\Re\frac{\theta}{T}-T_0\eta_s\right)dV$$

$$+\oint_{(\Sigma)}\left[\boldsymbol{\sigma}_n\cdot\mathbf{v}-\frac{H_*}{T}\theta^2+\frac{\theta}{T}\left(\lambda\frac{\partial\theta}{\partial n}+H_*\theta\right)-\rho_0\tilde{\mu}_\pi'\frac{\partial\pi_{mn}}{\partial t}\right]d\Sigma\,. \qquad (2.236)$$

Here, $\boldsymbol{\nabla}\theta\cdot\mathbf{n}=\dfrac{\partial\theta}{\partial n}$ and $\pi_{mn}=\boldsymbol{\pi}_m\cdot\mathbf{n}$.

The energy balance equation (2.236) makes it possible to prove the uniqueness of the solution.

Assume two distinct solutions $\mathbf{u}_1(\mathbf{r},t)$, $\theta_1(\mathbf{r},t)$, $\tilde{\mu}_{\pi1}'(\mathbf{r},t)$ and $\mathbf{u}_2(\mathbf{r},t)$, $\theta_2(\mathbf{r},t)$, $\tilde{\mu}_{\pi2}'(\mathbf{r},t)$ to satisfy Eqs. (2.114)–(2.116) and the appropriate boundary and initial conditions. Therefore, their difference $\mathbf{u} = \mathbf{u}_1 - \mathbf{u}_2$, $\theta = \theta_1 - \theta_2$, and $\tilde{\mu}'_\pi = \tilde{\mu}'_{\pi 1} - \tilde{\mu}'_{\pi 2}$ satisfies the homogeneous equations and the homogeneous boundary and initial conditions:

$$\forall \mathbf{r} \in (\Sigma): \ \hat{\boldsymbol{\sigma}}_* \cdot \mathbf{n} = 0, \quad \lambda \frac{\partial \theta}{\partial n} + H_* \theta = 0, \quad \tilde{\mu}'_\pi = 0, \qquad (2.237)$$

$$\text{at} \quad t = t_0: \ \mathbf{u} = 0, \quad \frac{\partial \mathbf{u}}{\partial t} = 0, \quad \theta = 0, \quad \tilde{\mu}'_\pi = 0. \qquad (2.238)$$

In view of the homogeneity of the equations and boundary conditions (2.237), from the equation of energy balance (2.236), we obtain

$$\frac{\partial \mathcal{E}_*}{\partial t} = -\int_{(V)} T_0 \eta_s dV - \oint_{(\Sigma)} \frac{H_*}{T} \theta^2 d\Sigma .$$

Since $\eta_s \geq 0$ and $\frac{H_*}{T} \geq 0$, the following inequality should hold:

$$\frac{\partial \mathcal{E}_*}{\partial t} \leq 0. \qquad (2.239)$$

The difference of solutions satisfies the zero initial conditions, and, therefore, $\mathcal{E}_*$ equals zero at the initial moment in time. Thus, from the inequality (2.239), it follows that the function $\mathcal{E}_*$ is either negative or zero: $\mathcal{E}_* \leq 0$. On the other hand, according to Eq. (2.235), we have $\mathcal{E}_* \geq 0$ since G, $K - \frac{2}{3}G - \frac{K^2 \alpha_\rho^2}{\rho_0 d_\rho}$, C_V, d_ρ, and χ_m are positive-definite.

The above two inequalities can be fulfilled only if $\mathcal{E}_* = 0$. Consequently, taking the formula (2.235) into account, we can write

$$\int_{(V)} \left[\rho_0 \mathbf{v}^2 + \left(\bar{K} - \frac{2}{3} G \right) I_1^2 + 2G I_2 + \rho_0 \frac{C_V}{T_0} \theta^2 \right.$$

$$\left. + \rho_0 \chi_m \left(\nabla \tilde{\mu}'_\pi \right)^2 + \frac{\rho_0}{d_\rho} (\tilde{\mu}'_\pi + \beta_{T\rho} \theta)^2 \right] dV = 0 .$$

Since $K - \frac{2}{3}G - \frac{K^2 \alpha_\rho^2}{\rho_0 d_\rho} > 0$, $G > 0$, $C_V > 0$, $d_\rho > 0$, and $\chi_m > 0$ and the relation in brackets is positive-definite, the last formula implies $\mathbf{v} = 0$, $\hat{e} = 0$, $\theta = 0$, $\tilde{\mu}'_\pi = 0$, and $\nabla \tilde{\mu}'_\pi = 0$. Using the constitutive equations (2.78) and (2.80), we also obtain $\rho_m = 0$ and $\boldsymbol{\pi}_m = 0$. So, $\mathbf{u}_1 = \mathbf{u}_2$, $\theta_1 = \theta_2$, and $\tilde{\mu}'_{\pi 1} = \tilde{\mu}'_{\pi 2}$. Therefore, the coupled initial boundary-value problem of the local gradient thermoelasticity has only one solution, which is exactly what we intended to demonstrate.

Theorem 2. If ε_0, μ_0, χ_E, σ_e are positive and the functions $\mathbf{u}(\mathbf{r}, t)$, $\theta(\mathbf{r}, t)$, and $\tilde{\mu}'_\pi(\mathbf{r}, t)$ are known, then for the body domain (V) and vacuum (V_v), separated by a smooth surface (Σ), there is only one set of functions $\{\mathbf{E}, \mathbf{H}, \mathbf{E}_v, \mathbf{H}_v\}$ such that

- $\forall \mathbf{r} \in (V) \cup (\Sigma)$ and $\forall \mathbf{r}_v \in (V_v) \cup (\Sigma)$: $\{\mathbf{E}, \mathbf{H}, \mathbf{E}_v, \mathbf{H}_v\} \in C^{(2)}$;
- $\forall \mathbf{r} \in (V)$, it satisfies the differential equations (2.20), (2.21), (2.25), (2.26) and $\forall \mathbf{r}_v \in (V_v)$, it satisfies Eqs. (2.217), respectively;
- $\forall \mathbf{r} \in (V) \cup (\Sigma)$, it satisfies the constitutive relations (2.17), (2.24), (2.79), (2.91) and $\forall \mathbf{r}_v \in (V_v) \cup (\Sigma)$, it satisfies the constitutive relations (2.218), respectively;
- $\forall \mathbf{r}, \mathbf{r}_v \in (\Sigma)$, it fulfils the jump conditions (2.219) and the initial conditions
  At $t = t_0$: $\mathbf{E} = \mathbf{E}_0$, $\mathbf{H} = \mathbf{H}_0$, $\mathbf{E}_v = \mathbf{E}_{v0}$, $\mathbf{H}_v = \mathbf{H}_{v0}$.

Proof. Assume the two sets of fields $\{\mathbf{E}_1, \mathbf{H}_1, \mathbf{E}_{v1}, \mathbf{H}_{v1}\}$ and $\{\mathbf{E}_2, \mathbf{H}_2, \mathbf{E}_{v2}, \mathbf{H}_{v2}\}$ to be the solutions to the above problem. The difference fields $\mathbf{E} = \mathbf{E}_1 - \mathbf{E}_2$, $\mathbf{H} = \mathbf{H}_1 - \mathbf{H}_2$, $\mathbf{E}_v = \mathbf{E}_{v1} - \mathbf{E}_{v2}$, and $\mathbf{H}_v = \mathbf{H}_{1v} - \mathbf{H}_{2v}$ satisfy the relations (2.20), (2.21), (2.25), (2.26), and (2.217), the trivial initial conditions, the constitutive relations (2.17), (2.24), (2.218), as well as

$$\boldsymbol{\Pi}_e = \rho_0 \chi_E \mathbf{E} \tag{2.240}$$

and the kinetic equation

$$\mathbf{J}_e = \sigma_e \mathbf{E}. \tag{2.241}$$

These functions satisfy the following energy balance equations for the electromagnetic field (see Eq. (2.34)):

$$\frac{\partial U_e}{\partial t} + \boldsymbol{\nabla} \cdot \mathbf{S}_e + \left(\mathbf{J}_e + \frac{\partial \boldsymbol{\Pi}_e}{\partial t} \right) \cdot \mathbf{E} = 0, \tag{2.242}$$

$$\frac{\partial U_{ev}}{\partial t} + \boldsymbol{\nabla} \cdot \mathbf{S}_{ev} = 0, \tag{2.243}$$

where

$$U_e = \frac{1}{2}\left(\varepsilon_0 \mathbf{E}^2 + \mu_0 \mathbf{H}^2\right), \quad \mathbf{S}_e = \mathbf{E}\times\mathbf{H}, \tag{2.244}$$

$$U_{ev} = \frac{1}{2}\left(\varepsilon_0 \mathbf{E}_v^2 + \mu_0 \mathbf{H}_v^2\right), \quad \mathbf{S}_{ev} = \mathbf{E}_v \times \mathbf{H}_v. \tag{2.245}$$

Substituting the formulae (2.240), (2.241), and (2.244) into Eq. (2.242), after some manipulations, we obtain

$$\frac{1}{2}\frac{\partial}{\partial t}\left(\varepsilon \mathbf{E}^2 + \mu_0 \mathbf{H}^2\right) + \nabla\cdot(\mathbf{E}\times\mathbf{H}) + \sigma_e \mathbf{E}\cdot\mathbf{E} = 0.$$

Here, $\varepsilon = \varepsilon_0 + \rho_0 \chi_E$. By integrating the obtained expression over the region (V), and using the divergence theorem, we get

$$\frac{1}{2}\int_{(V)} \frac{\partial}{\partial t}\left(\varepsilon \mathbf{E}^2 + \mu_0 \mathbf{H}^2\right) dV = -\int_{(\Sigma)} (\mathbf{E}\times\mathbf{H})\cdot\mathbf{n}\, d\Sigma - \sigma_e \int_{(V)} \mathbf{E}^2 dV. \tag{2.246}$$

Here, $\mathbf{n}$ is the unit exterior normal to (Σ). Substituting the formulae (2.245) into Eq. (2.243) and integrating the obtained result over the domain (V_v), we obtain

$$\frac{1}{2}\int_{(V_v)} \frac{\partial}{\partial t}\left(\varepsilon_0 \mathbf{E}_v^2 + \mu_0 \mathbf{H}_v^2\right) dV = \int_{(\Sigma)} \left(\mathbf{E}_v \times \mathbf{H}_v\right)\cdot\mathbf{n}\, d\Sigma. \tag{2.247}$$

Adding the expressions (2.246) and (2.247), we find that

$$\frac{\partial U_e^t}{\partial t} = -\int_{(\Sigma)} \left[(\mathbf{E}\times\mathbf{H}) - (\mathbf{E}_v \times \mathbf{H}_v)\right]\cdot\mathbf{n}\, d\Sigma - \sigma_e \int_{(V)} \mathbf{E}^2 dV, \tag{2.248}$$

where

$$U_e^t = \frac{1}{2}\left[\int_{(V)} \left(\varepsilon \mathbf{E}^2 + \mu_0 \mathbf{H}^2\right) dV + \int_{(V_v)} \left(\varepsilon_0 \mathbf{E}_v^2 + \mu_0 \mathbf{H}_v^2\right) dV\right] \geq 0. \tag{2.249}$$

In view of the jump conditions (2.219), we can write the expression (2.248) as follows:

$$\frac{\partial U_e^t}{\partial t} = -\int_{(\Sigma)} \mathbf{E}_s \cdot \mathbf{I}_s\, d\Sigma - \sigma_e \int_{(V)} \mathbf{E}^2 dV.$$

Since $\mathbf{I}_s = \sigma_e^s \mathbf{E}_s$, where $\mathbf{E}_s$ is the tangential component of vector of the electric field, we have

$$\frac{\partial U_e^t}{\partial t} = -\sigma_e^s \int_{(\Sigma)} \mathbf{E}_s^2 d\Sigma - \sigma_e \int_{(V)} \mathbf{E}^2 dV . \qquad (2.250)$$

From the formula (2.250), it follows that $\frac{\partial U_e^t}{\partial t} \le 0$, since σ_e and σ_e^s are positive. Thus, U_e^t is either a decreasing function or a constant. As the functions **E**, **H**, $\mathbf{E}_v$, $\mathbf{H}_v$ satisfy the trivial initial conditions at the initial time $t = t_0$, the function U_e^t is equal to zero at the initial moment in time. Hence, $U_e^t \le 0$. At the same time, as it follows from (2.249), the function U_e^t is positive-definite or equal to zero: $U_e^t \ge 0$. The last two inequalities hold only if $U_e^t = 0$. Thus,

$$U_e^t = \frac{1}{2}\left[\int_{(V)} \left(\varepsilon \mathbf{E}^2 + \mu_0 \mathbf{H}^2\right) dV + \int_{(V_v)} \left(\varepsilon_0 \mathbf{E}_v^2 + \mu_0 \mathbf{H}_v^2\right) dV\right] = 0 .$$

Since $\varepsilon_0 > 0$, $\mu_0 > 0$, $\varepsilon = \varepsilon_0 + \rho_0 \chi_E > 0$, from the above equation, we obtain $\mathbf{E} = \mathbf{E}_1 - \mathbf{E}_2 = 0$, $\mathbf{H} = \mathbf{H}_1 - \mathbf{H}_2 = 0$, $\mathbf{E}_v = \mathbf{E}_{v1} - \mathbf{E}_{v2} = 0$, and $\mathbf{H}_v = \mathbf{H}_{1v} - \mathbf{H}_{2v} = 0$. Thus, $\mathbf{E}_1 = \mathbf{E}_2$, $\mathbf{H}_1 = \mathbf{H}_2$, $\mathbf{E}_{v1} = \mathbf{E}_{v2}$, and $\mathbf{H}_{1v} = \mathbf{H}_{2v}$, which is exactly what we had to prove.

2.16 Reciprocity Theorem

In this section, we prove the work reciprocity theorem for ideal thermoelastic dielectrics. We consider two different stress–strain states of a dielectric solid, caused by two sets of external loading, namely, the mass forces $\mathbf{F}_*$ and $\mathbf{F}'_*$; the thermal sources $\Re$ and $\Re'$; the surface loadings $\boldsymbol{\sigma}_*$ and $\boldsymbol{\sigma}'_*$ on the surface (Σ_σ); the displacements $\mathbf{u}$ and $\mathbf{u}'$ on the surface (Σ_u); the surface electric charges $\boldsymbol{\Pi}_e \cdot \mathbf{n}$ and $\boldsymbol{\Pi}'_e \cdot \mathbf{n}$ on the surface (Σ_p); the electric potentials φ_e and φ'_e on the surface (Σ_φ); the disturbances of the temperature θ and θ' on the surface (Σ_θ); the heat fluxes $\mathbf{J}_q$ and $\mathbf{J}'_q$ on the surface (Σ_J); the vectors of local mass displacement $\boldsymbol{\pi}_m$ and $\boldsymbol{\pi}'_m$ on the surface (Σ_π); and the potentials $\tilde{\mu}'_\pi$ and $(\tilde{\mu}'_\pi)'$ on the surface (Σ_μ). Here, $(\Sigma_\sigma) \cup (\Sigma_u) = (\Sigma)$, $(\Sigma_\sigma) \cap (\Sigma_u) = \varnothing$, $(\Sigma_\theta) \cup (\Sigma_J) = (\Sigma)$, $(\Sigma_\theta) \cap (\Sigma_J) = \varnothing$,

$(\Sigma_\varphi)\bigcup(\Sigma_p)=(\Sigma)$, $(\Sigma_\varphi)\bigcap(\Sigma_p)=\varnothing$, $(\Sigma_\pi)\bigcup(\Sigma_\mu)=(\Sigma)$, $(\Sigma_\pi)\bigcap(\Sigma_\mu)=\varnothing$. As a consequence of such an external action, we have two states of the body that can be described by the stress tensors $\hat{\boldsymbol{\sigma}}_*$, $\hat{\boldsymbol{\sigma}}_*'$ and strain tensors $\hat{e}$, $\hat{e}'$; by disturbances of temperature θ, θ' and specific entropies s, s'; by specific densities of induced mass ρ_m, ρ_m' and modified potentials $\tilde{\mu}_\pi'$, $(\tilde{\mu}_\pi')'$; by the specific vectors of local mass displacement $\boldsymbol{\pi}_m$, $\boldsymbol{\pi}_m'$ and gradients of potentials $\boldsymbol{\nabla}\tilde{\mu}_\pi'$, $(\boldsymbol{\nabla}\tilde{\mu}_\pi')'$; as well as by the specific vectors of polarization $\boldsymbol{\pi}_e$, $\boldsymbol{\pi}_e'$ and the electric fields $\mathbf{E}$, $\mathbf{E}'$, correspondingly.

We apply a one-sided Laplace transform [Schiff 1999]

$$\mathfrak{L}\left[f(\mathbf{r},t)\right]=f^L(\mathbf{r},\zeta)=\int_0^\infty f(\mathbf{r},t)e^{-\zeta t}dt$$

to the foregoing equations of the local gradient theory of dielectrics. Here, $f(\mathbf{r},t)=\left\{\hat{\boldsymbol{\sigma}}_*, \hat{e}, \mathbf{F}_*, \mathbf{u}, \mathbf{B}, \mathbf{E}, \mathbf{D}, \mathbf{H}, \boldsymbol{\pi}_e, \boldsymbol{\pi}_m, \theta, \mu_\pi', \rho_m, \Re\right\}$, and ζ is a parameter of the Laplace transform.

Assume all the initial conditions for the perturbation of functions to be equal to zero. Let the body be an ideal dielectric. We also restrict ourselves to the linear approximation, in which the equation of motion (2.54) looks as follows:

$$\boldsymbol{\nabla}\cdot\hat{\boldsymbol{\sigma}}_* + \rho_0\mathbf{F}_* = \rho_0\frac{\partial^2\mathbf{u}}{\partial t^2}. \tag{2.251}$$

For the considered two systems of external loads, applying a Laplace transform to the linear momentum equation (2.251), we obtain

$$\boldsymbol{\nabla}\cdot\hat{\boldsymbol{\sigma}}_*^L + \rho_0\mathbf{F}_*^L = \rho_0\zeta^2\mathbf{u}^L, \tag{2.252}$$

$$\boldsymbol{\nabla}\cdot\hat{\boldsymbol{\sigma}}_*'^L + \rho_0\mathbf{F}_*'^L = \rho_0\zeta^2\mathbf{u}'^L. \tag{2.253}$$

Multiplying Eqs. (2.252) and (2.253) by the displacement vectors $\mathbf{u}'^L$ and $\mathbf{u}^L$, respectively, taking the difference between the obtained relations and integrating the result over the body volume (V), we obtain the following formula:

$$\int_{(V)}\left[\left(\boldsymbol{\nabla}\cdot\hat{\boldsymbol{\sigma}}_*^L\right)\cdot\mathbf{u}'^L + \rho_0\mathbf{F}_*^L\cdot\mathbf{u}'^L - \left(\boldsymbol{\nabla}\cdot\hat{\boldsymbol{\sigma}}_*'^L\right)\cdot\mathbf{u}^L - \rho_0\mathbf{F}_*'^L\cdot\mathbf{u}^L\right]dV = 0. \tag{2.254}$$

Making use of the relations

$$\left(\nabla\cdot\hat{\boldsymbol{\sigma}}_*^L\right)\cdot\mathbf{u}'^L=\nabla\cdot\left(\hat{\boldsymbol{\sigma}}_*^L\cdot\mathbf{u}'^L\right)-\hat{\boldsymbol{\sigma}}_*^L:\nabla\mathbf{u}'^L,$$

$$\left(\nabla\cdot\hat{\boldsymbol{\sigma}}_*'^L\right)\cdot\mathbf{u}^L=\nabla\cdot\left(\hat{\boldsymbol{\sigma}}_*'^L\cdot\mathbf{u}^L\right)-\hat{\boldsymbol{\sigma}}_*'^L:\nabla\mathbf{u}^L,$$

the strain–displacement relation (2.8) and the divergence theorem, from the integral equation (2.254), we arrive at

$$\int\limits_{(\Sigma)}\left(\boldsymbol{\sigma}_*^L\cdot\mathbf{u}'^L-\boldsymbol{\sigma}_*'^L\cdot\mathbf{u}^L\right)d\Sigma+\int\limits_{(V)}\rho_0\left(\mathbf{F}_*^L\cdot\mathbf{u}'^L-\mathbf{F}_*'^L\cdot\mathbf{u}^L\right)dV$$

$$=\int\limits_{(V)}\left(\hat{\boldsymbol{\sigma}}_*^L:\hat{e}'^L-\hat{\boldsymbol{\sigma}}_*'^L:\hat{e}^L\right)dV.$$

Here, $\boldsymbol{\sigma}_*^L=\hat{\boldsymbol{\sigma}}_*^L\cdot\mathbf{n}$ and $\boldsymbol{\sigma}_*'^L=\hat{\boldsymbol{\sigma}}_*'^L\cdot\mathbf{n}$.

Substituting the constitutive equation (2.76) into the right-hand side of the obtained equation leads to the following result:

$$\int\limits_{(\Sigma)}\left(\boldsymbol{\sigma}_*^L\cdot\mathbf{u}'^L-\boldsymbol{\sigma}_*'^L\cdot\mathbf{u}^L\right)d\Sigma+\int\limits_{(V)}\rho_0\left(\mathbf{F}_*^L\cdot\mathbf{u}'^L-\mathbf{F}_*'^L\cdot\mathbf{u}^L\right)dV$$

$$=K\alpha_T\int\limits_{(V)}\left(\theta'^Le^L-\theta^Le'^L\right)dV-K\alpha_\rho\int\limits_{(V)}\left(\rho_m^Le'^L-\rho_m'^Le^L\right)dV.\quad(2.255)$$

Let us return to the equation of entropy balance (2.11). Making use of constitutive equations (2.77) and (2.91), from Eq. (2.255), in linear approximation, we obtain the following heat equation for ideal dielectrics:

$$\rho_0C_V\frac{\partial\theta}{\partial t}=\lambda\Delta\theta-T_0K\alpha_T\frac{\partial e}{\partial t}-\rho_0T_0\beta_{T\rho}\frac{\partial\rho_m}{\partial t}+\rho_0\Re.\quad(2.256)$$

Applying a Laplace transform to Eq. (2.256), for two systems of external loads, we can write:

$$\rho_0C_V\zeta\theta^L=\lambda\Delta\theta^L-T_0K\alpha_T\zeta e^L-\rho_0T_0\beta_{T\rho}\zeta\rho_m^L+\rho_0\Re^L,\quad(2.257)$$

$$\rho_0C_V\zeta\theta'^L=\lambda\Delta\theta'^L-T_0K\alpha_T\zeta e'^L-\rho_0T_0\beta_{T\rho}\zeta\rho_m'^L+\rho_0\Re'^L.\quad(2.258)$$

Multiplying Eqs. (2.257) and (2.258) by the functions θ'^L and θ^L, respectively, subtracting them and integrating the obtained result over the area (V), we eventually find that

$$\lambda \int_{(V)} \left(\theta'^L \Delta\theta^L - \theta^L \Delta\theta'^L\right) dV - T_0 K \alpha_T \zeta \int_{(V)} \left(\theta'^L e^L - \theta^L e'^L\right) dV$$

$$-\rho_0 T_0 \beta_{T\rho} \zeta \int_{(V)} \left(\theta'^L \rho_m^L - \theta^L \rho_m'^L\right) dV + \rho_0 \int_{(V)} \left(\theta'^L \mathfrak{R}^L - \theta^L \mathfrak{R}'^L\right) dV = 0.$$

In the first integral of the formula obtained above, we take the following expressions into account

$$\theta'^L \Delta\theta^L - \theta^L \Delta\theta'^L = \boldsymbol{\nabla} \cdot \left(\theta'^L \boldsymbol{\nabla}\theta^L - \theta^L \boldsymbol{\nabla}\theta'^L\right).$$

After applying the divergence theorem, we can rewrite this relation as follows:

$$\frac{\lambda}{\zeta T_0} \int_{(\Sigma)} \left(\theta'^L \boldsymbol{\nabla}\theta^L - \theta^L \boldsymbol{\nabla}\theta'^L\right) \cdot \mathbf{n}\, d\Sigma - K \alpha_T \int_{(V)} \left(\theta'^L e^L - \theta^L e'^L\right) dV$$

$$-\rho_0 \beta_{T\rho} \int_{(V)} \left(\theta'^L \rho_m^L - \theta^L \rho_m'^L\right) dV + \frac{\rho_0}{\zeta T_0} \int_{(V)} \left(\theta'^L \mathfrak{R}^L - \theta^L \mathfrak{R}'^L\right) dV = 0. \tag{2.259}$$

We restrict ourselves to considering a quasi-static electric field and assume that $\frac{\partial \mathbf{A}}{\partial t} = 0$. In view of formula (2.173), we can write

$$\mathbf{E} = -\boldsymbol{\nabla}\varphi_e. \tag{2.260}$$

Applying a Laplace transform to Eq. (2.26) and taking the formulae (2.24) and (2.260) into account, we obtain

$$-\varepsilon_0 \nabla^2 \varphi_e^L + \boldsymbol{\nabla} \cdot \boldsymbol{\Pi}_e^L = 0, \tag{2.261}$$

$$-\varepsilon_0 \nabla^2 \varphi_e'^L + \boldsymbol{\nabla} \cdot \boldsymbol{\Pi}_e'^L = 0. \tag{2.262}$$

We multiply Eqs. (2.261) and (2.262) by the functions $\varphi_e'^L$ and φ_e^L. Proceeding in a similar manner, we obtain

$$-\varepsilon_0 \int_{(V)} \left[\boldsymbol{\nabla} \cdot \left(\boldsymbol{\nabla}\varphi_e^L\right) \varphi_e'^L - \boldsymbol{\nabla} \cdot \left(\boldsymbol{\nabla}\varphi_e'^L\right) \varphi_e^L\right] dV$$

$$= \int_{(V)} \left[\left(\boldsymbol{\nabla} \cdot \boldsymbol{\Pi}_e^L\right) \varphi_e'^L - \left(\boldsymbol{\nabla} \cdot \boldsymbol{\Pi}_e'^L\right) \varphi_e^L\right] dV. \tag{2.263}$$

Further, taking the following expressions into account

$$\boldsymbol{\nabla} \cdot \left(\boldsymbol{\nabla}\varphi_e^L\right) \varphi_e'^L = \boldsymbol{\nabla} \cdot \left[\left(\boldsymbol{\nabla}\varphi_e^L\right) \varphi_e'^L\right] - \left(\boldsymbol{\nabla}\varphi_e^L\right) \cdot \left(\boldsymbol{\nabla}\varphi_e'^L\right),$$

$$\nabla\cdot\left(\nabla\varphi_e'^L\right)\varphi_e^L=\nabla\cdot\left[\left(\nabla\varphi_e'^L\right)\varphi_e^L\right]-\left(\nabla\varphi_e'^L\right)\cdot\left(\nabla\varphi_e^L\right),$$

$$\left(\nabla\cdot\Pi_e^L\right)\varphi_e'^L=\nabla\cdot\left(\Pi_e^L\varphi_e'^L\right)-\Pi_e^L\cdot\left(\nabla\varphi_e'^L\right),$$

$$\left(\nabla\cdot\Pi_e'^L\right)\varphi_e^L=\nabla\cdot\left(\Pi_e'^L\varphi_e^L\right)-\Pi_e'^L\cdot\left(\nabla\varphi_e^L\right),$$

we get

$$\nabla\cdot\left(\nabla\varphi_e^L\right)\varphi_e'^L-\nabla\cdot\left(\nabla\varphi_e'^L\right)\varphi_e^L=\nabla\cdot\left[\left(\nabla\varphi_e^L\right)\varphi_e'^L\right]-\nabla\cdot\left[\left(\nabla\varphi_e'^L\right)\varphi_e^L\right],\tag{2.264}$$

$$\left(\nabla\cdot\Pi_e^L\right)\varphi_e'^L-\left(\nabla\cdot\Pi_e'^L\right)\varphi_e^L$$
$$=\nabla\cdot\left(\Pi_e^L\varphi_e'^L\right)-\nabla\cdot\left(\Pi_e'^L\varphi_e^L\right)+\Pi_e'^L\cdot\left(\nabla\varphi_e^L\right)-\Pi_e^L\cdot\left(\nabla\varphi_e'^L\right).\tag{2.265}$$

In view of the relations (2.264), (2.265) and the divergence theorem, Eq. (2.263) may be written as follows:

$$\varepsilon_0\int_{(\Sigma)}\left(\varphi_e'^L\,\nabla\,\varphi_e^L-\varphi_e^L\,\nabla\,\varphi_e'^L\right)\cdot\mathbf{n}\,d\Sigma+\int_{(\Sigma)}\left(\varphi_e'^L\Pi_e^L-\varphi_e^L\Pi_e'^L\right)\cdot\mathbf{n}\,d\Sigma$$
$$=\int_{(V)}\left(\Pi_e^L\cdot\nabla\varphi_e'^L-\Pi_e'^L\cdot\nabla\varphi_e^L\right)dV\tag{2.266}$$

or

$$\int_{(\Sigma)}\left(\varphi_e'^L\mathbf{D}^L-\varphi_e^L\mathbf{D}'^L\right)\cdot\mathbf{n}d\Sigma=\int_{(V)}\left(\Pi_e^L\cdot\nabla\varphi_e'^L-\Pi_e'^L\cdot\nabla\varphi_e^L\right)dV.\tag{2.267}$$

Combining Eqs. (2.255), (2.259), and (2.267) yields

$$\int_{(\Sigma)}\left[\boldsymbol{\sigma}_*^L\cdot\mathbf{u}'^L-\boldsymbol{\sigma}_*'^L\cdot\mathbf{u}^L+\left(\varphi_e'^L\mathbf{D}^L-\varphi_e^L\mathbf{D}'^L\right)\cdot\mathbf{n}\right]d\Sigma$$
$$-\frac{\lambda}{\zeta T_0}\int_{(\Sigma)}\left(\theta'^L\nabla\theta^L-\theta^L\nabla\theta'^L\right)\cdot\mathbf{n}d\Sigma+\rho_0\int_{(V)}\left(\mathbf{F}_*^L\cdot\mathbf{u}'^L-\mathbf{F}_*'^L\cdot\mathbf{u}^L\right)dV$$
$$-\frac{\rho_0}{\zeta T_0}\int_{(V)}\left(\theta'^L\Re^L-\theta^L\Re'^L\right)dV=\int_{(V)}\left[\Pi_e^L\cdot\nabla\varphi'^L-\Pi_e'^L\cdot\nabla\varphi^L\right]dV$$
$$-\int_{(V)}\left[K\alpha_\rho\left(\rho_m^Le'^L-\rho_m'^Le^L\right)+\rho_0\beta_{T\rho}\left(\theta'^L\rho_m^L-\theta^L\rho_m'^L\right)\right]dV.\tag{2.268}$$

We simplify the integrand in the right-hand side of Eq. (2.268). First, we transform the integrand in the last line of this equation. Using the constitutive relation (2.78), we can write

$$K\alpha_\rho e^L = -\rho_0 \tilde{\mu}_\pi'^L + \rho_0 d_\rho \rho_m^L - \rho_0 \beta_{T\rho} \theta^L, \tag{2.269}$$

$$K\alpha_\rho e'^L = -\rho_0 \left(\tilde{\mu}_\pi'\right)'^L + \rho_0 d_\rho \rho_m'^L - \rho_0 \beta_{T\rho} \theta'^L, \tag{2.270}$$

Substituting the expressions (2.269) and (2.270) into the integrand, we obtain the following formula:

$$K\alpha_\rho \left(\rho_m^L e'^L - \rho_m'^L e^L\right) + \rho_0 \beta_{T\rho} \left(\theta'^L \rho_m^L - \theta^L \rho_m'^L\right)$$
$$= \rho_0 \left[\rho_m'^L \tilde{\mu}_\pi'^L - \rho_m^L \left(\tilde{\mu}_\pi'\right)'^L\right]. \tag{2.271}$$

In view of the constitutive relations (2.79) and (2.80), it can be shown that the following expression is true for a quasi-static electric field:

$$\mathbf{\Pi}_e^L \cdot \nabla \varphi_e'^L - \mathbf{\Pi}_e'^L \cdot \nabla \varphi_e^L = \rho_0 \left[\left(\nabla \tilde{\mu}_\pi'\right)^L \cdot \boldsymbol{\pi}_m'^L - \left(\nabla \tilde{\mu}_\pi'\right)'^L \cdot \boldsymbol{\pi}_m^L\right]. \tag{2.272}$$

Using the relations (2.271) and (2.272), as well as the formula $\rho_m^L = -\nabla \cdot \boldsymbol{\pi}_m^L$ and $\rho_m'^L = -\nabla \cdot \boldsymbol{\pi}_m'^L$, we transform the right-hand side of Eq. (2.268) to obtain

$$\int_{(V)} \left[\mathbf{\Pi}_e^L \cdot \nabla \varphi_e'^L - \mathbf{\Pi}_e'^L \cdot \nabla \varphi_e^L - K\alpha_\rho \left(\rho_m^L e'^L - \rho_m'^L e^L\right)\right.$$
$$\left. + \rho_0 \beta_{T\rho} \left(\theta'^L \rho_m^L - \theta^L \rho_m'^L\right)\right] dV = -\rho_0 \int_{(V)} \nabla \cdot \left[\boldsymbol{\pi}_m^L \left(\tilde{\mu}_\pi'\right)'^L - \boldsymbol{\pi}_m'^L \tilde{\mu}_\pi'^L\right] dV. \tag{2.273}$$

Finally, substituting the expression (2.273) into (2.268) and taking the divergence theorem into account, we obtain the generalized reciprocity theorem in the transformed domain:

$$\zeta T_0 \left\{ \int_{(\Sigma)} \left[\boldsymbol{\sigma}_*^L \cdot \mathbf{u}'^L - \boldsymbol{\sigma}_*'^L \cdot \mathbf{u}^L + \left(\varphi_e'^L \mathbf{D}^L - \varphi_e^L \mathbf{D}'^L\right) \cdot \mathbf{n}\right.\right.$$
$$\left.\left. + \rho_0 \left(\boldsymbol{\pi}_m^L \left(\tilde{\mu}_\pi'\right)'^L - \boldsymbol{\pi}_m'^L \tilde{\mu}_\pi'^L\right) \cdot \mathbf{n}\right] d\Sigma + \rho_0 \int_{(V)} \left(\mathbf{F}_*^L \cdot \mathbf{u}'^L - \mathbf{F}_*'^L \cdot \mathbf{u}^L\right) dV \right\}$$
$$+ \lambda \int_{(\Sigma)} \left(\theta^L \nabla \theta'^L - \theta'^L \nabla \theta^L\right) \cdot \mathbf{n}\, d\Sigma + \rho_0 \int_{(V)} \left(\theta^L \Re'^L - \theta'^L \Re^L\right) dV = 0.$$

Inverting the Laplace transform yields the theorem of reciprocity of work in the desired form

$$T_0 \left\{ \int_{(\Sigma)} \left[\boldsymbol{\sigma}_* \odot \mathbf{u}' - \boldsymbol{\sigma}'_* \odot \mathbf{u} + \varphi'_e \bigcirc (\mathbf{D} \cdot \mathbf{n}) - \varphi_e \bigcirc (\mathbf{D}' \cdot \mathbf{n}) \right. \right.$$

$$\left. \left. + \rho_0 (\boldsymbol{\pi}_m \cdot \mathbf{n}) \bigcirc (\tilde{\mu}'_\pi)' - \rho_0 (\boldsymbol{\pi}'_m \cdot \mathbf{n}) \bigcirc \tilde{\mu}'_\pi \right] d\Sigma + \rho_0 \int_{(V)} (\mathbf{F}_* \odot \mathbf{u}' - \mathbf{F}'_* \odot \mathbf{u}) dV \right\}$$

$$+ \lambda \int_{(\Sigma)} \left[\theta * (\boldsymbol{\nabla}\theta' \cdot \mathbf{n}) - \theta' * (\boldsymbol{\nabla}\theta \cdot \mathbf{n}) \right] d\Sigma + \rho_0 \int_{(V)} (\theta * \Re' - \theta' * \Re) dV = 0 .$$

(2.274)

Here, we use the following notation to indicate the time convolutions:

$$\mathbf{f} \odot \mathbf{g} = \int_0^t \mathbf{f}(\mathbf{r}, t-\tau) \cdot \frac{\partial \mathbf{g}(\mathbf{r},\tau)}{\partial \tau} d\tau ,$$

$$f \bigcirc g = \int_0^t f(\mathbf{r}, t-\tau) \frac{\partial g(\mathbf{r},\tau)}{\partial \tau} d\tau ,$$

$$f * g = \int_0^t f(\mathbf{r}, t-\tau) g(\mathbf{r},\tau) d\tau .$$

Equation (2.274) corresponds to the reciprocity theorem generalized to non-stationary problems of the linear theory of local gradient electrothermoelasticity. It is worth noting that the occurrence of convolutions $\rho_0 (\boldsymbol{\pi}_m \cdot \mathbf{n}) \odot (\tilde{\mu}'_\pi)'$ and $\rho_0 (\boldsymbol{\pi}'_m \cdot \mathbf{n}) \odot \tilde{\mu}'_\pi$ in Eq. (2.274) is the result of taking into account the local mass displacement. In the absence of the local mass displacement effects, Eq. (2.274) yields the reciprocity relation of the classical thermopiezoelectricity obtained by Nowacki (1983).

For stationary processes, Eq. (2.274) can be simplified to the following form:

$$\int_{(\Sigma)} \left\{ \boldsymbol{\sigma}_* \cdot \mathbf{u}' - \boldsymbol{\sigma}'_* \cdot \mathbf{u} + (\varphi'_e \mathbf{D} - \varphi_e \mathbf{D}') \cdot \mathbf{n} + \rho_0 \left[(\tilde{\mu}'_\pi)' \boldsymbol{\pi}_m - \tilde{\mu}'_\pi \boldsymbol{\pi}'_m \right] \cdot \mathbf{n} \right\} d\Sigma$$

$$+\rho_0 \int_{(V)} \left[\mathbf{F}_* \cdot \mathbf{u}' - \mathbf{F}'_* \cdot \mathbf{u} + \beta_{T\rho}\left(\rho'_m\theta - \rho_m\theta'\right) + \frac{K\alpha_T}{\rho_0}\left(e'\theta - e\theta'\right) \right] dV = 0 .$$

(2.275)

In view of the expression

$$\beta_{T\rho}\left(\rho'_m\theta - \rho_m\theta'\right) + K\frac{\alpha_T}{\rho_0}\left(e'\theta - e\theta'\right) = s'\theta - s\theta' ,$$

in a stationary case, the generalized theorem of the reciprocity of work can be formulated as follows:

$$\int_{(\Sigma)} \left\{ \boldsymbol{\sigma}_* \cdot \mathbf{u}' - \boldsymbol{\sigma}'_* \cdot \mathbf{u} + \left(\varphi'_e \mathbf{D} - \varphi_e \mathbf{D}'\right) \cdot \mathbf{n} + \rho_0 \left[\left(\tilde{\mu}'_\pi\right)' \boldsymbol{\pi}_m - \tilde{\mu}'_\pi \boldsymbol{\pi}'_m \right] \cdot \mathbf{n} \right\} d\Sigma$$

$$+\rho_0 \int_{(V)} \left(\mathbf{F}_* \cdot \mathbf{u}' - \mathbf{F}'_* \cdot \mathbf{u} + s'\theta - s\theta'\right) dV = 0 . \qquad (2.276)$$

From Eq. (2.276), in the isothermal approximation, we obtain

$$\int_{(\Sigma)} \left\{ \boldsymbol{\sigma}_* \cdot \mathbf{u}' - \boldsymbol{\sigma}'_* \cdot \mathbf{u} + \left(\varphi'_e \mathbf{D} - \varphi_e \mathbf{D}'\right) \cdot \mathbf{n} + \rho_0 \left[\left(\tilde{\mu}'_\pi\right)' \boldsymbol{\pi}_m - \tilde{\mu}'_\pi \boldsymbol{\pi}'_m \right] \cdot \mathbf{n} \right\} d\Sigma$$

$$+\rho_0 \int_{(V)} \left(\mathbf{F}_* \cdot \mathbf{u}' - \mathbf{F}'_* \cdot \mathbf{u}\right) dV = 0 . \qquad (2.277)$$

2.17 Comparison of the Local Gradient Theory of Dielectrics with Generalized Theories

2.17.1 Constitutive Equations of Integral Type

Let us show that the consideration of the process of local mass displacement in a certain sense is equivalent to the application of integral constitutive relations of spatial type. To this end, we use the Eq. (2.96). Integrating this equation, we can define ρ_m as a functional of displacement **u**, temperature T, and electric field **E**. We demonstrate it by the example of an infinite medium. We represent the solution to Eq. (2.96) as follows (Vladimirov 1971):

$$\rho_m(\mathbf{r}) = \frac{1}{4\pi d_\rho} \int f(\mathbf{r}-\mathbf{r}') \left[K \frac{\alpha_\rho}{\rho_0} \Delta' e(\mathbf{r}') + \beta_{T\rho} \Delta' \theta(\mathbf{r}') + \frac{\chi_{Em}}{\chi_m} \nabla' \cdot \mathbf{E}(\mathbf{r}') \right] d\mathbf{r}', \tag{2.278}$$

where $f(\mathbf{r}) = e^{-\lambda_\mu r}/r$ is the solution to the equation $\Delta f - \lambda_\mu^2 f = -4\pi\delta(\mathbf{r})$; $\lambda_\mu^2 = (d_\rho \chi_m)^{-1}$; $e(\mathbf{r}') = \nabla' \cdot \mathbf{u}(\mathbf{r}')$, where the prime at the Laplace operator means that the derivative was taken by the coordinates with the prime.

Now we consider the integral

$$\begin{aligned}
&\int f(\mathbf{r}-\mathbf{r}')\Delta' g(\mathbf{r}')d\mathbf{r}' \\
&= \int \Delta' \left[f(\mathbf{r}-\mathbf{r}') g(\mathbf{r}') \right] d\mathbf{r}' - 2\int \nabla' f(\mathbf{r}-\mathbf{r}') \cdot \nabla' g(\mathbf{r}') d\mathbf{r}' \\
&\quad - \int g(\mathbf{r}')\Delta' f(\mathbf{r}-\mathbf{r}') d\mathbf{r}' = \oint \nabla' \left[f(\mathbf{r}-\mathbf{r}') g(\mathbf{r}') \right] \cdot \mathbf{n}' d\Sigma' \\
&\quad - 2\int \nabla' f(\mathbf{r}-\mathbf{r}') \cdot \nabla' g(\mathbf{r}') d\mathbf{r}' - \int g(\mathbf{r}') \Delta f(\mathbf{r}-\mathbf{r}') d\mathbf{r}',
\end{aligned} \tag{2.279}$$

Where $g(\mathbf{r}') = \{e(\mathbf{r}'), \theta(\mathbf{r}')\}$ and $\mathbf{n}'$ is the outward normal to the surface Σ'.

Let us choose a surface Σ' in the form of a sphere and direct its radius to infinity. The surface integral in the formula (2.279) over the sphere gives a zero contribution. As a result, we obtain

$$\begin{aligned}
&\int f(\mathbf{r}-\mathbf{r}')\Delta' g(\mathbf{r}')d\mathbf{r}' = -\int g(\mathbf{r}') \left\{ \left[\Delta f(\mathbf{r}-\mathbf{r}') - \lambda_\mu^2 f(\mathbf{r}-\mathbf{r}') \right] \right. \\
&\quad \left. + \lambda_\mu^2 f(\mathbf{r}-\mathbf{r}') \right\} d\mathbf{r}' - 2\int \nabla' f(\mathbf{r}-\mathbf{r}') \cdot \nabla' g(\mathbf{r}') d\mathbf{r}' \\
&= -\int g(\mathbf{r}') \left[-4\pi\delta(\mathbf{r}-\mathbf{r}') + \lambda_\mu^2 f(\mathbf{r}-\mathbf{r}') \right] d\mathbf{r}' - 2\int \nabla' f(\mathbf{r}-\mathbf{r}') \cdot \nabla' g(\mathbf{r}') d\mathbf{r}' \\
&= 4\pi g(\mathbf{r}) - \lambda_\mu^2 \int f(\mathbf{r}-\mathbf{r}') g(\mathbf{r}') d\mathbf{r}' - 2\int \nabla' f(\mathbf{r}-\mathbf{r}') \cdot \nabla' g(\mathbf{r}') d\mathbf{r}'.
\end{aligned} \tag{2.280}$$

In view of the relation (2.280), from the formula (2.278), we get

$$\rho_m(\mathbf{r}) = \frac{1}{d_\rho}\left[K\frac{\alpha_\rho}{\rho_0}e(\mathbf{r}) + \beta_{T\rho}\theta(\mathbf{r})\right]$$

$$-\frac{1}{2\pi d_\rho}\int \nabla' f(\mathbf{r}-\mathbf{r}')\cdot\left[K\frac{\alpha_\rho}{\rho_0}\nabla' e(\mathbf{r}') + \beta_{T\rho}\nabla'\theta(\mathbf{r}')\right]d\mathbf{r}'$$

$$-\frac{\lambda_\mu^2}{4\pi d_\rho}\int f(\mathbf{r}-\mathbf{r}')\left[K\frac{\alpha_\rho}{\rho_0}e(\mathbf{r}') + \beta_{T\rho}\theta(\mathbf{r}') - d_\rho\chi_{Em}\nabla'\cdot\mathbf{E}(\mathbf{r}')\right]d\mathbf{r}'. \tag{2.281}$$

Note that Eqs. (2.24), (2.26), and (2.30) yield

$$\nabla\cdot\mathbf{E} = \varepsilon_0^{-1}(\rho_e + \rho_{e\pi}). \tag{2.282}$$

Thus, taking this equation into account, we can write the expression (2.281) as follows:

$$\rho_m(\mathbf{r}) = \frac{1}{d_\rho}\left[K\frac{\alpha_\rho}{\rho_0}e(\mathbf{r}) + \beta_{T\rho}\theta(\mathbf{r})\right] - \frac{\lambda_\mu^2}{4\pi d_\rho}\int f(\mathbf{r}-\mathbf{r}')$$

$$\times\left\{K\frac{\alpha_\rho}{\rho_0}e(\mathbf{r}') + \beta_{T\rho}\theta(\mathbf{r}') - \frac{d_\rho\chi_{Em}}{\varepsilon_0}\left[\rho_e(\mathbf{r}') + \rho_{e\pi}(\mathbf{r}')\right]\right\}d\mathbf{r}'$$

$$-\frac{1}{2\pi d_\rho}\int \nabla' f(\mathbf{r}-\mathbf{r}')\cdot\left[K\frac{\alpha_\rho}{\rho_0}\nabla' e(\mathbf{r}') + \beta_{T\rho}\nabla'\theta(\mathbf{r}')\right]d\mathbf{r}'.$$

Finally, we substitute the formula (2.281) into constitutive equations (2.76)–(2.80). As a result, we obtain the nonlocal constitutive relations of integral type. For example, for ideal dielectrics, the constitutive relations for the stress tensor and entropy take the form

$$\hat{\sigma}_*(\mathbf{r}) = 2G\hat{e}(\mathbf{r}) + \left\langle\left(\bar{K} - \frac{2}{3}G\right)e(\mathbf{r})\right.$$

$$-K\bar{\alpha}_T\theta(\mathbf{r}) + \lambda_\mu^2\frac{K\alpha_\rho}{4\pi d_\rho}\int f(\mathbf{r}-\mathbf{r}')\left\{K\frac{\alpha_\rho}{\rho_0}e(\mathbf{r}')\right.$$

$$\left.+\beta_{T\rho}\theta(\mathbf{r}') - \frac{d_\rho\chi_{Em}}{\varepsilon_0}\left[\rho_e(\mathbf{r}') + \rho_{e\pi}(\mathbf{r}')\right]\right\}d\mathbf{r}'$$

$$\left.+\frac{K\alpha_\rho}{2\pi d_\rho}\int \nabla' f(\mathbf{r}-\mathbf{r}')\cdot\left[K\frac{\alpha_\rho}{\rho_0}\nabla' e(\mathbf{r}') + \beta_{T\rho}\nabla'\theta(\mathbf{r}')\right]d\mathbf{r}'\right\rangle\hat{\mathbf{I}}, \tag{2.283}$$

$$s(\mathbf{r}) = s_0 + \frac{\bar{C}_V}{T_0}\theta(\mathbf{r}) + K\frac{\bar{\alpha}_T}{\rho_0}e(\mathbf{r}) - \frac{\beta_{T\rho}\lambda_\mu^2}{4\pi d_\rho}\int f(\mathbf{r}-\mathbf{r}')$$

$$\times\left[K\frac{\alpha_\rho}{\rho_0}e(\mathbf{r}') + \beta_{T\rho}\theta(\mathbf{r}') - d_\rho\chi_{Em}\boldsymbol{\nabla}'\cdot\mathbf{E}(\mathbf{r}')\right]d\mathbf{r}'$$

$$-\frac{\beta_{T\rho}}{2\pi d_\rho}\int \boldsymbol{\nabla}'f(\mathbf{r}-\mathbf{r}')\cdot\left[K\frac{\alpha_\rho}{\rho_0}\boldsymbol{\nabla}'e(\mathbf{r}') + \beta_{T\rho}\boldsymbol{\nabla}'\theta(\mathbf{r}')\right]d\mathbf{r}'. \quad (2.284)$$

Here, the quantities $\bar{C}_V$ and $\bar{\alpha}_T$ are presented by the formulae (2.84).

Note that the expression for the polarization vector $\boldsymbol{\pi}_e$ has a similar structure. In such a way, we here showed that excluding the quantities that were introduced to describe the local mass displacement, we obtained the integral-type constitutive relations of the nonlocal theory of electrothermoelastic dielectrics. In this case, the governing set of equations (2.94), (2.95), and (2.97)–(2.99) becomes integro-differential.

2.17.2 Dependence of Constitutive Equations on Gradients of Strain Tensor, Temperature, and Electric Field

For simplicity, we restrict ourselves to a consideration of the ideal dielectrics. By substituting the formula (2.126) into the constitutive equation (2.79) for the polarization vector, we obtain

$$\boldsymbol{\pi}_e = \chi_E\mathbf{E} - \frac{\chi_{Em}}{\kappa_E\lambda_{\mu E}^2}\boldsymbol{\nabla}(\boldsymbol{\nabla}\cdot\mathbf{E}) + \underbrace{\chi_{Em}\frac{K\alpha_\rho}{\rho_0}\boldsymbol{\nabla}(\boldsymbol{\nabla}\cdot\mathbf{u}) + \chi_{Em}\beta_{T\rho}\boldsymbol{\nabla}\theta}_{\substack{\text{the body polarization caused by}\\ \text{the local mass displacement and by the}\\ \text{influence of fields of "non-electric" nature}}}\cdot \quad (2.285)$$

Here, the quantity $\lambda_{\mu E}$ is denoted by the formula (2.111). The relation (2.285) shows that the body polarization is caused not only by the electric field vector, but also by the spatial derivatives $\boldsymbol{\nabla}(\boldsymbol{\nabla}\cdot\mathbf{E})$, $\boldsymbol{\nabla}(\boldsymbol{\nabla}\cdot\mathbf{u})$, and $\boldsymbol{\nabla}\theta$. Therefore, the local gradient theory of dielectrics will describe the linear response of the polarization to the deformation gradient (flexoelectric effect) and to the temperature gradient (thermopolarization effect). Note that such type of relation for the polarization vector is obtained in the piezoelectric theory, which takes the higher-order gradients of the strain tensor (Mindlin 1972a) or higher electric quadrupoles (Kafadar 1971; Yang et al. 2004) into account. This allows us to assert that in a certain sense, taking into account the local mass displacement is equivalent to accounting for, in the theoretical description, the dependence of the local thermodynamic state of the body on the spatial gradients of strain, of electric field, and of temperature field.

Substitution of the formula (2.132) into the constitutive equation (2.76) yields

$$\hat{\boldsymbol{\sigma}}_* = 2G\hat{e} + \left[\left(K - \frac{2}{3}G\right)e - K\alpha_T\theta - K\alpha_\rho \frac{\chi_m - \kappa_E\chi_{Em}}{\kappa_E}\boldsymbol{\nabla}\cdot\mathbf{E}\right]\hat{\mathbf{I}}.$$

The above formula and the relation (2.285) show that the local gradient theory of dielectrics accommodates an electromechanical interaction in isotropic (including centrosymmetric) materials.

Problems

1. Starting with Eq. (2.47), give the detailed calculations leading to the relations (2.49)–(2.52).
2. Find the nonlinear explicit forms of constitutive relations for isotropic thermoelastic polarized media.
3. Construct the corresponding constitutive relations, taking the density of internal energy $u = u(\hat{e}, s, \boldsymbol{\pi}_e, \rho_m, \boldsymbol{\pi}_m)$ as a thermodynamic potential. Build an explicit form of linear constitutive relations for centrosymmetric cubic crystals.
4. Make detailed calculations leading to the relation (2.162).
5. Using the relations (2.172) and (2.173), formulate the boundary conditions in terms of scalar φ_u, φ_e and vector $\boldsymbol{\psi}_u$, **A** fields of the displacement vector and electromagnetic field vectors.
6. Using the expressions of differential operators in cylindrical coordinates (r, φ, z)

$$\boldsymbol{\nabla} a = \frac{\partial a}{\partial r}\mathbf{i}_r + \frac{1}{r}\frac{\partial a}{\partial \varphi}\mathbf{i}_\varphi + \frac{\partial a}{\partial z}\mathbf{i}_z,$$

$$\nabla\cdot\mathbf{a}=\frac{1}{r}\frac{\partial(ra_r)}{\partial r}+\frac{1}{r}\frac{\partial a_\varphi}{\partial\varphi}+\frac{\partial a_z}{\partial z},$$

$$\nabla\times\mathbf{a}=\left(\frac{1}{r}\frac{\partial a_z}{\partial\varphi}-\frac{\partial a_\varphi}{\partial z}\right)\mathbf{i}_r$$

$$+\left(\frac{\partial a_r}{\partial z}-\frac{\partial a_z}{\partial r}\right)\mathbf{i}_\varphi+\frac{1}{r}\left(\frac{\partial(ra_\varphi)}{\partial r}-\frac{\partial a_r}{\partial\varphi}\right)\mathbf{i}_z,$$

express the governing set of equations (2.153)–(2.155) of the elastic polarized solids in cylindrical coordinates.

7. Study the literature and obtain constitutive relations for the local gradient theory of non-ferromagnetic thermoelastic polarized binary solid solutions. **Instruction**: The local mass displacement of the subsystem of an admixture should be neglected.
8. Study the literature and construct a local gradient theory of a viscous incompressible polarized liquid.
9. Prove the theorem of the reciprocity of work for stationary processes. Make detailed calculations leading to the relation (2.276).

Chapter 3

Generalized Local Gradient Theories of Dielectrics

In Section 3.1, the local gradient theory of dielectrics is generalized for polarized media in which the mass flux and the density of the induced mass are tensors of second order. This theory enables one to describe the effect of the local mass displacement on the shear stresses in isotropic materials, including centrosymmetric cubic crystals.

In Section 3.2, an approach that accounts for the irreversibility and inertia of both the local mass displacement and electric polarization is proposed. It is shown that the dissipation and inertia of the mentioned processes impose the rheological constitutive relations for the heat flux and electric current as well as for vectors of the local mass displacement and polarization.

Section 3.3 present the constitutive relations of the local gradient electrothermomechanics of a rheological dielectric medium with a fading memory.

In Section 3.4, continuum–thermodynamic approach for the construction of a complete set of equations of the local gradient theory of non-ferromagnetic dielectric media with electric quadrupoles is proposed. In this case, the quadrupole polarization, the local mass displacement and its irreversibility are taken into account in order to obtain the general theory of dielectrics. Within the framework of this theory, the space of constitutive variables

Local Gradient Theory for Dielectrics: Fundamentals and Applications
Olha Hrytsyna and Vasyl Kondrat

ISBN 978-981-4800-62-4 (Hardcover), 978-1-003-00686-2 (eBook)
www.jennystanford.com

additionally includes the gradient of the electric field vector. It is shown that the electric quadrupole is a thermodynamic conjugate of the electric field gradient.

3.1 Local Gradient Theory of Thermoelastic Polarized Media: Tensor-Like Representation of Parameters Related to the Local Mass Displacement

It was shown in Section 2.13 that in view of the assumptions adopted in Chapter 2, the local mass displacement does not cause any shear stresses in polarized solids. Therefore, the tangential stresses in an isotropic body may occur only due to mechanical motion. Here we intend to obtain a set of relations of the local gradient theory of electrothermoelastic non-ferromagnetic polarized bodies, which takes into account the effect of the local mass displacement on the shear stresses. To this end, we assume that the parameters associated with the local mass displacement, namely, the density of the induced mass and the measure of the effect of the local mass displacement on the internal energy, are tensors of the second order.

3.1.1 Mass Balance Equation

Following Pidstryhach (1965), we characterize the mass density by a symmetric tensor of the second order $\hat{\boldsymbol{\rho}}$ and its flux by a tensor of the third order $\hat{\mathbf{J}}^{(3)}_{m*}$. Thus, we can write the mass balance equation in the integral form as follows:

$$\frac{d}{dt}\int_{(V)} \hat{\rho}\, dV = -\oint_{(\Sigma)} \hat{\mathbf{J}}^{(3)}_{m*} \cdot \mathbf{n}\, d\Sigma . \tag{3.1}$$

We represent the tensor of mass flux density $\hat{\mathbf{J}}^{(3)}_{m*}$ as

$$\hat{\mathbf{J}}^{(3)}_{m*} = \hat{\mathbf{J}}^{(3)}_{mc} + \hat{\mathbf{J}}^{(3)}_{ms} ,$$

where $\hat{\mathbf{J}}^{(3)}_{mc}$ is the convective component of mass flux and $\hat{\mathbf{J}}^{(3)}_{ms}$ is the mass flux related to the local mass displacement.

We define the constituent $\hat{\mathbf{J}}^{(3)}_{mc}$ as a dyadic product of the mass density tensor and of the velocity $\mathbf{v}_*$ of the convective transition of a small body element, i.e., $\hat{\mathbf{J}}_{mc}^{(3)} = \mathbf{v}_* \otimes \hat{\boldsymbol{\rho}}$ (Pidstryhach 1965). Introducing the tensor of the local mass displacement $\widehat{\boldsymbol{\Pi}}^{m(3)}$

$$\hat{\mathbf{J}}_{ms}^{(3)} = \frac{\partial \widehat{\boldsymbol{\Pi}}^{m(3)}}{\partial t}, \tag{3.2}$$

we obtain the following relation for the mass flux density:

$$\hat{\mathbf{J}}_{m*}^{(3)} = \mathbf{v}_* \otimes \hat{\boldsymbol{\rho}} + \frac{\partial \widehat{\boldsymbol{\Pi}}^{m(3)}}{\partial t} \tag{3.3}$$

In view of the divergence theorem and formula (3.3), the mass balance equation (3.1) can be written in the local form

$$\frac{\partial \hat{\boldsymbol{\rho}}}{\partial t} = -\boldsymbol{\nabla} \cdot \left(\mathbf{v}_* \otimes \hat{\boldsymbol{\rho}} + \frac{\partial \widehat{\boldsymbol{\Pi}}^{m(3)}}{\partial t} \right). \tag{3.4}$$

Through the relation

$$\mathbf{v} \otimes \hat{\boldsymbol{\rho}} = \mathbf{v}_* \otimes \hat{\boldsymbol{\rho}} + \frac{\partial \widehat{\boldsymbol{\Pi}}^{m(3)}}{\partial t} \tag{3.5}$$

we introduce the velocity vector **v** of the continuum of the mass centers of a small body element. Then, the equation of mass balance (3.4) takes the form

$$\frac{d\hat{\boldsymbol{\rho}}}{dt} + (\boldsymbol{\nabla} \cdot \mathbf{v})\hat{\boldsymbol{\rho}} = 0. \tag{3.6}$$

Note that Eq. (3.6) coincides with a result obtained by Pidstryhach (1965).

We define the average mass density by the formula $\rho \equiv \rho_{ii}$. Then, convolving Eq. (3.6), we can obtain a standard form of the mass balance equation (see Eq. (2.55)):

$$\frac{\partial \rho}{\partial t} + \boldsymbol{\nabla} \cdot (\rho \mathbf{v}) = 0. \tag{3.7}$$

3.1.2 Energy Balance Equation

Under the foregoing assumptions, the balance equation for the total energy of a “body–electromagnetic field” system can be written as follows:

$$\frac{d}{dt}\int\limits_{(V)}\left(\rho u+\frac{1}{2}\rho\mathbf{v}^2+U_e\right)dV=-\oint\limits_{(\Sigma)}\left[\rho\left(u+\frac{1}{2}\mathbf{v}^2\right)\mathbf{v}-\hat{\boldsymbol{\sigma}}\cdot\mathbf{v}+\mathbf{J}_q\right.$$

$$\left.+\mathbf{S}_e+\hat{\mathbf{J}}_m^{(3)}:\hat{\boldsymbol{\mu}}+\hat{\mathbf{J}}_{ms}^{(3)}:\hat{\boldsymbol{\mu}}^{\pi}\right]\cdot\mathbf{n}\,d\Sigma+\int\limits_{(V)}(\rho\mathbf{F}\cdot\mathbf{v}+\rho\Re)dV. \tag{3.8}$$

Here, $\hat{\boldsymbol{\mu}}$ is the tensor of the chemical potential (Pidstryhach 1965); $\hat{\boldsymbol{\mu}}^{\pi}$ is the tensor of the second order whose components characterize the change in the internal energy of the system caused by the local mass displacement; $\hat{\mathbf{J}}_m^{(3)}=(\mathbf{v}_*-\mathbf{v})\otimes\hat{\boldsymbol{\rho}}$ is the mass flux related to the tensor $\widehat{\boldsymbol{\Pi}}^{m(3)}$ of the local mass displacement by the formula

$$\hat{\mathbf{J}}_m^{(3)}=-\frac{\partial\widehat{\boldsymbol{\Pi}}^{m(3)}}{\partial t}. \tag{3.9}$$

Note that fluxes $\hat{\mathbf{J}}_m^{(3)}$ and $\widehat{\boldsymbol{\Pi}}^{m(3)}$ are symmetric with respect to the second and third indices.

If we write Eq. (3.8) in the local form and take into account the mass conservation law (3.7), the entropy balance equation (2.11), the balance law (2.36) for energy of the electromagnetic field as well as the formulae (3.2) and (3.9), we can obtain the following balance equation for the internal energy

$$\rho\frac{du}{dt}=\left[\hat{\boldsymbol{\sigma}}-\rho(\mathbf{E}_*\cdot\boldsymbol{\pi}_e)\hat{\mathbf{I}}\right]:(\boldsymbol{\nabla}\otimes\mathbf{v})+\rho T\frac{ds}{dt}+\rho\mathbf{E}_*\cdot\frac{d\boldsymbol{\pi}_e}{dt}$$

$$-\frac{\partial\left(\boldsymbol{\nabla}\cdot\widehat{\boldsymbol{\Pi}}^{m(3)}\right)}{\partial t}:\hat{\boldsymbol{\mu}}'^{\pi}-\frac{\partial\widehat{\boldsymbol{\Pi}}^{m(3)}}{\partial t}\overset{(3)}{\cdot}\widehat{\mathbf{M}}^{(3)}-\mathbf{J}_q\cdot\frac{\boldsymbol{\nabla}T}{T}+\mathbf{J}_e\cdot\mathbf{E}_*-T\eta_s$$

$$+\mathbf{v}\cdot\left(-\rho\frac{d\mathbf{v}}{dt}+\boldsymbol{\nabla}\cdot\left[\hat{\boldsymbol{\sigma}}-\rho(\mathbf{E}_*\cdot\boldsymbol{\pi}_e)\hat{\mathbf{I}}\right]+\mathbf{F}_e+\rho\mathbf{F}\right). \tag{3.10}$$

Here, $\hat{\boldsymbol{\mu}}'^{\pi}=\hat{\boldsymbol{\mu}}^{\pi}-\hat{\boldsymbol{\mu}}$, and $\widehat{\mathbf{M}}^{(3)}=(\boldsymbol{\nabla}\otimes\hat{\boldsymbol{\mu}}'^{\pi})^{\mathrm{T}(1,3)}$ is a tensor of the third order, which is an isomer of the tensor $\boldsymbol{\nabla}\otimes\hat{\boldsymbol{\mu}}'^{\pi}$ that was formed by transposition of the first and third base vectors; the dot means, as before, the convolution with respect to a pair of indices, and the number in parentheses above it indicates the number of such convolutions.

Using the relation

$$\hat{\boldsymbol{\rho}}^{m\pi} = -\nabla \cdot \widehat{\boldsymbol{\Pi}}^{m(3)}, \tag{3.11}$$

we bring into consideration the tensor of induced mass density. Since the tensor $\widehat{\boldsymbol{\Pi}}^{m(3)}$ is symmetric with respect to the transposition of the second and third indices, then $\hat{\boldsymbol{\rho}}^{m\pi}$ is a symmetric tensor. Note that differentiating formula (3.11) with respect to time and taking the relation (3.2) into account, we obtain the following balance equation of the induced mass

$$\frac{\partial \hat{\boldsymbol{\rho}}^{m\pi}}{\partial t} + \nabla \cdot \hat{\mathbf{J}}_{ms}^{(3)} = 0. \tag{3.12}$$

By the formulae $\rho_{m\pi} \equiv \hat{\boldsymbol{\rho}}^{m\pi} : \hat{\mathbf{I}} = \rho_{kk}^{m\pi}$ and $\boldsymbol{\Pi}_m \equiv \widehat{\boldsymbol{\Pi}}^{m(3)} : \hat{\mathbf{I}}$, we introduce the average density of the induced mass and vector $\boldsymbol{\Pi}_m$. By the convolution of the relation (3.11), we obtain a formula (2.40), which relates the average density of the induced mass $\rho_{m\pi}$ and the vector $\boldsymbol{\Pi}_m$.

Using the formulae

$$\hat{\boldsymbol{\pi}}^{m(3)} = \frac{1}{\rho} \widehat{\boldsymbol{\Pi}}^{m(3)}, \quad \hat{\boldsymbol{\rho}}^{m} = \frac{1}{\rho} \hat{\boldsymbol{\rho}}^{m\pi}, \tag{3.13}$$

we introduce the specific values of corresponding quantities.

Now, substituting formulae (3.11) and (3.13) into relation (3.10) and taking the mass balance equation (3.7) into account, we can write the internal energy balance equation in the form:

$$\rho \frac{du}{dt} = \hat{\boldsymbol{\sigma}}_* : (\nabla \otimes \mathbf{v}) + \rho T \frac{ds}{dt} + \rho \mathbf{E}_* \cdot \frac{d\boldsymbol{\pi}_e}{dt} + \rho \hat{\boldsymbol{\mu}}'^{\pi} : \frac{d\hat{\boldsymbol{\rho}}^{m}}{dt} - \rho \frac{d\hat{\boldsymbol{\pi}}^{m(3)}}{dt} \overset{(3)}{\cdot} \widehat{\mathbf{M}}^{(3)}$$

$$- \mathbf{J}_q \cdot \frac{\nabla T}{T} + \mathbf{J}_{e*} \cdot \mathbf{E}_* - T\eta_s + \mathbf{v} \cdot \left(-\rho \frac{d\mathbf{v}}{dt} + \nabla \cdot \hat{\boldsymbol{\sigma}}_* + \mathbf{F}_e + \rho \mathbf{F}_* \right), \tag{3.14}$$

where

$$\hat{\boldsymbol{\sigma}}_* = \hat{\boldsymbol{\sigma}} - \rho \left[\mathbf{E}_* \cdot \boldsymbol{\pi}_e + \hat{\boldsymbol{\rho}}^{m} : \hat{\boldsymbol{\mu}}'^{\pi} - \hat{\boldsymbol{\pi}}^{m(3)} \overset{(3)}{\cdot} (\nabla \otimes \hat{\boldsymbol{\mu}}'^{\pi})^{\mathrm{T}(1,3)} \right] \hat{\mathbf{I}}, \tag{3.15}$$

$$\mathbf{F}_* = \mathbf{F} + \hat{\boldsymbol{\rho}}^{m} : (\nabla \otimes \hat{\boldsymbol{\mu}}'^{\pi})^{\mathrm{T}(1,3)} - \hat{\boldsymbol{\pi}}^{m(3)} \overset{(3)}{\cdot} (\nabla \otimes \nabla \otimes \hat{\boldsymbol{\mu}}'^{\pi})^{\mathrm{T}(1,3)(2,4)}. \tag{3.16}$$

Here, the superscript T(1,3)(2,4) in the relation (3.16) indicates the isomer of the fourth-order tensor $\nabla\otimes\nabla\otimes\hat{\boldsymbol{\mu}}'^{\pi}$ formed by the transposition of the first and third indices, as well as of the second and fourth indices.

It is easy to show that the consequence of the invariance of the internal energy balance equation (3.14) with respect to the space translations is a symmetry of the stress tensor $\hat{\boldsymbol{\sigma}}_*$ and the momentum equation

$$\nabla\cdot\hat{\boldsymbol{\sigma}}_* + \mathbf{F}_e + \rho\mathbf{F}_* = \rho\frac{d\mathbf{v}}{dt}. \tag{3.17}$$

The latter equation coincides with Eq. (2.54), but here the stress tensor and an additional mass force are determined by formulae (3.15) and (3.16).

Taking Eq. (3.17) into account, we simplify the expression (3.14) to the form:

$$\rho\frac{du}{dt} = \hat{\boldsymbol{\sigma}}_* : \frac{d\hat{e}}{dt} + \rho T\frac{ds}{dt} + \rho\mathbf{E}_*\cdot\frac{d\boldsymbol{\pi}_e}{dt} + \rho\hat{\boldsymbol{\mu}}'^{\pi} : \frac{d\hat{\boldsymbol{\rho}}^{m}}{dt} - \rho\frac{d\hat{\boldsymbol{\pi}}^{m(3)}}{dt} \overset{(3)}{\cdot} \widehat{\mathbf{M}}^{(3)}$$

$$-\mathbf{J}_q\cdot\frac{\nabla T}{T} + \mathbf{J}_{e*}\cdot\mathbf{E}_* - T\eta_s. \tag{3.18}$$

We introduce the generalized Helmholtz free energy by the relation

$$f = u - Ts - \mathbf{E}_*\cdot\boldsymbol{\pi}_e + \hat{\boldsymbol{\pi}}^{m(3)} \overset{(3)}{\cdot} \widehat{\mathbf{M}}^{(3)}.$$

Based on Eq. (3.18), for the free energy *f*, we get

$$\rho\frac{df}{dt} = \hat{\boldsymbol{\sigma}}_* : \frac{d\hat{e}}{dt} - \rho s\frac{dT}{dt} - \rho\boldsymbol{\pi}_e\cdot\frac{d\mathbf{E}_*}{dt} + \rho\hat{\boldsymbol{\mu}}'^{\pi} : \frac{d\hat{\boldsymbol{\rho}}^{m}}{dt} + \rho\hat{\boldsymbol{\pi}}^{m(3)} \overset{(3)}{\cdot} \frac{d\widehat{\mathbf{M}}^{(3)}}{dt}$$

$$-\mathbf{J}_q\cdot\frac{\nabla T}{T} + \mathbf{J}_{e*}\cdot\mathbf{E}_* - T\eta_s. \tag{3.19}$$

3.1.3 Constitutive Equations

Assume the free energy to be determined by a scalar parameter *T*, a vector parameter $\mathbf{E}_*$, and by three tensor parameters $\hat{e}$, $\hat{\boldsymbol{\rho}}^{m}$, and $\widehat{\mathbf{M}}^{(3)}$. Then, based on Eq. (3.19), we obtain a differential one-form

$$df = \frac{1}{\rho}\hat{\boldsymbol{\sigma}}_* : d\hat{e} - s\,dT - \boldsymbol{\pi}_e\cdot d\mathbf{E}_* + \hat{\boldsymbol{\mu}}'^{\pi} : d\hat{\boldsymbol{\rho}}^{m} + \hat{\boldsymbol{\pi}}^{m(3)} \overset{(3)}{\cdot} d\widehat{\mathbf{M}}^{(3)}, \tag{3.20}$$

which corresponds to the generalized Gibbs equation and an expression for entropy production

$$\eta_s = \frac{1}{T}\left(-\mathbf{J}_q \cdot \frac{\nabla T}{T} + \mathbf{J}_{e*} \cdot \mathbf{E}_*\right). \tag{3.21}$$

Making use of the generalized Gibbs equation (3.20), we formulate the constitutive relations

$$\hat{\boldsymbol{\sigma}}_* = \rho \frac{\partial f}{\partial \hat{e}}\bigg|_{T,\mathbf{E}_*,\hat{\boldsymbol{\rho}}^m,\widehat{\mathbf{M}}^{(3)}}, \quad s = -\frac{\partial f}{\partial T}\bigg|_{\hat{e},\mathbf{E}_*,\hat{\boldsymbol{\rho}}^m,\widehat{\mathbf{M}}^{(3)}}, \quad \boldsymbol{\pi}_e = -\frac{\partial f}{\partial \mathbf{E}_*}\bigg|_{\hat{e},T,\hat{\boldsymbol{\rho}}^m,\widehat{\mathbf{M}}^{(3)}}, \tag{3.22}$$

$$\hat{\boldsymbol{\mu}}'^{\pi} = \frac{\partial f}{\partial \hat{\boldsymbol{\rho}}^m}\bigg|_{\hat{e},T,\mathbf{E}_*,\widehat{\mathbf{M}}^{(3)}}, \quad \hat{\boldsymbol{\pi}}^{m(3)} = \frac{\partial f}{\partial \widehat{\mathbf{M}}^{(3)}}\bigg|_{\hat{e},T,\mathbf{E}_*,\hat{\boldsymbol{\rho}}^m}. \tag{3.23}$$

From the expression for entropy production (3.21), we also write the following relations for fluxes

$$\mathbf{j}_i = \mathbf{j}_i(\mathbf{X}_1, \mathbf{X}_2; A), \quad i = 1, 2, \tag{3.24}$$

where

$$\mathbf{j}_1 = \mathbf{j}_q, \quad \mathbf{j}_2 = \mathbf{j}_{e*}, \quad \mathbf{X}_1 = -\nabla T/T, \quad \mathbf{X}_2 = \mathbf{E}_*.$$

According to formulae (3.22) and (3.23), along with the strain tensor, temperature, and electric field, the space of the constitutive parameters also contains a tensor of the third order, $\widehat{\mathbf{M}}^{(3)}$ (i.e., gradient of potential $\hat{\boldsymbol{\mu}}'^{\pi}$), and a tensor of the induced mass density, $\hat{\boldsymbol{\rho}}^m$ (i.e., divergence of the tensor of the local mass displacement $\widehat{\boldsymbol{\Pi}}^{m(3)}$). A specific tensor of the local mass displacement, $\hat{\boldsymbol{\pi}}^{m(3)}$, and a tensor of the modified chemical potential, $\hat{\boldsymbol{\mu}}'^{\pi}$, are conjugate to them.

A complete set of nonlinear equations of local gradient electrothermoelasticity of non-ferromagnetic polarized solids includes (i) balance equations (2.11), (3.6), (3.12), and (3.17); (ii) equations of electrodynamics containing Maxwell equations (2.20), (2.21), (2.25), (2.26) and constitutive relations for vectors of the electromagnetic field (2.17), (2.24); (iii) expression (3.21) for entropy production; (iv) geometric and physical relations (2.8), and (3.22)–(3.24). In the formulation of boundary-value problems of mathematical physics, this set should be complemented by corresponding boundary and initial conditions.

Assuming that the free energy is an analytic function of the variables $\hat{e}$, θ, $\mathbf{E}$, $\hat{\boldsymbol{\rho}}^m$, and $\widehat{\mathbf{M}}^{(3)}$, we can expand it in the Taylor series with respect to small perturbations of parameters concerning the natural state of the infinite anisotropic homogeneous medium. Let

$$\hat{e}=0, \quad \hat{\boldsymbol{\sigma}}_* =0, \quad T=T_0, \quad \boldsymbol{\pi}_e=0, \quad \mathbf{E}_*=0,$$

$$\hat{\boldsymbol{\rho}}^m=0, \quad \hat{\boldsymbol{\mu}}'^{\pi}=\hat{\boldsymbol{\mu}}_0'^{\pi}, \quad \widehat{\mathbf{M}}^{(3)}=0, \quad \hat{\boldsymbol{\pi}}^{m(3)}=0$$

in this state. In order to formulate the linear constitutive equations, in the expansion of the free energy f, we retain the summands to at most the second order

$$\begin{aligned} f = {} & f_0 - s_0\,\theta + \hat{\boldsymbol{\mu}}_0'^{\pi} : \hat{\boldsymbol{\rho}}^m + \frac{1}{2\rho_0}\hat{\mathbf{C}}^{(4)} \overset{(4)}{\cdot} (\hat{e}\otimes\hat{e}) + \frac{1}{2}\hat{d}^{\rho(4)} \overset{(4)}{\cdot} (\hat{\boldsymbol{\rho}}^m \otimes \hat{\boldsymbol{\rho}}^m) - \frac{C_V}{2T_0}\theta^2 \\ & - \frac{1}{\rho_0}\hat{\boldsymbol{\alpha}}^{\rho(4)} \overset{(4)}{\cdot} (\hat{e}\otimes\hat{\boldsymbol{\rho}}^m) - \frac{1}{\rho_0}\hat{\boldsymbol{\beta}} : \hat{e}\,\theta - \frac{1}{\rho_0}\hat{\mathbf{g}}^{(5)} \overset{(5)}{\cdot} (\hat{e}\otimes\widehat{\mathbf{M}}^{(3)}) - \boldsymbol{\beta}^E \cdot \mathbf{E}\,\theta - \hat{\boldsymbol{\beta}}^{T\rho} : \hat{\boldsymbol{\rho}}^m\,\theta \\ & - \frac{1}{\rho_0}\hat{\mathbf{f}}^{(3)} \overset{(3)}{\cdot} (\hat{e}\otimes\mathbf{E}) - \frac{1}{2}\hat{\boldsymbol{\chi}}^{E} \overset{(2)}{\cdot} (\mathbf{E}\otimes\mathbf{E}) - \hat{\boldsymbol{\gamma}}^{E(3)} \overset{(3)}{\cdot} (\mathbf{E}\otimes\hat{\boldsymbol{\rho}}^m) - \hat{\boldsymbol{\beta}}^{\mu(3)} \overset{(3)}{\cdot} \widehat{\mathbf{M}}^{(3)}\,\theta \\ & - \frac{1}{2}\hat{\boldsymbol{\chi}}^{m(6)} \overset{(6)}{\cdot} (\widehat{\mathbf{M}}^{(3)} \otimes \widehat{\mathbf{M}}^{(3)}) - \hat{\boldsymbol{\gamma}}^{\rho(5)} \overset{(5)}{\cdot} (\widehat{\mathbf{M}}^{(3)} \otimes \hat{\boldsymbol{\rho}}^m) + \hat{\boldsymbol{\chi}}^{Em(4)} \overset{(4)}{\cdot} (\mathbf{E}\otimes\widehat{\mathbf{M}}^{(3)}), \end{aligned} \tag{3.25}$$

where $\hat{\mathbf{C}}^{(4)}$ and $\hat{\boldsymbol{\alpha}}^{\rho(4)}$ are the fourth-order tensors of the elastic constants and of a volumetric expansion caused by the local mass displacement; $\hat{\boldsymbol{\beta}}$ is a tensor of thermal expansion; $\hat{\mathbf{f}}^{(3)}$, $\hat{\mathbf{g}}^{(3)}$, and $\hat{\boldsymbol{\beta}}^{\mu(3)}$ are the third-order tensors of piezoelectric, piezomass, and pyromass coefficients, respectively; C_V is the specific heat capacity; $\hat{\boldsymbol{\beta}}^{T\rho}$ and $\hat{d}^{\rho(4)}$ are the second- and fourth-order tensors of isothermal–isochoric coefficients of the dependency of entropy and potential $\hat{\boldsymbol{\mu}}'^{\pi}$ on the density of the induced mass; $\boldsymbol{\beta}^E$ is the pyroelectric coefficients; $\hat{\boldsymbol{\chi}}^{E}$ is the dielectric susceptibility tensor; $\hat{\boldsymbol{\chi}}^{m(6)}$ and $\hat{\boldsymbol{\chi}}^{Em(4)}$ are the sixth- and fourth-order tensors characterizing the dependence of vectors of the local mass displacement and the polarization on tensor $\widehat{\mathbf{M}}^{(3)}$; $\hat{\boldsymbol{\gamma}}^{\rho(5)}$ and $\hat{\boldsymbol{\gamma}}^{E(3)}$ are the third- and fifth-order tensors of the coefficients that characterize the dependence of the tensor $\hat{\boldsymbol{\mu}}'^{\pi}$ on its gradient (i.e., tensor $\widehat{\mathbf{M}}^{(3)}$) and electric field $\mathbf{E}$.

Using representation (3.25), in the linear approximation, Eqs. (3.22) and (3.23) for an anisotropic elastic medium can be written in the explicit form

$$\hat{\boldsymbol{\sigma}}_* = \hat{\mathbf{C}}^{(4)} : \hat{e} - \hat{\boldsymbol{\beta}}\theta - \hat{\boldsymbol{\rho}}^m : \hat{\boldsymbol{\alpha}}^{\rho(4)} - \hat{\mathbf{M}}^{(3)} \overset{(3)}{\cdot} \hat{\mathbf{g}}^{(5)} - \mathbf{E}\cdot\hat{\mathbf{f}}^{(3)}, \tag{3.26}$$

$$s = s_0 + \frac{C_V}{T_0}\theta + \hat{\boldsymbol{\beta}}^{T\rho} : \hat{\boldsymbol{\rho}}^m + \hat{\boldsymbol{\beta}}^{\mu(3)} \overset{(3)}{\cdot} \hat{\mathbf{M}}^{(3)} + \frac{1}{\rho_0}\hat{\boldsymbol{\beta}} : \hat{e} + \boldsymbol{\beta}^E \cdot \mathbf{E}, \tag{3.27}$$

$$\hat{\boldsymbol{\mu}}'^{\pi} = \hat{\boldsymbol{\mu}}_0'^{\pi} + \hat{d}^{\rho(4)} : \hat{\boldsymbol{\rho}}^m - \hat{\boldsymbol{\gamma}}^{\rho(5)} \overset{(3)}{\cdot} \hat{\mathbf{M}}^{(3)} - \frac{1}{\rho_0}\hat{\boldsymbol{\alpha}}^{\rho(4)} : \hat{e} - \hat{\boldsymbol{\beta}}^{T\rho}\theta - \hat{\boldsymbol{\gamma}}^{E(3)} \cdot \mathbf{E}, \tag{3.28}$$

$$\boldsymbol{\pi}_e = \hat{\boldsymbol{\chi}}^E \cdot \mathbf{E} - \hat{\mathbf{M}}^{(3)} \overset{(3)}{\cdot} \hat{\boldsymbol{\chi}}^{Em(4)} + \hat{\boldsymbol{\rho}}_m : \boldsymbol{\gamma}^{E(3)} + \frac{1}{\rho_0}\hat{\mathbf{f}}^{(3)} : \hat{e} + \boldsymbol{\beta}^E\theta, \tag{3.29}$$

$$\hat{\boldsymbol{\pi}}^{m(3)} = -\hat{\boldsymbol{\chi}}^{m(6)} \overset{(3)}{\cdot} \hat{\mathbf{M}}^{(3)} - \hat{\boldsymbol{\rho}}^m : \hat{\boldsymbol{\gamma}}^{\rho(5)} - \frac{1}{\rho_0}\hat{\mathbf{g}}^{(5)} : \hat{e} - \hat{\boldsymbol{\beta}}^{\mu(3)}\theta + \hat{\boldsymbol{\chi}}^{Em(4)} \cdot \mathbf{E}. \tag{3.30}$$

We can see that under the assumptions accepted within this subsection, the local mass displacement can cause the emergence of shear stresses not only in anisotropic materials but also in crystals with a high symmetry of the lattice structure.

3.1.4 Governing Equations for Isotropic Elastic Medium

Let us write down the governing set of equations for a mathematical model of an elastic medium. For simplicity, consider an isothermal approximation and restrict ourselves to the processes of deformation and the local mass displacement only. For isotropic materials, all tensors of odd orders vanish, while for the components of tensors of the fourth and sixth order, we get

$$\mathrm{C}_{ijkl} = C_1\delta_{ij}\delta_{kl} + C_2\delta_{ik}\delta_{jl} + C_3\delta_{il}\delta_{jk},$$

$$\alpha^{\rho}_{ijkl} = a_1\delta_{ij}\delta_{kl} + a_2\delta_{ik}\delta_{jl} + a_3\delta_{il}\delta_{jk},$$

$$d^{\rho}_{ijkl} = d_1\delta_{ij}\delta_{kl} + d_2\delta_{ik}\delta_{jl} + d_3\delta_{il}\delta_{jk},$$

$$\begin{aligned}\chi^m_{ijknpl} = {}& \chi^m_1\delta_{ij}\delta_{kn}\delta_{pl} + \chi^m_2\delta_{ij}\delta_{kp}\delta_{nl} + \chi^m_3\delta_{ij}\delta_{kl}\delta_{np} + \chi^m_4\delta_{ik}\delta_{jn}\delta_{pl} \\ & + \chi^m_5\delta_{ik}\delta_{jp}\delta_{nl} + \chi^m_6\delta_{ik}\delta_{jl}\delta_{np} + \chi^m_7\delta_{in}\delta_{jk}\delta_{pl} + \chi^m_8\delta_{in}\delta_{jp}\delta_{kl} \\ & + \chi^m_9\delta_{in}\delta_{jl}\delta_{kp} + \chi^m_{10}\delta_{ip}\delta_{jk}\delta_{nl}\end{aligned}$$

$$+\chi_{11}^{m}\delta_{ip}\delta_{jn}\delta_{kl}+\chi_{12}^{m}\delta_{ip}\delta_{jl}\delta_{kn}+\chi_{13}^{m}\delta_{il}\delta_{jk}\delta_{np}$$
$$+\chi_{14}^{m}\delta_{il}\delta_{jn}\delta_{kp}+\chi_{15}^{m}\delta_{il}\delta_{jp}\delta_{kn},$$

where C_i, a_i, d_i, and χ_j^m $(i=\overline{1,3}\,,\, j=\overline{1,15})$ are constants.

Since $\hat{\mathbf{C}}^{(4)}$, $\hat{\boldsymbol{\alpha}}^{\rho(4)}$, and $\hat{\boldsymbol{d}}^{\rho(4)}$ are symmetric about the transposition of the last two indices, the constitutive relations (3.26), (3.28), and (3.30) for isotropic materials take the form

$$\sigma_{*ij}=2Ge_{ij}-2G_{\rho e}\rho_{ij}^{m}+\left[\left(K-\frac{2}{3}G\right)e-\left(K_{\rho e}-\frac{2}{3}G_{\rho e}\right)\rho_m\right]\delta_{ij}, \quad (3.31)$$

$$\tilde{\mu}_{ij}^{\prime\pi}=2G_{\rho}\rho_{ij}^{m}-\frac{2}{\rho_0}G_{\rho e}e_{ij}+\left[\left(K_{\rho}-\frac{2}{3}G_{\rho}\right)\rho_m-\frac{1}{\rho_0}\left(K_{\rho e}-\frac{2}{3}G_{\rho e}\right)e\right]\delta_{ij}, \quad (3.32)$$

$$\pi_{ijk}=-\kappa_1\nabla_i\tilde{\mu}_{ll}^{\prime\pi}\delta_{jk}-\kappa_2\nabla_j\tilde{\mu}_{ll}^{\prime\pi}\delta_{ik}-\kappa_3\nabla_k\tilde{\mu}_{ll}^{\prime\pi}\delta_{ij}-\kappa_4\nabla_l\tilde{\mu}_{li}^{\prime\pi}\delta_{jk}$$
$$-\kappa_5\nabla_l\tilde{\mu}_{lj}^{\prime\pi}\delta_{ik}-\kappa_6\nabla_l\tilde{\mu}_{lk}^{\prime\pi}\delta_{ij}-\kappa_7\nabla_i\tilde{\mu}_{kj}^{\prime\pi}-\kappa_8\nabla_j\tilde{\mu}_{ki}^{\prime\pi}-\kappa_9\nabla_k\tilde{\mu}_{ij}^{\prime\pi}. \quad (3.33)$$

Here, σ_{*ij}, e_{ij}, $\tilde{\mu}_{ij}^{\prime\pi}$, ρ_{ij}^{m}, and π_{ijk} are the components of the tensors $\hat{\boldsymbol{\sigma}}_*$, $\hat{\boldsymbol{e}}$, $\hat{\tilde{\boldsymbol{\mu}}}^{\prime\pi}$, $\hat{\boldsymbol{\rho}}^{m}$, and $\hat{\boldsymbol{\pi}}^{m(3)}$, respectively, and the following notations are used:

$$\tilde{\mu}_{ij}^{\pi}=\mu_{ij}^{\prime\pi}-\mu_{0\,ij}^{\prime\pi},\quad \rho_m=\rho_{m\pi}/\rho_0,\quad G=(C_2+C_3)/2,\quad G_{\rho e}=(a_2+a_3)/2,$$
$$G_{\rho}=(d_2+d_3)/2,\quad K=C_1+(C_2+C_3)/3,\quad K_{\rho e}=a_1+(a_2+a_3)/3,$$
$$K_{\rho}=d_1+(d_2+d_3)/3,\quad \kappa_1=\chi_7^m,\quad \kappa_2=\chi_4^m,\quad \kappa_3=\chi_1^m,\quad \kappa_4=\chi_{10}^m+\chi_{13}^m,$$
$$\kappa_5=\chi_5^m+\chi_6^m,\quad \kappa_6=\chi_2^m+\chi_3^m,\quad \kappa_7=\chi_8^m+\chi_9^m,$$
$$\kappa_8=\chi_{11}^m+\chi_{14}^m,\quad \kappa_9=\chi_{12}^m+\chi_{15}^m.$$

Note that in contrast to the constitutive relation (2.76), the formula (3.31) describes the effect of the local mass displacement not only on the deformation of the compression and tension but also on the change in the shape of the body.

Since the tensor $\hat{\boldsymbol{\pi}}^{m(3)}$ is symmetric by the last two indices, $\kappa_1=\varsigma_1$, $\kappa_2=\kappa_3=\varsigma_2$, $\kappa_4=\varsigma_3$, $\kappa_5=\kappa_6=\varsigma_4$, $\kappa_7=\varsigma_5$, and $\kappa_8=\kappa_9=\varsigma_6$. Thus, within the framework of the constructed theory, we have 12 material constants for an isotropic material: K, G, K_ρ, G_ρ, $G_{\rho e}$, $K_{\rho e}$, ς_i, $i=\overline{1,6}$.

In terms of the displacement vector and tensor of the induced mass, the set of governing equations for isotropic medium is as follows:

$$G\Delta u_j + \left(K + \frac{1}{3}G\right)\nabla_j \nabla_l u_l - 2G_{\rho e}\nabla_l \rho_{lj}^m$$

$$-\left(K_{\rho e} - \frac{2}{3}G_{\rho e}\right)\nabla_j \rho_m + \rho_0 F_{*j} = \rho_0 \frac{\partial^2 u_j}{\partial t^2}, \tag{3.34}$$

$$\rho_{jk}^m = \left[W_1 \Delta\rho_m - \left(W_{1e} + \varsigma_3 \frac{2}{\rho_0} G_{\rho e}\right)\Delta(\nabla_l u_l)\right]\delta_{jk} + 2\varsigma_3 G_\rho \nabla_i \nabla_l \rho_{li}^m \delta_{jk}$$

$$+W_2 \nabla_j \nabla_k \rho_m + 2\varsigma_5 G_\rho \Delta\rho_{kj}^m + 2G_\rho(\varsigma_4 + \varsigma_6)\left[\nabla_j \nabla_l \rho_{lk}^m + \nabla_k \nabla_l \rho_{lj}^m\right]$$

$$-W_{2e}\nabla_k \nabla_j \nabla_l u_l - \frac{G_{\rho e}}{\rho_0}(\varsigma_4 + \varsigma_5 + \varsigma_6)\Delta(\nabla_j u_k + \nabla_k u_j), \tag{3.35}$$

where

$$W_1 = 3\varsigma_1 K_\rho + (\varsigma_3 + \varsigma_5)\left(K_\rho - \frac{2}{3}G_\rho\right),$$

$$W_{1e} = \frac{1}{\rho_0}\left[3\varsigma_1 K_{\rho e} + (\varsigma_3 + \varsigma_5)\left(K_{\rho e} - \frac{2}{3}G_{\rho e}\right)\right],$$

$$W_2 = 2\left[3\varsigma_2 K_\rho + (\varsigma_4 + \varsigma_6)\left(K_\rho - \frac{2}{3}G_\rho\right)\right],$$

$$W_{2e} = \frac{2}{\rho_0}\left[3\varsigma_2 K_{\rho e} + (\varsigma_4 + \varsigma_6)\left(K_{\rho e} - \frac{2}{3}G_{\rho e}\right)\right].$$

3.2 Local Gradient Theory of Dielectrics Taking into Account the Inertia and Irreversibility of Local Mass Displacement and Polarization

In the previous chapter, a complete set of relations for the local gradient theory of deformation of thermoelastic dielectrics was formulated, neglecting the dissipation and inertia of the local mass displacement. As a result, the mechanical and thermal equations and equations of electromagnetic field were dynamic, while the equation for local mass displacement is static. Therefore, the electromagnetic, mechanical, and thermal fields are dynamically uncoupled with the local mass displacement. Such an approximation does not allow us to investigate the transitional modes of the formation of near-surface inhomogeneity of physical and mechanical fields. This approximation may also be insufficient and even unacceptable for the study of the perturbation of electrothermomechanical processes by shock loads, as well as for the description of acoustic and electromagnetic emission caused by the formation of surfaces, etc.

For an appropriate description of the mentioned phenomena, a complete set of equations of the generalized local gradient theory of polarized solids will be obtained in this section, taking the inertia and dissipation of polarization and local mass displacement into account.

3.2.1 Inertia of Local Mass Displacement and Polarization

In the total energy balance equation, we took into account the kinetic energies of polarization (Maugin 1977, 1988) and local mass displacement (Hrytsyna 2011; Kondrat and Hrytsyna 2012c), which enables us to describe the inertia of the above processes. Following Maugin (1988), we relate this inertia with some scalar quantities d_m and d_E. Here we introduce the kinetic energy of the local mass displacement similar to the kinetic energy of polarization introduced by Maugin (1977).

We assume that the total energy $\mathcal{E}$ of the "solid–electromagnetic field" system is the sum of internal energy ρu, kinetic energy of the mass center $\frac{1}{2}\rho\mathbf{v}^2$, energy of the electromagnetic field U_e, polarization kinetic energy $\frac{1}{2}\rho d_E\left(\frac{d\boldsymbol{\pi}_e}{dt}\right)^2$, and kinetic energy of the local mass displacement $\frac{1}{2}\rho d_m\left(\frac{d\boldsymbol{\pi}_m}{dt}\right)^2$, that is

$$\mathcal{E} = \rho u + U_e + \frac{1}{2}\rho \mathbf{v}^2 + \frac{1}{2}\rho d_E \left(\frac{d\boldsymbol{\pi}_e}{dt}\right)^2 + \frac{1}{2}\rho d_m \left(\frac{d\boldsymbol{\pi}_m}{dt}\right)^2. \quad (3.36)$$

In this case, the law of total energy conservation can be given in an integral form:

$$\frac{d}{dt}\int_{(V)} \mathcal{E} dV = -\oint_{(\Sigma)} \left[\rho\left(u + \frac{1}{2}\mathbf{v}^2 + \frac{d_E}{2}\left(\frac{d\boldsymbol{\pi}_e}{dt}\right)^2 + \frac{d_m}{2}\left(\frac{d\boldsymbol{\pi}_m}{dt}\right)^2\right)\mathbf{v} - \hat{\boldsymbol{\sigma}}\cdot\mathbf{v} \right.$$

$$\left. + \mathbf{S}_e + \mathbf{J}_q + \mu\mathbf{J}_m + \mu_\pi \mathbf{J}_{ms} \right]\cdot \mathbf{n}\, d\Sigma + \int_{(V)} (\rho\mathbf{F}\cdot\mathbf{v} + \rho\Re)\, dV$$

or in a local form:

$$\frac{\partial \mathcal{E}}{\partial t} = -\nabla\cdot\left[\rho\left(u + \frac{1}{2}\mathbf{v}^2 + \frac{d_E}{2}\left(\frac{d\boldsymbol{\pi}_e}{dt}\right)^2 + \frac{d_m}{2}\left(\frac{d\boldsymbol{\pi}_m}{dt}\right)^2\right)\mathbf{v} - \hat{\boldsymbol{\sigma}}\cdot\mathbf{v} \right.$$

$$\left. + \mathbf{S}_e + \mathbf{J}_q + \mu\mathbf{J}_m + \mu_\pi \mathbf{J}_{ms} \right] + \rho\mathbf{F}\cdot\mathbf{v} + \rho\Re. \quad (3.37)$$

Using Eq. (3.37), representation (3.36), the entropy balance equation (2.11) and the electromagnetic energy balance equation (2.36), formulae (2.37), (2.46), and (2.48), we can obtain, after some algebra, the following formulation for the conservation energy law:

$$\rho\frac{du}{dt} = \hat{\boldsymbol{\sigma}}_* : (\nabla\otimes\mathbf{v}) + \rho T\frac{ds}{dt} + \rho\mathbf{E}_*\cdot\frac{d\boldsymbol{\pi}_e}{dt} + \rho\mu'_\pi\frac{d\rho_m}{dt} - \rho\nabla\mu'_\pi\cdot\frac{d\boldsymbol{\pi}_m}{dt}$$

$$-\left[\frac{\partial\rho}{\partial t} + \nabla\cdot(\rho\mathbf{v})\right]\left[u + \frac{1}{2}\mathbf{v}^2 + \frac{1}{2}d_E\left(\frac{d\boldsymbol{\pi}_e}{dt}\right)^2\right.$$

$$\left. + \frac{1}{2}d_m\left(\frac{d\boldsymbol{\pi}_m}{dt}\right)^2 - Ts - \boldsymbol{\pi}_e\cdot\mathbf{E}_* - \rho_m\mu'_\pi + \nabla\mu'_\pi\cdot\boldsymbol{\pi}_m\right]$$

$$-\rho d_E\frac{d^2\boldsymbol{\pi}_e}{dt^2}\cdot\frac{d\boldsymbol{\pi}_e}{dt} - \rho d_m\frac{d^2\boldsymbol{\pi}_m}{dt^2}\cdot\frac{d\boldsymbol{\pi}_m}{dt} + \mathbf{J}_{e*}\cdot\mathbf{E}_* - \mathbf{J}_q\cdot\frac{\nabla T}{T} - T\eta_s$$

$$+\mathbf{v}\cdot\left(-\rho\frac{d\mathbf{v}}{dt} + \nabla\cdot\hat{\boldsymbol{\sigma}}_* + \mathbf{F}_e + \rho\mathbf{F}_*\right).$$

From the requirement that the conservation law of free energy should be invariant with respect to space translations, in a manner similar to Section 2.7, we obtain the laws of conservation of mass and linear momentum

$$\frac{\partial \rho}{\partial t} + \nabla \cdot (\rho \mathbf{v}) = 0, \quad (3.38)$$

$$\nabla \cdot \hat{\boldsymbol{\sigma}}_* + \mathbf{F}_e + \rho \mathbf{F}_* = \rho \frac{d\mathbf{v}}{dt}. \quad (3.39)$$

In view of Eqs. (3.38) and (3.39), the law of energy conservation takes the form

$$\rho \frac{du}{dt} = \hat{\boldsymbol{\sigma}}_* : (\nabla \otimes \mathbf{v}) + \rho T \frac{ds}{dt} + \rho \mathbf{E}_* \cdot \frac{d\boldsymbol{\pi}_e}{dt} + \rho \mu'_\pi \frac{d\rho_m}{dt} - \rho \nabla \mu'_\pi \cdot \frac{d\boldsymbol{\pi}_m}{dt}$$

$$-\rho d_E \frac{d^2\boldsymbol{\pi}_e}{dt^2} \cdot \frac{d\boldsymbol{\pi}_e}{dt} - \rho d_m \frac{d^2\boldsymbol{\pi}_m}{dt^2} \cdot \frac{d\boldsymbol{\pi}_m}{dt} + \mathbf{J}_{e*} \cdot \mathbf{E}_* - \mathbf{J}_q \cdot \frac{\nabla T}{T} - T\eta_s. \quad (3.40)$$

Here, $\hat{\boldsymbol{\sigma}}_*$, $\mathbf{F}_*$, and $\mathbf{F}_e$ are determined according to formulae (2.50)–(2.52).

3.2.2 Irreversibility of the Local Mass Displacement and Polarization

It is known that reversible processes do not actually occur being only idealizations of actual processes. Therefore, in this section, we take the non-reversibility of the processes of polarization and local mass displacement into account. To this end, we split $\mathbf{E}_*$ and $\nabla \mu'_\pi$ into reversible $\mathbf{E}^r_*$, $(\nabla \mu'_\pi)^r$ and dissipative $\mathbf{E}^i_*$, $(\nabla \mu'_\pi)^i$ parts as follows:

$$\mathbf{E}_* = \mathbf{E}^r_* + \mathbf{E}^i_*, \; \nabla \mu'_\pi = (\nabla \mu'_\pi)^r + (\nabla \mu'_\pi)^i. \quad (3.41)$$

In view of these formulae, Eq. (3.40) becomes

$$\rho \frac{du}{dt} = \hat{\boldsymbol{\sigma}}_* : (\nabla \otimes \mathbf{v}) + \rho T \frac{ds}{dt} + \rho \mathbf{E}^r_* \cdot \frac{d\boldsymbol{\pi}_e}{dt} + \rho \mu'_\pi \frac{d\rho_m}{dt} - \rho (\nabla \mu'_\pi)^r \cdot \frac{d\boldsymbol{\pi}_m}{dt}$$

$$-\rho d_E \frac{d^2\boldsymbol{\pi}_e}{dt^2} \cdot \frac{d\boldsymbol{\pi}_e}{dt} - \rho d_m \frac{d^2\boldsymbol{\pi}_m}{dt^2} \cdot \frac{d\boldsymbol{\pi}_m}{dt}$$

$$+\mathbf{J}_{e*} \cdot \mathbf{E}_* - \mathbf{J}_q \cdot \frac{\nabla T}{T} + \rho \mathbf{E}^i_* \cdot \frac{d\boldsymbol{\pi}_e}{dt} - \rho (\nabla \mu'_\pi)^i \cdot \frac{d\boldsymbol{\pi}_m}{dt} - T\eta_s. \quad (3.42)$$

The Voigt classical theory of dielectrics uses a notion of macroscopic (i.e., mean) electric field. However, the field of each particle of dielectric bodies can be influenced not only by the mean macroscopic field, but also by the fields of its neighboring particles. Consequently, the macroscopic electric field and the local field in dielectric media exhibit some difference. The nonlinear theory of dielectrics developed by Toupin (1956, 1963) takes such a difference into account.

Following Toupin, we take into account the difference between the local and macroscopic fields. To this end, we assume that the local thermodynamic state of a dielectric body is determined not by the reversible part of the macroscopic fields $\mathbf{E}_*^r$ and $(\nabla\mu_\pi')^r$, but by their local values $\mathbf{E}^L$ and $(\nabla\mu_\pi')^L$, which, in general, are different from the mean values. Therefore, we rewrite the balance equation (3.42) of internal energy as follows:

$$\rho\frac{du}{dt}=\hat{\boldsymbol{\sigma}}_*:(\nabla\otimes\mathbf{v})+\rho T\frac{ds}{dt}+\rho\mathbf{E}^L\cdot\frac{d\boldsymbol{\pi}_e}{dt}+\rho\mu_\pi'\frac{d\rho_m}{dt}-\rho(\nabla\mu_\pi')^L\cdot\frac{d\boldsymbol{\pi}_m}{dt}$$

$$+\rho\left(\mathbf{E}_*^r-\mathbf{E}^L-d_E\frac{d^2\boldsymbol{\pi}_e}{dt^2}\right)\cdot\frac{d\boldsymbol{\pi}_e}{dt}+\rho\left((\nabla\mu_\pi')^L-(\nabla\mu_\pi')^r-d_m\frac{d^2\boldsymbol{\pi}_m}{dt^2}\right)\cdot\frac{d\boldsymbol{\pi}_m}{dt}$$

$$+\mathbf{J}_{e*}\cdot\mathbf{E}_*-\mathbf{J}_q\cdot\frac{\nabla T}{T}+\rho\mathbf{E}_*^i\cdot\frac{d\boldsymbol{\pi}_e}{dt}-\rho(\nabla\mu_\pi')^i\cdot\frac{d\boldsymbol{\pi}_m}{dt}-T\eta_s. \quad (3.43)$$

Now, the free energy can be introduced through the following Legendre transform:

$$f=u-Ts-\mathbf{E}^L\cdot\mathbf{p}+(\nabla\mu_\pi')^L\cdot\boldsymbol{\pi}_m.$$

Then, using Eq. (3.43), we can write:

$$\rho\frac{df}{dt}=\hat{\boldsymbol{\sigma}}_*:\frac{d\hat{e}}{dt}-\rho s\frac{dT}{dt}-\rho\boldsymbol{\pi}_e\cdot\frac{d\mathbf{E}^L}{dt}+\rho\mu_\pi'\frac{d\rho_m}{dt}+\rho\boldsymbol{\pi}_m\cdot\frac{d(\nabla\mu_\pi')^L}{dt}$$

$$+\rho\left(\mathbf{E}_*^r-\mathbf{E}^L-d_E\frac{d^2\boldsymbol{\pi}_e}{dt^2}\right)\cdot\frac{d\boldsymbol{\pi}_e}{dt}+\rho\left((\nabla\mu_\pi')^L-(\nabla\mu_\pi')^r-d_m\frac{d^2\boldsymbol{\pi}_m}{dt^2}\right)\cdot\frac{d\boldsymbol{\pi}_m}{dt}$$

$$+\mathbf{J}_{e*}\cdot\mathbf{E}_*-\mathbf{J}_q\cdot\frac{\nabla T}{T}+\rho\mathbf{E}_*^i\cdot\frac{d\boldsymbol{\pi}_e}{dt}-\rho(\nabla\mu_\pi')^i\cdot\frac{d\boldsymbol{\pi}_m}{dt}-T\eta_s. \quad (3.44)$$

Since the summands in the last two lines of Eq. (3.44) do not depend on the rates of change in the strain tensor, temperature, induced mass density, and the local values of both the electric field and gradient of modified chemical potential

$$\frac{d\hat{e}}{dt},\ \frac{dT}{dt},\ \frac{d\mathbf{E}^L}{dt},\ \frac{d\rho_m}{dt},\text{ and }\frac{d(\nabla\mu_\pi')^L}{dt},$$

we get the generalized Gibbs equation

$$df = \frac{1}{\rho}\hat{\boldsymbol{\sigma}}_* : d\hat{e} - s dT - \boldsymbol{\pi}_e \cdot d\mathbf{E}^L + \mu'_\pi d\rho_m + \boldsymbol{\pi}_m \cdot d(\nabla \mu'_\pi)^L, \quad (3.45)$$

the relation for the entropy production

$$\eta_s = -\mathbf{J}_q \cdot \frac{\nabla T}{T^2} + \mathbf{J}_{e*} \cdot \frac{\mathbf{E}_*}{T} + \rho \frac{d\boldsymbol{\pi}_e}{dt} \cdot \frac{\mathbf{E}_*^i}{T} - \rho \frac{d\boldsymbol{\pi}_m}{dt} \cdot \frac{(\nabla \mu'_\pi)^i}{T} \quad (3.46)$$

and two differential equations

$$\mathbf{E}_*^r - \mathbf{E}^L = d_E \frac{d^2 \boldsymbol{\pi}_e}{dt^2}, \quad (3.47)$$

$$(\nabla \mu'_\pi)^L - (\nabla \mu'_\pi)^r = d_m \frac{d^2 \boldsymbol{\pi}_m}{dt^2}, \quad (3.48)$$

relating the specific polarization $\boldsymbol{\pi}_e$ with the vectors of the local electric field $\mathbf{E}^L$ and the reversible component $\mathbf{E}_*^r$ of the macroscopic electric field $\mathbf{E}_*$, as well as the vector of the local mass displacement $\boldsymbol{\pi}_m$ with the local field $(\nabla \mu'_\pi)^L$ and reversible part $(\nabla \mu'_\pi)^r$ of gradient of the modified chemical potential $\nabla \mu'_\pi$.

Note that in order to obtain formulae (3.47) and (3.48), it is taken into account that summands

$$\rho\left(\mathbf{E}_*^r - \mathbf{E}^L - d_E \frac{d^2 \boldsymbol{\pi}_e}{dt^2}\right) \cdot \frac{d\boldsymbol{\pi}_e}{dt} \quad \text{and}$$

$$\rho\left((\nabla \mu'_\pi)^L - (\nabla \mu'_\pi)^r - d_m \frac{d^2 \boldsymbol{\pi}_m}{dt^2}\right) \cdot \frac{d\boldsymbol{\pi}_m}{dt},$$

that contain equilibrium values $\mathbf{E}_*^r$, $\mathbf{E}^L$, $(\nabla \mu'_\pi)^r$, and $(\nabla \mu'_\pi)^L$, cannot be included into the expression for entropy production.

Also note that Eq. (3.47) is equivalent to "the equation of intramolecular force balance" obtained by Maugin (1977, 1988) who took the inertia of dielectric polarization into account. If a material constant d_E vanishes ($d_E = 0$), this equation yields the so-called "equation of molecular equilibrium" (Toupin 1956, 1963).

3.2.3 Constitutive Equations

Since parameters T, ρ_m, $\mathbf{E}^L$, $(\nabla \mu'_\pi)^L$, and $\hat{e}$ are independent, the Gibbs equation (3.45) yields the following state laws:

$$\hat{\boldsymbol{\sigma}}_* = \rho \frac{\partial f}{\partial \hat{e}}\bigg|_{T,\,\mathbf{E}^L,\,\rho_m,\,(\boldsymbol{\nabla}\mu'_\pi)^L}, \qquad s = -\frac{\partial f}{\partial T}\bigg|_{\hat{e},\,\mathbf{E}^L,\,\rho_m,\,(\boldsymbol{\nabla}\mu'_\pi)^L}, \tag{3.49}$$

$$\boldsymbol{\pi}_e = -\frac{\partial f}{\partial \mathbf{E}^L}\bigg|_{\hat{e},\,T,\,\rho_m,\,(\boldsymbol{\nabla}\mu'_\pi)^L}, \qquad \boldsymbol{\pi}_m = \frac{\partial f}{\partial (\boldsymbol{\nabla}\mu'_\pi)^L}\bigg|_{\hat{e},\,T,\,\mathbf{E}^L,\,\rho_m}, \tag{3.50}$$

$$\mu'_\pi = \frac{\partial f}{\partial \rho_m}\bigg|_{\hat{e},\,T,\mathbf{E}^L,(\boldsymbol{\nabla}\mu'_\pi)^L}. \tag{3.51}$$

We should write Eqs. (3.49)–(3.51) in an explicit form. Within the framework of the linear theory for homogeneous isotropic materials, the free energy density f can be written in the following quadratic form:

$$f = f_0 - s_0\,\theta + \mu'_{\pi 0}\rho_m - \frac{C_V}{2T_0}\theta^2 + \frac{1}{2\rho_0}\left(K - \frac{2}{3}G\right) I_1^2 + \frac{G}{\rho_0} I_2$$

$$+ \frac{d_\rho}{2}\rho_m^2 - \frac{K\alpha_T}{\rho_0} I_1\theta - \frac{K\alpha_\rho}{\rho_0} I_1\,\rho_m - \beta_{T\rho}\,\rho_m\theta - \frac{1}{2}\chi_m(\boldsymbol{\nabla}\mu'_\pi)^L\cdot(\boldsymbol{\nabla}\mu'_\pi)^L$$

$$-\frac{1}{2}\chi_E\,\mathbf{E}_*^L\cdot\mathbf{E}_*^L + \chi_{Em}\,\mathbf{E}_*^L\cdot(\boldsymbol{\nabla}\mu'_\pi)^L. \tag{3.52}$$

Using the formulae (3.49)–(3.51) and (3.52) for isotropic dielectric materials, we obtain the following nonlocal linear constitutive relations in an explicit form:

$$\hat{\boldsymbol{\sigma}}_* = 2G\hat{e} + \left[\left(K - \frac{2}{3}G\right)e - K(\alpha_T\theta + \alpha_\rho\rho_m)\right]\hat{\mathbf{I}}, \tag{3.53}$$

$$s = s_0 + \frac{C_V}{T_0}\theta + \frac{1}{\rho_0}K\alpha_T\,e + \beta_{T\rho}\,\rho_m, \tag{3.54}$$

$$\mu'_\pi = \mu'_{\pi 0} + d_\rho\rho_m - \beta_{T\rho}\theta - \frac{1}{\rho_0}K\alpha_\rho\,e, \tag{3.55}$$

$$\boldsymbol{\pi}_e = \chi_E\mathbf{E}^L - \chi_{Em}(\boldsymbol{\nabla}\mu'_\pi)^L, \tag{3.56}$$

$$\boldsymbol{\pi}_m = -\chi_m(\boldsymbol{\nabla}\mu'_\pi)^L + \chi_{Em}\mathbf{E}^L. \tag{3.57}$$

Taking Eqs. (3.47) and (3.48) into account, we rewrite the state laws (3.56) and (3.57) as follows:

$$\boldsymbol{\pi}_e + \chi_E d_E\frac{\partial^2\boldsymbol{\pi}_e}{\partial t^2} + \chi_{Em}d_m\frac{\partial^2\boldsymbol{\pi}_m}{\partial t^2} = \chi_E\mathbf{E}^r - \chi_{Em}(\boldsymbol{\nabla}\mu'_\pi)^r, \tag{3.58}$$

$$\boldsymbol{\pi}_m + \chi_m d_m \frac{\partial^2 \boldsymbol{\pi}_m}{\partial t^2} + \chi_{Em} d_E \frac{\partial^2 \boldsymbol{\pi}_e}{\partial t^2} = -\chi_m (\boldsymbol{\nabla}\mu'_\pi)^r + \chi_{Em}\mathbf{E}^r . \quad (3.59)$$

Note that within a linear approximation, $\mathbf{E}_* = \mathbf{E}$ and $\frac{d\ldots}{dt} = \frac{\partial\ldots}{\partial t}$.

We obtain kinetic relations based on the relation (3.46) for entropy production. Let us assume that thermodynamic fluxes

$$\mathbf{j}_1 = \mathbf{J}_q, \quad \mathbf{j}_2 = \mathbf{J}_{e*}, \quad \mathbf{j}_3 = \rho\frac{d\boldsymbol{\pi}_e}{dt}, \quad \mathbf{j}_4 = \rho\frac{d\boldsymbol{\pi}_m}{dt}$$

are linear functions of the following thermodynamic forces:

$$\mathbf{X}_1 = -\frac{\boldsymbol{\nabla}T}{T}, \quad \mathbf{X}_2 = \mathbf{E}_*, \quad \mathbf{X}_3 = \mathbf{E}_*^i, \quad \mathbf{X}_4 = -(\boldsymbol{\nabla}\mu'_\pi)^i,$$

that is

$$\mathbf{j}_i = \mathbf{j}_i(\mathbf{X}_1,\mathbf{X}_2,\mathbf{X}_3,\mathbf{X}_4;A), \quad i=\overline{1,4}. \quad (3.60)$$

Here, $\mathbf{j}_i = \mathbf{j}_i(0,0,0,0;A), \quad i=\overline{1,4}$.

For a linear approximation, Eqs. (3.60) take the form

$$\mathbf{J}_q = -\frac{L_{11}}{T_0}\boldsymbol{\nabla}T + L_{12}\mathbf{E} + L_{13}\mathbf{E}^i - L_{14}(\boldsymbol{\nabla}\mu'_\pi)^i, \quad (3.61)$$

$$\mathbf{J}_e = -\frac{L_{21}}{T_0}\boldsymbol{\nabla}T + L_{22}\mathbf{E} + L_{23}\mathbf{E}^i - L_{24}(\boldsymbol{\nabla}\mu'_\pi)^i, \quad (3.62)$$

$$\rho_0\frac{\partial\boldsymbol{\pi}_e}{\partial t} = -\frac{L_{31}}{T_0}\boldsymbol{\nabla}T + L_{32}\mathbf{E} + L_{33}\mathbf{E}^i - L_{34}(\boldsymbol{\nabla}\mu'_\pi)^i, \quad (3.63)$$

$$\rho_0\frac{\partial\boldsymbol{\pi}_m}{\partial t} = -\frac{L_{41}}{T_0}\boldsymbol{\nabla}T + L_{42}\mathbf{E} + L_{43}\mathbf{E}^i - L_{44}(\boldsymbol{\nabla}\mu'_\pi)^i, \quad (3.64)$$

where L_{ij} $(i,j=\overline{1,4})$ are constant kinetic coefficients.

Taking expressions (3.41) into account, from the state laws (3.58), (3.59) and kinetic equations (3.61)–(3.64), we exclude irreversible terms $\mathbf{E}^i$, $(\boldsymbol{\nabla}\mu'_\pi)^i$ and reversible terms $\mathbf{E}^r$, $(\boldsymbol{\nabla}\mu'_\pi)^r$ of vectors $\mathbf{E}$ and $\boldsymbol{\nabla}\mu'_\pi$. As a result, we obtain the following relations for the vectors of heat flux and electric current density:

$$\mathbf{J}_q = -\mathfrak{L}_1^T\boldsymbol{\nabla}T + \mathfrak{L}_1^E\mathbf{E} + \mathfrak{L}_1^\mu\boldsymbol{\nabla}\mu'_\pi + \mathfrak{L}_1^p\boldsymbol{\pi}_e + \mathfrak{L}_1^\pi\boldsymbol{\pi}_m - d_E\mathfrak{L}_1^d\frac{\partial^2\boldsymbol{\pi}_e}{\partial t^2} + d_m\mathfrak{L}_1^\mu\frac{\partial^2\boldsymbol{\pi}_m}{\partial t^2}, \quad (3.65)$$

$$\mathbf{J}_e = -\mathfrak{L}_2^T \nabla T + \mathfrak{L}_2^E \mathbf{E} + \mathfrak{L}_2^\mu \nabla \mu_\pi' + \mathfrak{L}_2^p \boldsymbol{\pi}_e + \mathfrak{L}_2^\pi \boldsymbol{\pi}_m - d_E \mathfrak{L}_2^d \frac{\partial^2 \boldsymbol{\pi}_e}{\partial t^2} + d_m \mathfrak{L}_2^\mu \frac{\partial^2 \boldsymbol{\pi}_m}{\partial t^2}, \tag{3.66}$$

and the rheological constitutive relations for vectors of polarization and local mass displacement:

$$d_E \mathfrak{L}_3^d \frac{\partial^2 \boldsymbol{\pi}_e}{\partial t^2} + \rho_0 \frac{\partial \boldsymbol{\pi}_e}{\partial t} - \mathfrak{L}_3^p \boldsymbol{\pi}_e - \mathfrak{L}_3^\pi \boldsymbol{\pi}_m - d_m \mathfrak{L}_3^\mu \frac{\partial^2 \boldsymbol{\pi}_m}{\partial t^2}$$
$$= -\mathfrak{L}_3^T \nabla T + \mathfrak{L}_3^E \mathbf{E} + \mathfrak{L}_3^\mu \nabla \mu_\pi', \tag{3.67}$$

$$-d_m \mathfrak{L}_4^\mu \frac{\partial^2 \boldsymbol{\pi}_m}{\partial t^2} + \rho_0 \frac{\partial \boldsymbol{\pi}_m}{\partial t} - \mathfrak{L}_4^\pi \boldsymbol{\pi}_m - \mathfrak{L}_4^p \boldsymbol{\pi}_e + d_E \mathfrak{L}_4^d \frac{\partial^2 \boldsymbol{\pi}_e}{\partial t^2}$$
$$= -\mathfrak{L}_4^T \nabla T + \mathfrak{L}_4^E \mathbf{E} + \mathfrak{L}_4^\mu \nabla \mu_\pi'. \tag{3.68}$$

In the expressions (3.65)–(3.68), the coefficients $\mathfrak{L}_i^T$, $\mathfrak{L}_i^E$, $\mathfrak{L}_i^\mu$, $\mathfrak{L}_i^d$, $\mathfrak{L}_i^p$, and $\mathfrak{L}_i^\pi$ $(i = \overline{1,4})$ are given by the formulae

$$\mathfrak{L}_i^T = \frac{L_{i1}}{T_0}, \quad \mathfrak{L}_i^E = L_{i2} + L_{i3}, \quad \mathfrak{L}_i^\mu = -L_{i4}, \quad \mathfrak{L}_i^d = L_{i3}, \tag{3.69}$$

$$\mathfrak{L}_i^p = L_{i4}\bar{\chi}_{Em} - L_{i3}\bar{\chi}_m, \quad \mathfrak{L}_i^\pi = L_{i3}\bar{\chi}_{Em} - L_{i4}\bar{\chi}_E, \quad i = \overline{1,4}, \tag{3.70}$$

where

$$\bar{\chi}_E = \frac{\chi_E}{\chi_E \chi_m - \chi_{Em}^2}, \quad \bar{\chi}_m = \frac{\chi_m}{\chi_E \chi_m - \chi_{Em}^2}, \quad \bar{\chi}_{Em} = \frac{\chi_{Em}}{\chi_E \chi_m - \chi_{Em}^2}.$$

Due to simple transformations, we can rewrite relations (3.65)–(3.68) as follows:

$$\mathsf{L}(\mathbf{J}_q) = -\underline{\mathsf{L}}_{1T}(\nabla T) + \underline{\mathsf{L}}_{1E}(\mathbf{E}) + \underline{\mathsf{L}}_{1\mu}(\nabla \mu_\pi'), \tag{3.71}$$

$$\mathsf{L}(\mathbf{J}_e) = -\underline{\mathsf{L}}_{2T}(\nabla T) + \underline{\mathsf{L}}_{2E}(\mathbf{E}) + \underline{\mathsf{L}}_{2\mu}(\nabla \mu_\pi'), \tag{3.72}$$

$$\mathsf{L}(\boldsymbol{\pi}_e) = -\underline{\mathsf{L}}_{3T}(\nabla T) + \underline{\mathsf{L}}_{3E}(\mathbf{E}) + \underline{\mathsf{L}}_{3\mu}(\nabla \mu_\pi'), \tag{3.73}$$

$$\mathsf{L}(\boldsymbol{\pi}_m) = -\underline{\mathsf{L}}_{4T}(\nabla T) + \underline{\mathsf{L}}_{4E}(\mathbf{E}) + \underline{\mathsf{L}}_{4\mu}(\nabla \mu_\pi'). \tag{3.74}$$

Here, operators L, $\underline{\mathsf{L}}_{iT}$, $\underline{\mathsf{L}}_{iE}$, $\underline{\mathsf{L}}_{i\mu}$, $i = \overline{1, 4}$, are given by the formulae:

$$\mathsf{L} = \left(\rho_0 \frac{\partial}{\partial t} - \mathsf{L}_{4\pi}\right)\left(\rho_0 \frac{\partial}{\partial t} - \mathsf{L}_{3p}\right) - \mathsf{L}_{4p}\mathsf{L}_{3\pi},$$

$$\mathsf{L}_{ip} = \mathfrak{L}_i^p - d_E \mathfrak{L}_i^d \frac{\partial^2}{\partial t^2}, \quad \mathsf{L}_{i\pi} = \mathfrak{L}_i^\pi + d_m \mathfrak{L}_i^\mu \frac{\partial^2}{\partial t^2}, \quad i = \overline{1, 4},$$

$$\underline{\mathbf{L}}_{j\alpha} = \mathfrak{L}_j^{\alpha}\mathbf{L} + \mathbf{L}_{jp}\underline{\mathbf{L}}_{3\alpha} + \mathbf{L}_{j\pi}\underline{\mathbf{L}}_{4\alpha}, \; j = 1, 2,$$

$$\underline{\mathbf{L}}_{3\alpha} = \mathfrak{L}_4^{\alpha}\mathbf{L}_{3\pi} + \mathfrak{L}_3^{\alpha}\left(\rho_0\frac{\partial}{\partial t} - \mathbf{L}_{4\pi}\right),$$

$$\underline{\mathbf{L}}_{4\alpha} = \mathfrak{L}_3^{\alpha}\mathbf{L}_{4\mathbf{p}} + \mathfrak{L}_4^{\alpha}\left(\rho_0\frac{\partial}{\partial t} - \mathbf{L}_{3\mathbf{p}}\right), \; \alpha = \{\mathbf{T}, \mathbf{E}, \mu\}. \tag{3.75}$$

By comparing Eqs. (3.67), (3.68) and Eqs. (2.74), (2.75) as well as Eqs. (3.65), (3.66) and Eqs. (2.91), we can see that due to accounting for the inertia and irreversibility of polarization and local mass displacement, not only the constitutive relations for the specific polarization $\boldsymbol{\pi}_e$ and the vector of local mass displacement $\boldsymbol{\pi}_m$ become rheological, but also the expressions for the heat flux $\mathbf{J}_q$ and electric current $\mathbf{J}_e$. Note that summands $d_E\mathfrak{L}_i^d\frac{\partial^2\boldsymbol{\pi}_e}{\partial t^2}$ and $d_m\mathfrak{L}_i^{\mu}\frac{\partial^2\boldsymbol{\pi}_m}{\partial t^2}$ ($i = 3, 4$) in the left-hand parts of Eqs. (3.67) and (3.68) appear due to the inertia of the processes of polarization and local mass displacement being taken into account while the summands $\rho_0\frac{\partial\boldsymbol{\pi}_e}{\partial t}$, $\rho_0\frac{\partial\boldsymbol{\pi}_m}{\partial t}$, $\mathfrak{L}_i^p\boldsymbol{\pi}_e$, $\mathfrak{L}_i^{\pi}\boldsymbol{\pi}_m$, and $\mathfrak{L}_i^T\nabla T$ ($i = 3, 4$) appear due to considering the irreversibility of the mentioned processes. According to Eq. (3.67), a body can be polarized not only in the electric field but also in a gradient of temperature as well as in a gradient of the modified chemical potential, i.e., $\nabla\mu_{\pi}'$. Therefore, the rheological relation (3.67) should describe the surface polarization as well as the linear response of the dielectric polarization to the temperature gradient (i.e., so-called thermopolarization effect).

3.2.4 Constitutive Equations for Ideal Dielectrics

Since there are no conduction and convective currents in ideal dielectrics, in order to satisfy the condition $\mathbf{J}_e = 0$, we assume

$$L_{21} = L_{12} = 0, \quad L_{22} = 0, \quad L_{32} = L_{23} = 0, \quad L_{42} = L_{24} = 0.$$

In view of this assumption, formulae (3.69) and (3.70) become

$$\mathfrak{L}_1^T = \frac{L_{11}}{T_0}, \; \mathfrak{L}_2^T = 0, \; \mathfrak{L}_3^T = \frac{L_{13}}{T_0} = \frac{\mathfrak{L}_1^E}{T_0}, \; \mathfrak{L}_4^T = \frac{L_{41}}{T_0} = -\frac{\mathfrak{L}_1^{\mu}}{T_0}, \tag{3.76}$$

$$\mathfrak{L}_1^E = L_{13}\,,\ \mathfrak{L}_2^E = 0\,,\ \mathfrak{L}_3^E = L_{33}\,,\ \mathfrak{L}_4^E = L_{43}\,, \tag{3.77}$$

$$\mathfrak{L}_1^\mu = -L_{14}\,,\ \mathfrak{L}_2^\mu = 0\,,\ \mathfrak{L}_3^\mu = -L_{34} = -\mathfrak{L}_4^E\,,\ \mathfrak{L}_4^\mu = -L_{44}\,, \tag{3.78}$$

$$\mathfrak{L}_1^d = L_{13} = \mathfrak{L}_1^E,\ \mathfrak{L}_2^d = 0\,,\ \mathfrak{L}_3^d = L_{33} = \mathfrak{L}_3^E,\ \mathfrak{L}_4^d = L_{43} = \mathfrak{L}_4^E, \tag{3.79}$$

$$\mathfrak{L}_1^p = -\mathfrak{L}_1^\mu \bar{\chi}_{Em} - \mathfrak{L}_1^E \bar{\chi}_m,\ \mathfrak{L}_2^p = 0,$$

$$\mathfrak{L}_3^p = \mathfrak{L}_4^E \bar{\chi}_{Em} - \mathfrak{L}_3^E \bar{\chi}_m,\ \mathfrak{L}_4^p = -\mathfrak{L}_4^\mu \bar{\chi}_{Em} - \mathfrak{L}_4^E \bar{\chi}_m\,, \tag{3.80}$$

$$\mathfrak{L}_1^\pi = \mathfrak{L}_1^E \bar{\chi}_{Em} + \mathfrak{L}_1^\mu \bar{\chi}_E\,,\ \mathfrak{L}_2^\pi = 0\,,$$

$$\mathfrak{L}_3^\pi = \mathfrak{L}_3^E \bar{\chi}_{Em} - \mathfrak{L}_4^E \bar{\chi}_E\,,\ \mathfrak{L}_4^\pi = \mathfrak{L}_4^E \bar{\chi}_{Em} + \mathfrak{L}_4^\mu \bar{\chi}_E\,. \tag{3.81}$$

Therefore, the constitutive relations (3.65)–(3.68) for ideal dielectrics are as follows:

$$\mathbf{J}_q = -\mathfrak{L}_1^T \nabla T + \mathfrak{L}_1^E \mathbf{E} + \mathfrak{L}_1^\mu \nabla \mu_\pi' - \left(\mathfrak{L}_1^\mu \bar{\chi}_{Em} + \mathfrak{L}_1^E \bar{\chi}_m\right)\boldsymbol{\pi}_e - d_E \mathfrak{L}_1^E \frac{\partial^2 \boldsymbol{\pi}_e}{\partial t^2}$$

$$+\left(\mathfrak{L}_1^E \bar{\chi}_{Em} + \mathfrak{L}_1^\mu \bar{\chi}_E\right)\boldsymbol{\pi}_m + d_m \mathfrak{L}_1^\mu \frac{\partial^2 \boldsymbol{\pi}_m}{\partial t^2}\,, \tag{3.82}$$

$$d_E \mathfrak{L}_3^d \frac{\partial^2 \boldsymbol{\pi}_e}{\partial t^2} + \rho_0 \frac{\partial \boldsymbol{\pi}_e}{\partial t} - \left(\mathfrak{L}_4^E \bar{\chi}_{Em} - \mathfrak{L}_3^E \bar{\chi}_m\right)\boldsymbol{\pi}_e = -\frac{\mathfrak{L}_1^E}{T_0}\nabla T + \mathfrak{L}_3^E \mathbf{E}$$

$$-\mathfrak{L}_4^E \nabla \mu_\pi' + \left(\mathfrak{L}_3^E \bar{\chi}_{Em} - \mathfrak{L}_4^E \bar{\chi}_E\right)\boldsymbol{\pi}_m - d_m \mathfrak{L}_4^E \frac{\partial^2 \boldsymbol{\pi}_m}{\partial t^2}\,, \tag{3.83}$$

$$-d_m \mathfrak{L}_4^\mu \frac{\partial^2 \boldsymbol{\pi}_m}{\partial t^2} + \rho_0 \frac{\partial \boldsymbol{\pi}_m}{\partial t} - \left(\mathfrak{L}_4^E \bar{\chi}_{Em} + \mathfrak{L}_4^\mu \bar{\chi}_E\right)\boldsymbol{\pi}_m = \frac{\mathfrak{L}_1^\mu}{T_0}\nabla T + \mathfrak{L}_4^E \mathbf{E}$$

$$+\mathfrak{L}_4^\mu \nabla \mu_\pi' - \left(\mathfrak{L}_4^\mu \bar{\chi}_{Em} + \mathfrak{L}_4^E \bar{\chi}_m\right)\boldsymbol{\pi}_e - d_E \mathfrak{L}_4^E \frac{\partial^2 \boldsymbol{\pi}_e}{\partial t^2}\,. \tag{3.84}$$

Note that the formula (3.82) generalizes the Fourier heat conduction law for ideal dielectric media, taking into account inertia and dissipation of processes of local mass displacement and polarization.

3.2.5 Governing Equations for Ideal Dielectrics

The constitutive equations (2.17), (2.24), (3.53)–(3.55), and (3.71)–(3.74); the balance equations (2.11), (2.41), (3.39); the Maxwell equations (2.20), (2.21), (2.25), (2.26); formulae (2.7), (2.48); and the strain–displacement relation (2.8) comprise a fundamental set of relations of linear generalized local gradient theory of deformable non-ferromagnetic polarized solids in which account is taken of the inertia and the irreversibility of local mass displacement and polarization.

In this subsection, we restrict ourselves to the consideration of ideal dielectrics for which $\rho_e = 0$ and $\mathbf{J}_e = 0$.

Let us write the set of governing equations for perturbations of the following functions: $\mathbf{u}, \mathbf{E}, \mathbf{B}, \theta$ and ρ_m. Substituting the constitutive relations (2.17), (2.24), (3.53)–(3.55), (3.71)–(3.74) and geometric relations (2.7), (2.8) into the balance equations (2.11), (2.41), (3.39) and into the Maxwell equations (2.20), (2.21), (2.25), (2.26), we obtain five vector differential equations and two scalar ones:

$$\rho_0 \frac{\partial^2 \mathbf{u}}{\partial t^2} = \left(K + \frac{1}{3} G \right) \nabla(\nabla \cdot \mathbf{u}) + G \Delta \mathbf{u} - K \alpha_T \nabla \theta - K \alpha_\rho \nabla \rho_m + \rho_0 \mathbf{F}, \quad (3.85)$$

$$\rho_0 C_V \frac{\partial \theta}{\partial t} + T_0 K \alpha_T \frac{\partial (\nabla \cdot \mathbf{u})}{\partial t} + \rho_0 T_0 \bar{\beta}_{T\rho} \frac{\partial \rho_m}{\partial t} - \varepsilon_0 a_{TE} \frac{\partial (\nabla \cdot \mathbf{E})}{\partial t} = \bar{\lambda} \Delta \theta + \rho_0 \Re. \quad (3.86)$$

$$\Delta \rho_m - \frac{\bar{\chi}_E}{d_\rho} \rho_m + \frac{\rho_0}{d_\rho \mathfrak{L}_4^\mu} \left(1 - a_{\mu E} \mathfrak{L}_4^E \right) \frac{\partial \rho_m}{\partial t} - \frac{d_m}{d_\rho} \frac{\partial^2 \rho_m}{\partial t^2} = \frac{\bar{\beta}_{T\rho}}{d_\rho} \Delta \theta$$

$$+ K \frac{\alpha_\rho}{\rho_0 d_\rho} \Delta (\nabla \cdot \mathbf{E}) + \frac{\varepsilon_0}{\rho_0 d_\rho} \left[\rho_0 a_{\mu E} \frac{\partial (\nabla \cdot \mathbf{E})}{\partial t} - \bar{\chi}_{Em} (\nabla \cdot \mathbf{E}) \right], \quad (3.87)$$

$$\nabla \times \mathbf{E} = -\frac{\partial \mathbf{B}}{\partial t}, \quad (3.88)$$

$$\nabla \cdot \mathbf{B} = 0, \quad (3.89)$$

$$\left(1 + \frac{\varepsilon_0}{\rho_0} \bar{\chi}_m \right) \nabla \cdot \mathbf{E} + \frac{\varepsilon_0}{\mathfrak{L}_3^E} \frac{\partial (\nabla \cdot \mathbf{E})}{\partial t} + d_E \frac{\varepsilon_0}{\rho_0} \frac{\partial^2 (\nabla \cdot \mathbf{E})}{\partial t^2}$$

$$= \frac{a_{TE}}{T_0} \Delta \theta + \bar{\chi}_{Em} \rho_m - \rho_0 a_{\mu E} \frac{\partial \rho_m}{\partial t}, \quad (3.90)$$

$$\frac{1}{\mu_0} \nabla \times \mathsf{L}(\mathbf{B}) = (\varepsilon_0 \mathsf{L} + \rho_0 \underline{\mathsf{L}}_{3E}) \left(\frac{\partial \mathbf{E}}{\partial t} \right) + d_\rho \rho_0 \underline{\mathsf{L}}_{3\mu} \left(\frac{\partial (\nabla \rho_m)}{\partial t} \right)$$

$$- K \alpha_\rho \underline{\mathsf{L}}_{3\mu} \left(\frac{\partial [\nabla (\nabla \cdot \mathbf{u})]}{\partial t} \right) - \rho_0 (\underline{\mathsf{L}}_{3T} - \beta_{T\rho} \underline{\mathsf{L}}_{3\mu}) \left(\frac{\partial (\nabla \theta)}{\partial t} \right). \quad (3.91)$$

Here, the following notations were used:

$$\bar{\lambda} = \mathfrak{L}_1^T + \frac{\left(\mathfrak{L}_1^\mu\right)^2}{T_0 \mathfrak{L}_4^\mu} - \frac{\left(\mathfrak{L}_1^E \mathfrak{L}_4^\mu - \mathfrak{L}_4^E \mathfrak{L}_1^\mu\right)^2}{T_0 \mathfrak{L}_4^\mu \left[\mathfrak{L}_4^\mu \mathfrak{L}_3^E + \left(\mathfrak{L}_4^E\right)^2\right]},\ \bar{\beta}_{T\rho} = \beta_{T\rho} - \frac{\mathfrak{L}_3^E \mathfrak{L}_1^\mu + \mathfrak{L}_1^E \mathfrak{L}_4^E}{T_0\left[\mathfrak{L}_3^E \mathfrak{L}_4^\mu + \left(\mathfrak{L}_4^E\right)^2\right]},$$

$$a_{\mu E} = \frac{\mathfrak{L}_4^E}{\mathfrak{L}_4^\mu \bar{\mathfrak{L}}_3^E} = \frac{\mathfrak{L}_4^E}{\mathfrak{L}_3^E \mathfrak{L}_4^\mu + \left(\mathfrak{L}_4^E\right)^2},\ a_{TE} = \frac{\bar{\mathfrak{L}}_1^E}{\bar{\mathfrak{L}}_3^E} = \frac{\mathfrak{L}_1^E \mathfrak{L}_4^\mu - \mathfrak{L}_4^E \mathfrak{L}_1^\mu}{\mathfrak{L}_3^E \mathfrak{L}_4^\mu + \left(\mathfrak{L}_4^E\right)^2},$$

$$\bar{\mathfrak{L}}_1^E = \mathfrak{L}_1^E - \mathfrak{L}_4^E \frac{\mathfrak{L}_1^\mu}{\mathfrak{L}_4^\mu},\ \bar{\mathfrak{L}}_3^E = \mathfrak{L}_3^E + \frac{\left(\mathfrak{L}_4^E\right)^2}{\mathfrak{L}_4^\mu}.$$

Now we can elaborate on the operators $\mathbf{L}$, $\underline{\mathbf{L}}_{3T}$, $\underline{\mathbf{L}}_{3E}$, $\underline{\mathbf{L}}_{3\mu}$, $\mathbf{L}_{3\pi}$, $\mathbf{L}_{4\mu}$, $\mathbf{L}_{3p}$, and $\mathbf{L}_{4p}$ in Eq. (3.91). From formulae (3.73) and (3.76)–(3.81) for ideal dielectrics, we get

$$\mathbf{L} = \left(\rho_0 \frac{\partial}{\partial t} - \mathbf{L}_{4\pi}\right)\left(\rho_0 \frac{\partial}{\partial t} - \mathbf{L}_{3p}\right) - \mathbf{L}_{4p}\mathbf{L}_{3\pi}, \tag{3.92}$$

$$\mathbf{L}_{3p} = \mathfrak{L}_4^E \bar{\chi}_{Em} - \mathfrak{L}_3^E \bar{\chi}_m - d_E \mathfrak{L}_3^E \frac{\partial^2}{\partial t^2},\ \mathbf{L}_{4p} = -\mathfrak{L}_4^\mu \bar{\chi}_{Em} - \mathfrak{L}_4^E \bar{\chi}_m - d_E \mathfrak{L}_4^E \frac{\partial^2}{\partial t^2}, \tag{3.93}$$

$$\mathbf{L}_{3\pi} = \mathfrak{L}_3^E \bar{\chi}_{Em} - \mathfrak{L}_4^E \bar{\chi}_E - d_m \mathfrak{L}_4^E \frac{\partial^2}{\partial t^2},\ \mathbf{L}_{4\pi} = \mathfrak{L}_4^E \bar{\chi}_{Em} + \mathfrak{L}_4^\mu \bar{\chi}_E + d_m \mathfrak{L}_4^\mu \frac{\partial^2}{\partial t^2}, \tag{3.94}$$

$$\underline{\mathbf{L}}_{3T} = -\frac{\mathfrak{L}_1^\mu}{T_0}\mathbf{L}_{3\pi} + \frac{\mathfrak{L}_1^E}{T_0}\left(\rho_0 \frac{\partial}{\partial t} - \mathbf{L}_{4\pi}\right),\ \underline{\mathbf{L}}_{3E} = \mathfrak{L}_4^E \mathbf{L}_{3\pi} + \mathfrak{L}_3^E\left(\rho_0 \frac{\partial}{\partial t} - \mathbf{L}_{4\pi}\right), \tag{3.95}$$

$$\underline{\mathbf{L}}_{3\mu} = \mathfrak{L}_4^\mu \mathbf{L}_{3\pi} - \mathfrak{L}_4^E\left(\rho_0 \frac{\partial}{\partial t} - \mathbf{L}_{4\pi}\right). \tag{3.96}$$

If we choose the modified chemical potential $\tilde{\mu}'_\pi$ as a key function, then the set of governing equations appears as follows:

$$\rho_0 \frac{\partial^2 \mathbf{u}}{\partial t^2} = \left(\bar{K} + \frac{1}{3}G\right)\nabla(\nabla \cdot \mathbf{u}) + G\Delta\mathbf{u} - K\bar{\alpha}_T \nabla\theta - K\frac{\alpha_\rho}{d_\rho}\nabla\tilde{\mu}'_\pi + \rho_0\mathbf{F}, \tag{3.97}$$

$$\rho_0 \bar{\bar{C}}_V \frac{\partial \theta}{\partial t} + T_0 K \bar{\bar{\alpha}}_T \frac{\partial(\nabla \cdot \mathbf{u})}{\partial t} + \rho_0 T_0 \frac{\bar{\beta}_{T\rho}}{d_\rho} \frac{\partial \tilde{\mu}'_\pi}{\partial t} - \varepsilon_0\, a_{TE} \frac{\partial(\nabla \cdot \mathbf{E})}{\partial t} = \bar{\lambda} \Delta \theta + \rho_0 \Re, \tag{3.98}$$

$$\Delta \tilde{\mu}'_\pi - \bar{\lambda}_\mu^2 \tilde{\mu}'_\pi + \frac{\rho_0}{d_\rho \mathfrak{L}_4^\mu} \frac{\partial \tilde{\mu}'_\pi}{\partial t} - \frac{d_m}{d_\rho} \frac{\partial^2 \tilde{\mu}'_\pi}{\partial t^2} = -\frac{\mathfrak{L}_1^\mu}{T_0 \mathfrak{L}_4^\mu} \Delta \theta$$
$$+ \left[\bar{\lambda}_\mu^2 - \frac{\rho_0}{d_\rho \mathfrak{L}_4^\mu} \frac{\partial}{\partial t} + \frac{d_m}{d_\rho} \frac{\partial^2}{\partial t^2} \right] \left(K \frac{\alpha_\rho}{\rho_0} \nabla \cdot \mathbf{u} + \beta_{T\rho} \theta \right)$$
$$- \left[\frac{\mathfrak{L}_4^E}{\mathfrak{L}_4^\mu} + \frac{\varepsilon_0}{\rho_0} \left(\bar{\chi}_{Em} + \bar{\chi}_m \frac{\mathfrak{L}_4^E}{\mathfrak{L}_4^\mu} \right) + d_E \frac{\varepsilon_0}{\rho_0} \frac{\mathfrak{L}_4^E}{\mathfrak{L}_4^\mu} \frac{\partial^2}{\partial t^2} \right] (\nabla \cdot \mathbf{E}), \tag{3.99}$$

$$\nabla \times \mathbf{E} = -\frac{\partial \mathbf{B}}{\partial t}, \tag{3.100}$$

$$\nabla \cdot \mathbf{B} = 0, \tag{3.101}$$

$$\left(1 + \frac{\varepsilon_0}{\rho_0} \bar{\chi}_m \right) \nabla \cdot \mathbf{E} + \frac{\varepsilon_0}{\bar{\mathfrak{L}}_3^E} \frac{\partial(\nabla \cdot \mathbf{E})}{\partial t} + d_E \frac{\varepsilon_0}{\rho_0} \frac{\partial^2 (\nabla \cdot \mathbf{E})}{\partial t^2}$$
$$= \frac{a_{TE}}{T_0} \Delta \theta + \frac{1}{d_\rho} \left(\bar{\chi}_{Em} - \rho_0 a_{\mu E} \frac{\partial}{\partial t} \right) \left(\tilde{\mu}'_\pi + K \frac{\alpha_\rho}{\rho_0} \nabla \cdot \mathbf{u} + \beta_{T\rho} \theta \right), \tag{3.102}$$

$$\frac{1}{\mu_0} \nabla \times \mathbf{L}(\mathbf{B}) = (\varepsilon_0 \mathbf{L} + \rho_0 \underline{\mathbf{L}}_{3E}) \left(\frac{\partial \mathbf{E}}{\partial t} \right) - \rho_0 \underline{\mathbf{L}}_{3T} \left(\frac{\partial (\nabla \theta)}{\partial t} \right) + \rho_0 \underline{\mathbf{L}}_{3\mu} \left(\frac{\partial \nabla \tilde{\mu}'_\pi}{\partial t} \right). \tag{3.103}$$

Here,

$$\bar{\lambda}_\mu^2 = \frac{1}{d_\rho} \left(\bar{\chi}_E + \bar{\chi}_{Em} \frac{\mathfrak{L}_4^E}{\mathfrak{L}_4^\mu} \right), \quad \bar{\bar{\alpha}}_T = \alpha_T + \bar{\beta}_{T\rho} \frac{\alpha_\rho}{d_\rho},$$

$$\bar{\bar{C}}_V = C_V + T_0 \frac{\bar{\beta}_{T\rho} \beta_{T\rho}}{d_\rho}. \tag{3.104}$$

For the isothermal approximation, the set of equations (3.85)–(3.91) can be written in the form

$$\rho_0 \frac{\partial^2 \mathbf{u}}{\partial t^2} = \left(K + \frac{1}{3} G \right) \nabla (\nabla \cdot \mathbf{u}) + G \Delta \mathbf{u} - K \alpha_\rho \nabla \rho_m + \rho_0 \mathbf{F}, \tag{3.105}$$

$$\Delta\rho_m - \frac{\bar{\chi}_E}{d_\rho}\rho_m + \frac{\rho_0}{d_\rho \mathfrak{L}_4^\mu}\left(1 - a_{\mu E}\mathfrak{L}_4^E\right)\frac{\partial \rho_m}{\partial t} - \frac{d_m}{d_\rho}\frac{\partial^2 \rho_m}{\partial t^2}$$

$$= K\frac{\alpha_\rho}{\rho_0 d_\rho}\Delta(\mathbf{\nabla}\cdot\mathbf{u}) + \frac{\varepsilon_0}{\rho_0 d_\rho}\left[\rho_0 a_{\mu E}\frac{\partial(\mathbf{\nabla}\cdot\mathbf{E})}{\partial t} - \bar{\chi}_{Em}\mathbf{\nabla}\cdot\mathbf{E}\right], \quad (3.106)$$

$$\mathbf{\nabla}\times\mathbf{E} = -\frac{\partial \mathbf{B}}{\partial t}, \quad (3.107)$$

$$\mathbf{\nabla}\cdot\mathbf{B} = 0, \quad (3.108)$$

$$\left(1 + \bar{\chi}_m\frac{\varepsilon_0}{\rho_0}\right)\mathbf{\nabla}\cdot\mathbf{E} + \frac{\varepsilon_0}{\bar{\mathfrak{L}}_3^E}\frac{\partial(\mathbf{\nabla}\cdot\mathbf{E})}{\partial t} + d_E\frac{\varepsilon_0}{\rho_0}\frac{\partial^2(\mathbf{\nabla}\cdot\mathbf{E})}{\partial t^2}$$

$$= \bar{\chi}_{Em}\rho_m - \rho_0 a_{\mu E}\frac{\partial \rho_m}{\partial t}, \quad (3.109)$$

$$\frac{1}{\mu_0}\mathbf{\nabla}\times\mathsf{L}(\mathbf{B}) = (\varepsilon_0\mathsf{L} + \rho_0\underline{\mathsf{L}}_{3E})\left(\frac{\partial \mathbf{E}}{\partial t}\right)$$

$$+ d_\rho\rho_0\underline{\mathsf{L}}_{3\mu}\left(\frac{\partial(\mathbf{\nabla}\rho_m)}{\partial t}\right) - K\alpha_\rho\underline{\mathsf{L}}_{3\mu}\left(\frac{\partial[\mathbf{\nabla}(\mathbf{\nabla}\cdot\mathbf{u})]}{\partial t}\right). \quad (3.110)$$

In this case, the set of equations (3.97)–(3.103) is as follows:

$$\rho_0\frac{\partial^2\mathbf{u}}{\partial t^2} = \left(\bar{K} + \frac{1}{3}G\right)\mathbf{\nabla}(\mathbf{\nabla}\cdot\mathbf{u}) + G\Delta\mathbf{u} - K\frac{\alpha_\rho}{d_\rho}\mathbf{\nabla}\tilde{\mu}'_\pi + \rho_0\mathbf{F}, \quad (3.111)$$

$$\Delta\tilde{\mu}'_\pi - \left(\bar{\lambda}_\mu^2 - \frac{\rho_0}{d_\rho\mathfrak{L}_4^\mu}\frac{\partial}{\partial t} + \frac{d_m}{d_\rho}\frac{\partial^2}{\partial t^2}\right)\left(\tilde{\mu}'_\pi + K\frac{\alpha_\rho}{\rho_0}\mathbf{\nabla}\cdot\mathbf{u}\right)$$

$$= -\left[\frac{\mathfrak{L}_4^E}{\mathfrak{L}_4^\mu} + \frac{\varepsilon_0}{\rho_0}\left(\bar{\chi}_{Em} + \bar{\chi}_m\frac{\mathfrak{L}_4^E}{\mathfrak{L}_4^\mu}\right) + d_E\frac{\varepsilon_0}{\rho_0}\frac{\mathfrak{L}_4^E}{\mathfrak{L}_4^\mu}\frac{\partial^2}{\partial t^2}\right](\mathbf{\nabla}\cdot\mathbf{E}), \quad (3.112)$$

$$\mathbf{\nabla}\times\mathbf{E} = -\frac{\partial \mathbf{B}}{\partial t}, \quad (3.113)$$

$$\mathbf{\nabla}\cdot\mathbf{B} = 0, \quad (3.114)$$

$$\left(1+\frac{\varepsilon_0}{\rho_0}\bar{\chi}_m\right)\boldsymbol{\nabla}\cdot\mathbf{E}+\frac{\varepsilon_0}{\bar{\mathfrak{L}}_3^E}\frac{\partial(\boldsymbol{\nabla}\cdot\mathbf{E})}{\partial t}+d_E\frac{\varepsilon_0}{\rho_0}\frac{\partial^2(\boldsymbol{\nabla}\cdot\mathbf{E})}{\partial t^2}$$

$$=\frac{1}{d_\rho}\left(\bar{\chi}_{Em}-\rho_0 a_{\mu E}\frac{\partial}{\partial t}\right)\left(\tilde{\mu}_\pi'+K\frac{\alpha_\rho}{\rho_0}\boldsymbol{\nabla}\cdot\mathbf{u}\right), \tag{3.115}$$

$$\frac{1}{\mu_0}\boldsymbol{\nabla}\times\mathbf{L}(\mathbf{B})=(\varepsilon_0\mathbf{L}+\rho_0\underline{\mathbf{L}}_{3\mathrm{E}})\left(\frac{\partial\mathbf{E}}{\partial t}\right)+\rho_0\underline{\mathbf{L}}_{3\mu}\left(\frac{\partial\boldsymbol{\nabla}\tilde{\mu}_\pi'}{\partial t}\right). \tag{3.116}$$

Evidently, accounting for the dissipation and inertia of both the local mass displacement and the electric polarization results in (i) increment in the order of electrodynamic equations that now contain higher time derivatives; (ii) formation of dynamic equations for specific density of the induced mass and modified chemical potential (see Eqs. (3.87) and (3.99)). Note that the equations of motion (3.85) and (3.97) coincide with Eqs. (2.94) and (2.102) obtained in the previous chapter within the framework of mathematical model for dielectric continua with neglecting the irreversibility and inertia of polarization and local mass displacement. Equations (3.86), (3.87), (3.98), and (3.99), besides renormalizing their coefficients, contain the terms that are proportional to the rates of change in an electric field gradient $\nabla\cdot\mathbf{E}$. Thus, the set of equations (3.85)–(3.91) and (3.97)–(3.103) can be effectively used to analyze:

- Transitional modes of the formation of near-surface inhomogeneity in physical and mechanical fields,
- Spectra of signals of acoustic emission caused by the surface formation,
- High-frequency mechanical and electromagnetic waves,
- Stress and strain fields as well as polarization of bodies under the loadings with high-distribution gradients.

The sets of governing equations (3.85)–(3.91) and (3.97)–(3.103) are rather complicated. However, in some cases, for example, in the case of slowly changeable processes, it is possible to neglect the inertia of polarization and local mass displacement and take only the irreversibility of these processes into account. A reverse situation may arise if the influence of dissipation of the processes of polarization and local mass displacement is insignificant, but the inertia of these processes should be taken into account. In such cases, a simplified set of equations can be employed. Further, we consider two versions of the simplified theory and write down the corresponding sets of field equations for isotropic bodies.

3.2.6 Governing Equations when Neglecting the Inertia of Polarization and Local Mass Displacement

Let us write down the relations of the local gradient theory of dielectrics neglecting the inertia of both the local mass displacement and polarization. We obtain a corresponding set of relations assuming $d_E = 0$ and $d_m = 0$ in Eqs. (3.85)–(3.91) and (3.97)–(3.103). Then, the set of governing equations (3.85)–(3.91) can be simplified to the following form:

$$\rho_0 \frac{\partial^2 \mathbf{u}}{\partial t^2} = \left(K + \frac{1}{3} G \right) \nabla(\nabla \cdot \mathbf{u}) + G \Delta \mathbf{u} - K\alpha_T \nabla\theta - K\alpha_\rho \nabla\rho_m + \rho_0 \mathbf{F},$$

$$\rho_0 C_V \frac{\partial \theta}{\partial t} + T_0 K \alpha_T \frac{\partial(\nabla \cdot \mathbf{u})}{\partial t} + \rho_0 T_0 \bar{\beta}_{T\rho} \frac{\partial \rho_m}{\partial t} - \varepsilon_0\, a_{TE} \frac{\partial(\nabla \cdot \mathbf{E})}{\partial t} = \bar{\lambda} \Delta \theta + \rho_0 \Re,$$

$$\Delta \rho_m - \frac{\bar{\chi}_E}{d_\rho} \rho_m + \frac{\rho_0}{d_\rho \mathfrak{L}_4^\mu} \left(1 - a_{\mu E} \mathfrak{L}_4^E\right) \frac{\partial \rho_m}{\partial t} = K \frac{\alpha_\rho}{\rho_0 d_\rho} \Delta(\nabla \cdot \mathbf{u})$$

$$+ \frac{\bar{\beta}_{T\rho}}{d_\rho} \Delta\theta + \frac{\varepsilon_0}{\rho_0 d_\rho} \left[\rho_0 a_{\mu E} \frac{\partial(\nabla \cdot \mathbf{E})}{\partial t} - \bar{\chi}_{Em} \nabla \cdot \mathbf{E} \right],$$

$$\nabla \times \mathbf{E} = -\frac{\partial \mathbf{B}}{\partial t}, \quad \nabla \cdot \mathbf{B} = 0,$$

$$\left(1 + \frac{\varepsilon_0}{\rho_0} \bar{\chi}_m \right) \nabla \cdot \mathbf{E} + \frac{\varepsilon_0}{\mathfrak{L}_3^E} \frac{\partial(\nabla \cdot \mathbf{E})}{\partial t} = \frac{a_{TE}}{T_0} \Delta\theta + \bar{\chi}_{Em} \rho_m - \rho_0 a_{\mu E} \frac{\partial \rho_m}{\partial t},$$

$$\frac{1}{\mu_0} \nabla \times \mathsf{L}(\mathbf{B}) = (\varepsilon_0 \mathsf{L} + \rho_0 \underline{\mathsf{L}}_{3E}) \left(\frac{\partial \mathbf{E}}{\partial t} \right) + d_\rho \rho_0 \underline{\mathsf{L}}_{3\mu} \left(\frac{\partial(\nabla \rho_m)}{\partial t} \right)$$

$$- K\alpha_\rho \underline{\mathsf{L}}_{3\mu} \left(\frac{\partial [\nabla(\nabla \cdot \mathbf{u})]}{\partial t} \right) - \rho_0 \left(\underline{\mathsf{L}}_{3T} - \beta_{T\rho} \underline{\mathsf{L}}_{3\mu} \right) \left(\frac{\partial(\nabla\theta)}{\partial t} \right),$$

where operators L, $\underline{\mathsf{L}}_{3T}$, $\underline{\mathsf{L}}_{3E}$, and $\underline{\mathsf{L}}_{3\mu}$ are defined by the formulae

$$\mathsf{L} = \rho_0^2 \frac{\partial^2}{\partial t^2} + \rho_0 \left(\mathfrak{L}_3^E \bar{\chi}_m - \mathfrak{L}_4^\mu \bar{\chi}_E - 2\mathfrak{L}_4^E \bar{\chi}_{Em} \right) \frac{\partial}{\partial t} + \mathfrak{L} \left[\left(\bar{\chi}_{Em} \right)^2 - \bar{\chi}_m \bar{\chi}_E \right], \tag{3.117}$$

$$\underline{\mathsf{L}}_{3T} = \frac{1}{T_0} \left[\rho_0 \mathfrak{L}_1^E \frac{\partial}{\partial t} + \left(\mathfrak{L}_1^\mu \mathfrak{L}_4^E - \mathfrak{L}_1^E \mathfrak{L}_4^\mu \right) \bar{\chi}_E - \left(\mathfrak{L}_1^E \mathfrak{L}_4^E + \mathfrak{L}_1^\mu \mathfrak{L}_3^E \right) \bar{\chi}_{Em} \right], \tag{3.118}$$

$$\underline{\mathbf{L}}_{3\mathrm{E}} = -\mathfrak{L}\bar{\chi}_E + \rho_0 \mathfrak{L}_3^E \frac{\partial}{\partial t}, \ \underline{\mathbf{L}}_{3\mu} = \mathfrak{L}\bar{\chi}_{Em} - \rho_0 \mathfrak{L}_4^E \frac{\partial}{\partial t}. \tag{3.119}$$

Here, $\mathfrak{L} = \mathfrak{L}_3^E \mathfrak{L}_4^\mu + \left(\mathfrak{L}_4^E\right)^2.$

The set of equations (3.97)–(3.103) can also be simplified to the form:

$$\rho_0 \frac{\partial^2 \mathbf{u}}{\partial t^2} = \left(\bar{K} + \frac{1}{3}G\right)\nabla(\nabla\cdot\mathbf{u}) + G\Delta\mathbf{u} - K\bar{\alpha}_T\nabla\theta - K\frac{\alpha_\rho}{d_\rho}\nabla\tilde{\mu}'_\pi + \rho_0\mathbf{F}, \tag{3.120}$$

$$\rho_0\bar{\bar{C}}_V\frac{\partial\theta}{\partial t} + T_0K\bar{\bar{\alpha}}_T\frac{\partial(\nabla\cdot\mathbf{u})}{\partial t} + \rho_0T_0\frac{\bar{\beta}_{T\rho}}{d_\rho}\frac{\partial\tilde{\mu}'_\pi}{\partial t} - \varepsilon_0\, a_{TE}\frac{\partial(\nabla\cdot\mathbf{E})}{\partial t} = \bar{\lambda}\Delta\theta + \rho_0\Re, \tag{3.121}$$

$$\Delta\tilde{\mu}'_\pi - \bar{\lambda}_\mu^2\tilde{\mu}'_\pi + \frac{\rho_0}{d_\rho\mathfrak{L}_4^\mu}\frac{\partial\tilde{\mu}'_\pi}{\partial t} = -\frac{\mathfrak{L}_1^\mu}{T_0\mathfrak{L}_4^\mu}\Delta\theta + \left[\bar{\lambda}_\mu^2 - \frac{\rho_0}{d_\rho\mathfrak{L}_4^\mu}\frac{\partial}{\partial t}\right]\left(K\frac{\alpha_\rho}{\rho_0}\nabla\cdot\mathbf{u} + \beta_{T\rho}\theta\right) - \left[\frac{\mathfrak{L}_4^E}{\mathfrak{L}_4^\mu} + \frac{\varepsilon_0}{\rho_0}\left(\bar{\chi}_{Em} + \bar{\chi}_m\frac{\mathfrak{L}_4^E}{\mathfrak{L}_4^\mu}\right)\right]\nabla\cdot\mathbf{E}, \tag{3.122}$$

$$\nabla\times\mathbf{E} = -\frac{\partial\mathbf{B}}{\partial t}, \tag{3.123}$$

$$\nabla\cdot\mathbf{B} = 0, \tag{3.124}$$

$$\left(1 + \frac{\varepsilon_0}{\rho_0}\bar{\chi}_m\right)\nabla\cdot\mathbf{E} + \frac{\varepsilon_0}{\bar{\mathfrak{L}}_3^E}\frac{\partial(\nabla\cdot\mathbf{E})}{\partial t} = \frac{a_{TE}}{T_0}\Delta\theta + \frac{1}{d_\rho}\left(\bar{\chi}_{Em} - \rho_0 a_{\mu E}\frac{\partial}{\partial t}\right)\left(\tilde{\mu}'_\pi + K\frac{\alpha_\rho}{\rho_0}\nabla\cdot\mathbf{u} + \beta_{T\rho}\theta\right), \tag{3.125}$$

$$\frac{1}{\mu_0}\nabla\times\mathbf{L}(\mathbf{B}) = (\varepsilon_0\mathbf{L} + \rho_0\underline{\mathbf{L}}_{3\mathrm{E}})\left(\frac{\partial\mathbf{E}}{\partial t}\right) - \rho_0\underline{\mathbf{L}}_{3\mathrm{T}}\left(\frac{\partial(\nabla\theta)}{\partial t}\right) + \rho_0\underline{\mathbf{L}}_{3\mu}\left(\frac{\partial(\nabla\tilde{\mu}'_\pi)}{\partial t}\right). \tag{3.126}$$

Note that knowing the functions **u**, **E**, **B**, θ, and $\tilde{\mu}'_\pi$ (or ρ_m) and using the state laws (3.53) and stress–displacement relations (2.8), we can determine the components of the stress tensor. In order to find the polarization and the local mass displacement, we have the following rheological equations:

$$\rho_0 \frac{\partial \boldsymbol{\pi}_e}{\partial t} - \left(\mathfrak{L}_4^E \bar{\chi}_{Em} - \mathfrak{L}_3^E \bar{\chi}_m\right)\boldsymbol{\pi}_e = -\frac{\mathfrak{L}_1^E}{T_0}\nabla\theta + \mathfrak{L}_3^E \mathbf{E} - \mathfrak{L}_4^E \nabla\mu'_\pi + \left(\mathfrak{L}_3^E \bar{\chi}_{Em} - \mathfrak{L}_4^E \bar{\chi}_E\right)\boldsymbol{\pi}_m,$$

$$\rho_0 \frac{\partial \boldsymbol{\pi}_m}{\partial t} - \left(\mathfrak{L}_4^E \bar{\chi}_{Em} + \mathfrak{L}_4^\mu \bar{\chi}_E\right)\boldsymbol{\pi}_m = \frac{\mathfrak{L}_1^\mu}{T_0}\nabla\theta + \mathfrak{L}_4^E \mathbf{E} + \mathfrak{L}_4^\mu \nabla\mu'_\pi - \left(\mathfrak{L}_4^\mu \bar{\chi}_{Em} + \mathfrak{L}_4^E \bar{\chi}_m\right)\boldsymbol{\pi}_e.$$

If the initial conditions impose vectors $\boldsymbol{\pi}_m$ and $\boldsymbol{\pi}_e$ to be zeros, then these equations can be written in the integral form

$$\boldsymbol{\pi}_e(\mathbf{r},t) = \frac{\mathfrak{L}_3^E}{\rho_0}\int_0^t \exp\left(-\frac{t-t'}{\tau_p}\right)\left[\mathbf{E}(\mathbf{r},t') - \frac{\mathfrak{L}_1^E}{T_0 \mathfrak{L}_3^E}\nabla\theta(\mathbf{r},t') - \frac{\mathfrak{L}_4^E}{\mathfrak{L}_3^E}\nabla\mu'_\pi(\mathbf{r},t') + \left(\bar{\chi}_{Em} - \frac{\mathfrak{L}_4^E}{\mathfrak{L}_3^E}\bar{\chi}_E\right)\boldsymbol{\pi}_m(\mathbf{r},t')\right]dt', \quad (3.127)$$

$$\boldsymbol{\pi}_m(\mathbf{r},t) = \frac{\mathfrak{L}_4^\mu}{\rho_0}\int_0^t \exp\left(-\frac{t-t'}{\tau_\pi}\right)\left[\nabla\mu'_\pi(\mathbf{r},t') + \frac{\mathfrak{L}_1^\mu}{T_0 \mathfrak{L}_4^\mu}\nabla\theta(\mathbf{r},t') + \frac{\mathfrak{L}_4^E}{\mathfrak{L}_4^\mu}\mathbf{E}(\mathbf{r},t') - \left(\bar{\chi}_{Em} + \frac{\mathfrak{L}_4^E}{\mathfrak{L}_4^\mu}\bar{\chi}_m\right)\boldsymbol{\pi}_e(\mathbf{r},t')\right]dt'. \quad (3.128)$$

Here, τ_π and τ_p are the times of relaxation introduced through the formulae

$$\tau_\pi = \frac{\rho_0}{\mathfrak{L}_4^E \bar{\chi}_{Em} + \mathfrak{L}_4^\mu \bar{\chi}_E}, \quad \tau_p = \frac{\rho_0}{\mathfrak{L}_4^E \bar{\chi}_{Em} - \mathfrak{L}_3^E \bar{\chi}_m}.$$

If the vectors of polarization and local mass displacement are known, then using the relation

$$\mathbf{J}_q = -\mathfrak{L}_1^T \nabla\theta + \mathfrak{L}_1^E \mathbf{E} + \mathfrak{L}_1^\mu \nabla\mu'_\pi$$
$$-\left(\mathfrak{L}_1^\mu \bar{\chi}_{Em} + \mathfrak{L}_1^E \bar{\chi}_m\right)\boldsymbol{\pi}_e + \left(\mathfrak{L}_1^E \bar{\chi}_{Em} + \mathfrak{L}_1^\mu \bar{\chi}_E\right)\boldsymbol{\pi}_m, \quad (3.129)$$

one can calculate the vector of the heat flux density.

3.2.7 Governing Equations when Neglecting the Dissipation of Polarization and Local Mass Displacement

Let us derive the set of governing equations accounting for the inertia of polarization and of the local mass displacement but with no irreversibility. Assume that the dissipative parts of vectors $\mathbf{E}_*$ and $\nabla\mu'_\pi$ are equal to zero. Under this assumption, the following coefficients should vanish in the constitutive relations (3.66), (3.67):

$$L_{13} = L_{14} = L_{23} = L_{24} = 0, \quad L_{3i} = 0, \quad L_{4i} = 0 \quad (i = \overline{1,4}). \quad (3.130)$$

In this case, Eqs. (3.47) and (3.48) become

$$\mathbf{E}_* - \mathbf{E}^L = d_E \frac{d^2\boldsymbol{\pi}_e}{dt^2}, \quad (3.131)$$

$$(\nabla\mu'_\pi)^L - \nabla\mu'_\pi = d_m \frac{d^2\boldsymbol{\pi}_m}{dt^2}. \quad (3.132)$$

Consequently, taking the formulae (3.131) and (3.132) into account, the linear constitutive equations for vectors of polarization and local mass displacement appear as

$$\boldsymbol{\pi}_e + \chi_E d_E \frac{\partial^2\boldsymbol{\pi}_e}{\partial t^2} + \chi_{Em} d_m \frac{\partial^2\boldsymbol{\pi}_m}{\partial t^2} = \chi_E \mathbf{E} - \chi_{Em} \nabla\mu'_\pi, \quad (3.133)$$

$$\boldsymbol{\pi}_m + \chi_m d_m \frac{\partial^2\boldsymbol{\pi}_m}{\partial t^2} + \chi_{Em} d_E \frac{\partial^2\boldsymbol{\pi}_e}{\partial t^2} = -\chi_m \nabla\mu'_\pi + \chi_{Em} \mathbf{E}, \quad (3.134)$$

or, in the compact form,

$$\mathsf{L}(\boldsymbol{\pi}_e) = \underline{\mathsf{L}}_{3\mathrm{E}}(\mathbf{E}) - \chi_{Em} \nabla\mu'_\pi,$$

$$\mathsf{L}(\boldsymbol{\pi}_m) = \chi_{Em}\mathbf{E} - \underline{\mathsf{L}}_{4\mu}(\nabla\mu'_\pi).$$

Here, operators L, $\underline{\mathsf{L}}_{3\mathrm{E}}$, and $\underline{\mathsf{L}}_{4\mu}$ are determined through the formulae:

$$\mathsf{L} = \left(1 + \chi_E d_E \frac{\partial^2}{\partial t^2}\right)\left(1 + \chi_m d_m \frac{\partial^2}{\partial t^2}\right) - \chi_{Em}^2 d_m d_E \frac{\partial^4}{\partial t^4}, \quad (3.135)$$

$$\underline{\mathbf{L}}_{3\mathrm{E}} = \chi_E + \left(\chi_E \chi_m - \chi_{Em}^2\right) d_m \frac{\partial^2}{\partial t^2}, \tag{3.136}$$

$$\underline{\mathbf{L}}_{4\mu} = \chi_m + \left(\chi_E \chi_m - \chi_{Em}^2\right) d_E \frac{\partial^2}{\partial t^2}. \tag{3.137}$$

The state laws for the stress tensor, entropy, and potential μ'_π are of the form (3.53)–(3.55).

Now the heat flux and electric current are as follows:

$$\mathbf{J}_q = -\frac{L_{11}}{T_0}\nabla T + L_{12}\mathbf{E}, \quad \mathbf{J}_e = -\frac{L_{21}}{T_0}\nabla T + L_{22}\mathbf{E}.$$

Hence, for ideal dielectrics, we obtain

$$\mathbf{J}_q = -\mathfrak{L}_1^T \nabla\theta,$$

where $$\mathfrak{L}_1^T \equiv \lambda.$$

Thus, neglecting the dissipation of electric polarization and local mass displacement, the set of governing equations (3.85)–(3.91) can be written as:

$$\rho_0 \frac{\partial^2 \mathbf{u}}{\partial t^2} = \left(K + \frac{1}{3}G\right)\nabla(\nabla\cdot\mathbf{u}) + G\Delta\mathbf{u} - K\alpha_T\nabla\theta - K\alpha_\rho\nabla\rho_m + \rho_0\mathbf{F}, \tag{3.138}$$

$$\rho_0 C_V \frac{\partial\theta}{\partial t} + KT_0\alpha_T \frac{\partial(\nabla\cdot\mathbf{u})}{\partial t} + \rho_0 T_0 \beta_{T\rho}\frac{\partial\rho_m}{\partial t} = \lambda\Delta\theta + \rho_0\Re, \tag{3.139}$$

$$\begin{aligned}
&\Delta\rho_m - \lambda_\mu^2 \rho_m - \lambda_\mu^2\left(\chi_E d_E + \chi_m d_m\right)\frac{\partial^2\rho_m}{\partial t^2} \\
&+ \frac{\left(\chi_m\chi_E - \chi_{Em}^2\right)}{\chi_m} d_E \frac{\partial^2(\Delta\rho_m)}{\partial t^2} - \lambda_\mu^2\left(\chi_E\chi_m - \chi_{Em}^2\right) d_E d_m \frac{\partial^4\rho_m}{\partial t^4} \\
&= \lambda_\mu^2\left[\chi_m + \left(\chi_m\chi_E - \chi_{Em}^2\right) d_E \frac{\partial^2}{\partial t^2}\right]\left(\beta_{T\rho}\Delta\theta + \frac{K\alpha_\rho}{\rho_0}\Delta(\nabla\cdot\mathbf{u})\right) \\
&+ \lambda_\mu^2 \chi_{Em}\, \nabla\cdot\mathbf{E},
\end{aligned} \tag{3.140}$$

$$\nabla\times\mathbf{E} = -\frac{\partial\mathbf{B}}{\partial t}, \tag{3.141}$$

$$\nabla\cdot\mathbf{B} = 0, \tag{3.142}$$

$$(\varepsilon_0 \mathsf{L} + \rho_0 \underline{\mathsf{L}}_{3E})(\nabla \cdot \mathbf{E}) - \chi_{Em}\rho_0 d_\rho \Delta\rho_m$$
$$= -\beta_{T\rho}\chi_{Em}\rho_0\Delta\theta - K\alpha_\rho\chi_{Em}\Delta(\nabla\cdot\mathbf{u}), \tag{3.143}$$

$$\frac{1}{\mu_0}\mathsf{L}(\nabla\times\mathbf{B}) = (\varepsilon_0 \mathsf{L} + \rho_0 \underline{\mathsf{L}}_{3E})\left(\frac{\partial \mathbf{E}}{\partial t}\right)$$
$$-\rho_0\chi_{Em}\left[d_\rho\frac{\partial(\nabla\rho_m)}{\partial t} - K\frac{\alpha_\rho}{\rho_0}\frac{\partial(\nabla(\nabla\cdot\mathbf{u}))}{\partial t} - \beta_{T\rho}\frac{\partial(\nabla\theta)}{\partial t}\right]. \tag{3.144}$$

In this case, the set of equations (3.97)–(3.103) becomes

$$\rho_0\frac{\partial^2\mathbf{u}}{\partial t^2} = \left(\bar{K} + \frac{1}{3}G\right)\nabla(\nabla\cdot\mathbf{u}) + G\Delta\mathbf{u} - K\bar{\alpha}_T\nabla\theta - K\frac{\alpha_\rho}{d_\rho}\nabla\tilde{\mu}'_\pi + \rho_0\mathbf{F}, \tag{3.145}$$

$$\rho_0\bar{C}_V\frac{\partial\theta}{\partial t} + KT_0\bar{\alpha}_T\frac{\partial(\nabla\cdot\mathbf{u})}{\partial t} + \rho_0 T_0\frac{\beta_{T\rho}}{d_\rho}\frac{\partial\tilde{\mu}'_\pi}{\partial t} = \lambda\Delta\theta + \rho_0\Re, \tag{3.146}$$

$$\Delta\tilde{\mu}'_\pi - \lambda_\mu^2\tilde{\mu}'_\pi - \lambda_\mu^2(\chi_E d_E + \chi_m d_m)\frac{\partial^2\tilde{\mu}'_\pi}{\partial t^2}$$
$$+\frac{(\chi_m\chi_E - \chi_{Em}^2)}{\chi_m}d_E\frac{\partial^2(\Delta\tilde{\mu}'_\pi)}{\partial t^2} - \lambda_\mu^2(\chi_E\chi_m - \chi_{Em}^2)d_E d_m\frac{\partial^4\tilde{\mu}'_\pi}{\partial t^4}$$
$$= \lambda_\mu^2\left[1 + (\chi_E d_E + \chi_m d_m)\frac{\partial^2}{\partial t^2} + (\chi_E\chi_m - \chi_{Em}^2)d_E d_m\frac{\partial^4}{\partial t^4}\right]$$
$$\times\left(\beta_{T\rho}\theta + \frac{K\alpha_\rho}{\rho_0}\nabla\cdot\mathbf{u}\right) + \frac{\chi_{Em}}{\chi_m}\nabla\cdot\mathbf{E}, \tag{3.147}$$

$$\nabla\times\mathbf{E} = -\frac{\partial\mathbf{B}}{\partial t}, \tag{3.148}$$

$$\nabla\cdot\mathbf{B} = 0, \tag{3.149}$$

$$(\varepsilon_0 \mathsf{L} + \rho_0 \underline{\mathsf{L}}_{3E})(\nabla\cdot\mathbf{E}) - \chi_{Em}\rho_0\Delta\mu'_\pi = 0, \tag{3.150}$$

$$\frac{1}{\mu_0}\mathsf{L}(\nabla\times\mathbf{B}) = (\varepsilon_0 \mathsf{L} + \rho_0 \underline{\mathsf{L}}_{3E})\left(\frac{\partial\mathbf{E}}{\partial t}\right) - \rho_0\chi_{Em}\frac{\partial(\nabla\tilde{\mu}'_\pi)}{\partial t}. \tag{3.151}$$

Hence, comparing the obtained equations with their counterparts obtained in the second chapter, we can see that the consideration of the inertia of polarization and the local mass displacement manifested in increasing the order of the equations of electrodynamics and in the presence of "dynamic" terms in Eqs. (3.140) and (3.147) for the density of induced mass and for the modified chemical potential.

3.2.8 Generalized Lorentz Gauge Condition

Using the formulae (2.172)–(2.174), we represent vectors of the displacement **u** and mass forces **F** as the sum of their potential and solenoidal parts, as well as vectors of the electric field **E** and magnetic induction **B** through the scalar φ_e and vector **A** potentials. We also introduce the potential φ_{em} defined by the relation

$$\varphi_{em} = \frac{\rho_0}{\varepsilon}\left[\left(\frac{\varepsilon_0}{\rho_0}\mathsf{L} + \underline{\mathsf{L}}_{3\mathrm{E}}\right)\varphi_e + \underline{\mathsf{L}}_{3\mathrm{T}}\theta - \underline{\mathsf{L}}_{3\mu}\tilde{\mu}'_{\pi}\right]. \tag{3.152}$$

If the Lorentz gauge condition is modified as below

$$\mathsf{L}(\nabla\cdot\mathbf{A}) + \varepsilon\mu_0\frac{\partial\varphi_{em}}{\partial t} = 0, \tag{3.153}$$

then the electrodynamic equations (3.102) and (3.103) can be reduced to two uncoupled differential equatype equations for scalar φ_{em} and vector **A** potentials, namely

$$\mathsf{L}(\Delta\,\mathbf{A}) - \varepsilon_0\mu_0\left(\mathsf{L} + \frac{\rho_0}{\varepsilon_0}\underline{\mathsf{L}}_{3\mathrm{E}}\right)\left(\frac{\partial^2\mathbf{A}}{\partial t^2}\right) = 0, \tag{3.154}$$

$$\mathsf{L}(\Delta\,\varphi_{em}) - \varepsilon_0\mu_0\left(\mathsf{L} + \frac{\rho_0}{\varepsilon_0}\underline{\mathsf{L}}_{3\mathrm{E}}\right)\left(\frac{\partial^2\varphi_{em}}{\partial t^2}\right) = 0. \tag{3.155}$$

In Eqs. (3.152)–(3.155), the operators **L**, $\underline{\mathsf{L}}_{3\mathrm{T}}$, $\underline{\mathsf{L}}_{3\mathrm{E}}$, and $\underline{\mathsf{L}}_{3\mu}$ are given by the formulae (3.92), (3.95), and (3.96).

If the inertia of polarization, as well as the dissipation of the processes of local mass displacement and polarization, can be neglected, then the potential φ_{em} coincides with the potential $\varphi_{e\mu}$ introduced in the previous chapter by the formula (2.193).

Substituting the expressions (2.172)–(2.174) into Eqs. (3.97)–(3.99), in order to find the functions φ_u, $\boldsymbol{\psi}_u$, θ and $\tilde{\mu}'_{\pi}$, we obtain the following equations:

$$G\Delta\boldsymbol{\psi}_u + \rho_0\boldsymbol{\Psi} - \rho_0\frac{\partial^2\boldsymbol{\psi}_u}{\partial t^2} = 0, \tag{3.156}$$

$$\left(\bar{K}+\frac{4}{3}G\right)\Delta\varphi_u+\rho_0\Phi-\rho_0\frac{\partial^2\varphi_u}{\partial t^2}=K\frac{\alpha_\rho}{d_\rho}\tilde{\mu}'_\pi+K\bar{\alpha}_T\theta\,,\quad(3.157)$$

$$\left(\frac{\varepsilon_0}{\rho_0}\mathbf{L}+\underline{\mathbf{L}}_{3\mathrm{E}}\right)\mathbf{L}\left[\rho_0\bar{\bar{C}}_V\frac{\partial\theta}{\partial t}+T_0K\bar{\bar{\alpha}}_T\frac{\partial\Delta\varphi_u}{\partial t}+\rho_0T_0\frac{\bar{\beta}_{T\rho}}{d_\rho}\frac{\partial\tilde{\mu}'_\pi}{\partial t}\right.$$

$$\left.-\bar{\lambda}\Delta\theta-\rho_0\Re\right]+\varepsilon_0 a_{TE}\left(-\underline{\mathbf{L}}_{3\mathrm{T}}\mathbf{L}\frac{\partial\Delta\theta}{\partial t}+\underline{\mathbf{L}}_{3\mu}\mathbf{L}\frac{\partial\Delta\tilde{\mu}'_\pi}{\partial t}\right)=0.\quad(3.158)$$

$$\mathbf{L}\left\{\left(\frac{\varepsilon_0}{\rho_0}\mathbf{L}+\underline{\mathbf{L}}_{3\mathrm{E}}\right)\left(\Delta-\bar{\lambda}_\mu^2+\frac{\rho_0}{d_\rho\mathfrak{L}_4^\mu}\frac{\partial}{\partial t}-\frac{d_m}{d_\rho}\frac{\partial^2}{\partial t^2}\right)\tilde{\mu}'_\pi\right.$$

$$+\underline{\mathbf{L}}_{3\mu}\left(L+d_E\frac{\varepsilon_0}{\rho_0}\frac{\mathfrak{L}_4^E}{\mathfrak{L}_4^\mu}\frac{\partial^2}{\partial t^2}\right)\Delta\tilde{\mu}'_\pi$$

$$+\left[\frac{\mathfrak{L}_1^\mu}{T_0\mathfrak{L}_4^\mu}\left(\frac{\varepsilon_0}{\rho_0}\mathbf{L}+\underline{\mathbf{L}}_{3\mathrm{E}}\right)+\underline{\mathbf{L}}_{3\mathrm{T}}\left(L+d_E\frac{\varepsilon_0}{\rho_0}\frac{\mathfrak{L}_4^E}{\mathfrak{L}_4^\mu}\frac{\partial^2}{\partial t^2}\right)\right]\Delta\theta$$

$$\left.-\left(\frac{\varepsilon_0}{\rho_0}\mathbf{L}+\underline{\mathbf{L}}_{3\mathrm{E}}\right)\left(\bar{\lambda}_\mu^2-\frac{\rho_0}{d_\rho\mathfrak{L}_4^\mu}\frac{\partial}{\partial t}+\frac{d_m}{d_\rho}\frac{\partial^2}{\partial t^2}\right)\left(K\frac{\alpha_\mu}{\rho_0}\Delta\varphi_u+\beta_{T\rho}\theta\right)\right\}=0.$$

(3.159)

Here,

$$L=\frac{\mathfrak{L}_4^E}{\mathfrak{L}_4^\mu}+\frac{\varepsilon_0}{\rho_0}\left(\bar{\chi}_{Em}+\bar{\chi}_m\frac{\mathfrak{L}_4^E}{\mathfrak{L}_4^\mu}\right).$$

Comparing relations (3.156), (3.157) and their counterparts (2.175), (2.176), we can see that the inertia and dissipation of the electric polarization and local mass displacement do not influence the potentials $\boldsymbol{\psi}_u$ and φ_u of the displacement vector. However, in contrast to (2.178), Eq. (3.159) contains dynamic terms due to the inertia and dissipation of these processes. Equations (3.154) and (3.155) for a vector function **A** and for a modified electric potential φ_{em} have also changed in comparison with Eqs. (2.195) and (2.196) obtained in the previous chapter.

The set of equations (3.154)–(3.159) can be solved sequentially, namely, we may find functions φ_u, θ, and $\tilde{\mu}'_\pi$ as solutions of

Eqs. (3.157)–(3.159), and then determine the potentials **A** and φ_{em} as solutions to homogeneous Eqs. (3.154) and (3.155). If the functions θ, φ_{em}, and $\tilde{\mu}'_\pi$ are known, then in order to find the electric potential φ_e, we have a differential equation (3.152). Note that for vector potentials $\mathbf{\Psi}_u$ and **A**, we obtain homogeneous equations that are not connected either with each other or with other equations of this set.

3.3 Rheological Medium with Fading Memory

As a result of representation of the electric field vector and the gradient of a modified chemical potential μ'_π as the sum of their reversible and irreversible parts, we obtain time-nonlocal constitutive relations of the convolution type with an exponential kernel of relaxation (Kondrat and Hrytsyna 2009a). However, such kernels of relaxation do not always appropriately describe the behaviors of the observed irreversible processes.

Therefore, in this section, the local gradient theory of dielectrics is extended to the rheological dielectric media with a fading memory.

3.3.1 Energy Balance Equation

A complete set of equations of electrothermoelastic polarized non-ferromagnetic bodies with account for the rheological properties of a material includes field equations, as well as constitutive and geometric relations. For a rheological medium, the field equations that correspond to the Maxwell equations and to the fundamental physical laws of conservation of mass, momentum, entropy, and energy are the same as for a mathematical model of thermoelastic non-ferromagnetic polarized medium presented in Chapter 2. In order to obtain constitutive relations that take into account the rheological properties of a medium, we use the conservation law of free energy (2.64). For a rheological medium, we should remember that the generalized Helmholtz free energy is a functional of instantaneous values

$$\Lambda(t)=\left(\hat{e}(t),\, T(t),\, \rho_m(t), \mathbf{E}_*, \nabla\mu'_\pi(t)\right)$$

and histories of the variations in the strain tensor $\hat{e}$, the temperature T, the specific density of induced mass ρ_m, the electric field $\mathbf{E}_*$, and the gradient of the modified chemical potential μ'_π

$$\Lambda_t(\tau)=\left(\hat{e}_t(\tau),T_t(\tau),\rho_{mt}(\tau),\mathbf{E}_{*t}(\tau),\nabla\mu'_{\pi t}(\tau)\right),$$

that is, $f=\breve{f}=f\left[\Lambda(t),\Lambda_t(\tau)\right]$. In this case,

$$\frac{d\breve{f}}{dt}=\frac{\partial\breve{f}}{\partial\Lambda}\frac{d\Lambda}{dt}+\delta\breve{f}\left[\Lambda(t),\,\Lambda_t(\tau)\middle|\frac{d\Lambda_t(\tau)}{dt}\right],$$

where $\delta\breve{f}\left[\Lambda(t),\,\Lambda_t(\tau)\middle|\frac{d\Lambda_t(\tau)}{dt}\right]$ is the Frechet derivative (Cheney 2010).

Thus, the equation of energy conservation (2.64) can be written in the form

$$\rho\left(s+\frac{\partial\breve{f}}{\partial T}\right)\frac{dT}{dt}+\rho\left(-\frac{1}{\rho}\hat{\boldsymbol{\sigma}}_*+\frac{\partial\breve{f}}{\partial\hat{e}}\right):\frac{d\hat{e}}{dt}+\rho\left(\boldsymbol{\pi}_e+\frac{\partial\breve{f}}{\partial\mathbf{E}_*}\right)\cdot\frac{d\mathbf{E}_*}{dt}$$

$$+\rho\left(-\mu'_\pi+\frac{\partial\breve{f}}{\partial\rho_m}\right)\frac{d\rho_m}{dt}+\rho\left(-\boldsymbol{\pi}_m+\frac{\partial\breve{f}}{\partial\nabla\mu'_\pi}\right)\cdot\frac{d\nabla\mu'_\pi}{dt}$$

$$-\delta\breve{f}\left[\Lambda(t),\,\Lambda_t(\tau)\middle|\frac{d\Lambda_t(\tau)}{dt}\right]-\mathbf{J}_q\cdot\frac{\nabla T}{T}-T\eta_s$$

$$-\mathbf{v}\cdot\left(\rho\frac{d\mathbf{v}}{dt}-\nabla\cdot\hat{\boldsymbol{\sigma}}_*-\mathbf{F}_e-\rho\mathbf{F}_*\right)=0. \qquad (3.160)$$

3.3.2 Constitutive Equations

Since the quantities $\frac{dT}{dt}$, $\frac{d\hat{e}}{dt}$, $\frac{d\mathbf{E}_*}{dt}$, $\frac{d\rho_m}{dt}$, and $\frac{d\nabla\mu'_\pi}{dt}$ are independent, basing on Eq. (3.160), we get:

the constitutive relations

$$\hat{\boldsymbol{\sigma}}_*=\rho\frac{\partial\breve{f}}{\partial\hat{e}},\; s=-\frac{\partial\breve{f}}{\partial T},\; \mu'_\pi=\frac{\partial\breve{f}}{\partial\rho_m},\; \boldsymbol{\pi}_e=-\frac{\partial\breve{f}}{\partial\mathbf{E}_*},\; \boldsymbol{\pi}_m=\frac{\partial\breve{f}}{\partial\nabla\mu'_\pi},$$

the expression for entropy production

$$\eta_s = -\frac{1}{T}\delta\breve{f} - \mathbf{J}_q \cdot \frac{\nabla T}{T^2} \tag{3.161}$$

as well as the equation of motion

$$\rho\frac{d\mathbf{v}}{dt} = \nabla \cdot \hat{\boldsymbol{\sigma}}_* + \mathbf{F}_e + \rho\mathbf{F}_* .$$

Choosing an appropriate representation of the specific free energy $\breve{f}$, for an isotropic material, we can specify the linear constitutive relations as follows:

$$\hat{\boldsymbol{\sigma}}_* = \int_{-\infty}^{t} K_1^{\sigma}(t-t')\frac{\partial\hat{e}(t')}{\partial t'}dt' + \int_{-\infty}^{t} K_2^{\sigma}(t-t')\frac{\partial e(t')}{\partial t'}dt'\hat{\mathbf{I}}$$

$$+ \int_{-\infty}^{t} K_s^{\sigma}(t-t')\frac{\partial T(t')}{\partial t'}dt'\hat{\mathbf{I}} + \int_{-\infty}^{t} K_\mu^{\sigma}(t-t')\frac{\partial \rho_m(t')}{\partial t'}dt'\hat{\mathbf{I}}, \tag{3.162}$$

$$s - s_0 = -\int_{-\infty}^{t} K_s^{s}(t-t')\frac{\partial T(t')}{\partial t'}dt' - \int_{-\infty}^{t} K_\mu^{s}(t-t')\frac{\partial \rho_m(t')}{\partial t'}dt'$$

$$-\int_{-\infty}^{t} K_\sigma^{s}(t-t')\frac{\partial e(t')}{\partial t'}dt', \tag{3.163}$$

$$\mu'_\pi - \mu'_{\pi 0} = \int_{-\infty}^{t} K_\mu^{\mu}(t-t')\frac{\partial \rho_m(t')}{\partial t'}dt' + \int_{-\infty}^{t} K_s^{\mu}(t-t')\frac{\partial T(t')}{\partial t'}dt'$$

$$+\int_{-\infty}^{t} K_\sigma^{\mu}(t-t')\frac{\partial e(t')}{\partial t'}dt', \tag{3.164}$$

$$\boldsymbol{\pi}_e = -\int_{-\infty}^{t} K_p^{p}(t-t')\frac{\partial \mathbf{E}_*(t')}{\partial t'}dt' - \int_{-\infty}^{t} K_\pi^{p}(t-t')\frac{\partial \nabla\mu'_\pi(t')}{\partial t'}dt', \tag{3.165}$$

$$\boldsymbol{\pi}_m = \int_{-\infty}^{t} K_\pi^{\pi}(t-t')\frac{\partial \nabla\mu'_\pi(t')}{\partial t'}dt' + \int_{-\infty}^{t} K_p^{\pi}(t-t')\frac{\partial \mathbf{E}_*(t')}{\partial t'}dt'. \tag{3.166}$$

The relaxation kernels $K_i^{\sigma}(t-t')$, $K_s^{\sigma}(t-t')$, $K_\mu^{\sigma}(t-t')$, $K_w^{s}(t-t')$, $K_w^{\mu}(t-t')$, $K_b^{p}(t-t')$, $K_b^{\pi}(t-t')$, where i = 1, 2, $w \in \{\sigma, s, \mu\}$, and

$b \in \{p, \pi\}$, in the relations (3.162)–(3.166) should correspond to the principle of fading memory (Truesdell and Noll 1992). Hence, it follows that

$$K_{\beta}^{\alpha}(t-t') = K_{\beta 0}^{\alpha} + K_{\beta 1}^{\alpha}(t-t'), \quad \lim_{t-t' \to 0} K_{\beta 1}^{\alpha}(t-t') = 0. \quad (3.167)$$

Here, the kernels $K_{\beta 0}^{\alpha}$ do not depend on time and characterize an instantaneous reaction of the body.

Since $\eta_s \geq 0$, and $\breve{f}$ does not depend on the temperature gradient, we obtain: $-\delta \breve{f} \geq 0$. Then, from the expression (3.161) for entropy production, it follows that $-\mathbf{J}_q \cdot \nabla T / T^2 \geq 0$. The heat flux $\mathbf{J}_q$ is caused by the action of the thermodynamic force, i.e.,

$$\mathbf{J}_q = \mathbf{J}_q\,(\nabla T/T). \quad (3.168)$$

Let us assume that this dependence is functional. Taking the relations (3.167) and (3.168) into account, in the linear approximation, we obtain

$$\mathbf{J}_q = -\lambda_0 \nabla T + \int_{-\infty}^{t} \lambda_1(t-t') \frac{\partial \nabla T(t')}{\partial t'} dt'.$$

The relaxation kernels $K_i^{\sigma}(t-t')$, $K_s^{\sigma}(t-t')$, $K_{\mu}^{\sigma}(t-t')$, $K_w^{s}(t-t')$, $K_w^{\mu}(t-t')$, $K_b^{p}(t-t')$, $K_b^{\pi}(t-t')$ (i = 1, 2, $w \in \{\sigma, s, \mu\}$ $b \in \{p, \pi\}$) and $\lambda_1(t-t')$ are the material characteristics that should be determined by laboratory experiments. However, in contrast to the constitutive relations (3.127) and (3.128), these kernels may be not only exponential.

3.4 Local Gradient Theory of Dielectrics with Electric Quadrupoles

3.4.1 Electromagnetic Field Equations

Note that while constructing the governing equations in Chapter 2, only electric dipoles were taken into account. Here, we consider the contribution of electric dipoles and quadrupoles to the polarization current as well as the irreversibility of the local mass displacement. Based on this, we construct a more general theory of dielectrics.

In this section, we assume that the body polarization $\mathbf{\Pi}_e$ is caused by a change over time of both the dipole $\mathbf{P}$ and the quadrupole $\widehat{\mathbf{Q}}$ electric moments, namely (Fedorchenko 1988)

$$\mathbf{\Pi}_e = \mathbf{P} - \frac{1}{6}\nabla\cdot\widehat{\mathbf{Q}}. \tag{3.169}$$

Using the formulae (3.169) and (2.24), for an electric induction, we obtain the following relation:

$$\mathbf{D} = \varepsilon_0\mathbf{E} + \mathbf{P} - \frac{1}{6}\nabla\cdot\widehat{\mathbf{Q}}. \tag{3.170}$$

Let us return to the conservation law (2.34) for the energy of electromagnetic field. Note that in the co-moving frame, the vectors $\mathbf{E}$, $\mathbf{P}$, $\mathbf{J}_e$ and the tensor $\widehat{\mathbf{Q}}$ are transformed according to the relations:

$$\mathbf{E}_* = \mathbf{E} + \mathbf{v}\times\mathbf{B}, \quad \mathbf{P}_* = \mathbf{P}, \quad \widehat{\mathbf{Q}}_* = \widehat{\mathbf{Q}}, \quad \mathbf{J}_{e*} = \mathbf{J}_e + \rho_e\mathbf{v}. \tag{3.171}$$

Here, $\mathbf{E}_*$, $\mathbf{J}_{e*}$, $\mathbf{P}_*$, $\widehat{\mathbf{Q}}_*$ denote the electric field, density of electric current, dipole and quadrupole electric moments, respectively, referred to the co-moving frame and $\mathbf{E}$, $\mathbf{J}_e$, $\mathbf{P}$, $\widehat{\mathbf{Q}}$ are corresponding quantities, in a fixed frame.

Substituting relations (3.171) and (3.169) into Eq. (2.34) and taking the identity

$$\mathbf{a}\cdot(\mathbf{b}\times\mathbf{c}) = \mathbf{b}\cdot(\mathbf{c}\times\mathbf{a}) = \mathbf{c}\cdot(\mathbf{a}\times\mathbf{b})$$

into account, the balance equation for energy of the electromagnetic field can be reduced to the following form:

$$\frac{\partial U_e}{\partial t} + \nabla\cdot\mathbf{S}_e + \left(\mathbf{J}_{e*} + \frac{\partial\mathbf{\Pi}_e}{\partial t}\right)\cdot\mathbf{E}_*$$

$$+\mathbf{v}\cdot\left[\rho_e\mathbf{E}_* + \left(\mathbf{J}_{e*} + \frac{\partial\mathbf{\Pi}_e}{\partial t}\right)\times\mathbf{B}\right] = 0. \tag{3.172}$$

Taking the formula (3.169) into account, the last term in the first line of the above formula can be presented as follows:

$$\left(\mathbf{J}_{e*} + \frac{\partial\mathbf{\Pi}_e}{\partial t}\right)\cdot\mathbf{E}_* = \left(\mathbf{J}_{e*} + \frac{\partial\mathbf{P}}{\partial t}\right)\cdot\mathbf{E}_* - \frac{1}{6}\frac{\partial(\nabla\cdot\widehat{\mathbf{Q}})}{\partial t}\cdot\mathbf{E}_*. \tag{3.173}$$

In view of the identity

$$\nabla\cdot(\widehat{\mathbf{A}}\cdot\mathbf{a}) = (\nabla\cdot\widehat{\mathbf{A}})\cdot\mathbf{a} + \widehat{\mathbf{A}}^{\mathrm{T}}:(\nabla\otimes\mathbf{a}),$$

we can write

$$\frac{\partial(\nabla\cdot\widehat{\mathbf{Q}})}{\partial t}\cdot\mathbf{E}_*=\nabla\cdot\left(\frac{\partial\widehat{\mathbf{Q}}}{\partial t}\right)\cdot\mathbf{E}_*=\nabla\cdot\left(\frac{\partial\widehat{\mathbf{Q}}}{\partial t}\cdot\mathbf{E}_*\right)-\frac{\partial\widehat{\mathbf{Q}}^{\mathrm{T}}}{\partial t}:(\nabla\otimes\mathbf{E}_*). \quad (3.174)$$

Substituting Eqs. (3.173) and (3.174) into (3.172), after some algebra, the balance equation for the energy of an electromagnetic field can be rewritten as follows:

$$\frac{\partial U_e}{\partial t}+\nabla\cdot\mathbf{S}_e+\left(\mathbf{J}_{e*}+\frac{\partial\mathbf{P}}{\partial t}\right)\cdot\mathbf{E}_*+\frac{1}{6}\frac{\partial\widehat{\mathbf{Q}}}{\partial t}:(\nabla\otimes\mathbf{E}_*)$$

$$+\mathbf{v}\cdot\left[\rho_e\mathbf{E}_*+\left(\mathbf{J}_{e*}+\frac{\partial\boldsymbol{\Pi}_e}{\partial t}\right)\times\mathbf{B}\right]-\frac{1}{6}\nabla\cdot\left(\frac{\partial\widehat{\mathbf{Q}}}{\partial t}\cdot\mathbf{E}_*\right)=0. \quad (3.175)$$

Here, we take into consideration that $\widehat{\mathbf{Q}}$ is a symmetric tensor (Fedorchenko 1988).

Now, we can integrate Eq. (3.175) over the area (V). Using the divergence theorem, the last term in this equation can be transformed into the surface integral to obtain:

$$-\frac{1}{6}\oint_{(\Sigma)}\mathbf{n}\cdot\left(\frac{\partial\widehat{\mathbf{Q}}}{\partial t}\cdot\mathbf{E}_*\right)d\Sigma. \quad (3.176)$$

If the surface (Σ) is taken outside the body, then $\widehat{\mathbf{Q}}=0$ on this surface, that is, the surface integral (3.176) vanishes. Thus, finally, the conservation law (3.175) of the energy of an electromagnetic field can be rewritten as follows:

$$\frac{\partial U_e}{\partial t}+\nabla\cdot\mathbf{S}_e+\left(\mathbf{J}_{e*}+\frac{\partial\mathbf{P}}{\partial t}\right)\cdot\mathbf{E}_*+\frac{1}{6}\frac{\partial\widehat{\mathbf{Q}}}{\partial t}:(\nabla\otimes\mathbf{E}_*)$$

$$+\mathbf{v}\cdot\left[\rho_e\mathbf{E}_*+\left(\mathbf{J}_{e*}+\frac{\partial\boldsymbol{\Pi}_e}{\partial t}\right)\times\mathbf{B}\right]=0. \quad (3.177)$$

3.4.2 Energy Balance Equation

Making use of the divergence theorem as well as of the entropy balance equation (2.11), of the law of conservation of electromagnetic field energy (3.177), and of the formulae (2.37), (2.40), and (2.46), from the integral relation (2.43) after some calculations in the spirit of those in Section 2.6, we can write the following formulation for the conservation energy law:

$$\rho\frac{du}{dt}=\hat{\boldsymbol{\sigma}}_*:(\boldsymbol{\nabla}\otimes\mathbf{v})+\rho T\frac{ds}{dt}+\rho\mathbf{E}_*\cdot\frac{d\mathbf{p}}{dt}$$

$$+\rho(\boldsymbol{\nabla}\otimes\mathbf{E}_*)\cdot\frac{d\hat{\mathbf{q}}}{dt}+\rho\mu'_\pi\frac{d\rho_m}{dt}-\rho\,\boldsymbol{\nabla}\,\mu'_\pi\cdot\frac{d\boldsymbol{\pi}_m}{dt}$$

$$-\left[\frac{\partial\rho}{\partial t}+\boldsymbol{\nabla}\cdot(\rho\mathbf{v})\right]\left[u+\frac{1}{2}\mathbf{v}^2-Ts-\mathbf{p}\cdot\mathbf{E}_*-\hat{\mathbf{q}}:(\boldsymbol{\nabla}\otimes\mathbf{E}_*)-\rho_m\mu'_\pi+\boldsymbol{\nabla}\mu'_\pi\cdot\boldsymbol{\pi}_m\right]$$

$$+\mathbf{J}_{e*}\cdot\mathbf{E}_*-\mathbf{J}_q\cdot\frac{\boldsymbol{\nabla}T}{T}-T\eta_s+\mathbf{v}\cdot\left(-\rho\frac{d\mathbf{v}}{dt}+\boldsymbol{\nabla}\cdot\hat{\boldsymbol{\sigma}}_*+\mathbf{F}_e+\rho\mathbf{F}_*\right),\quad(3.178)$$

where

$$\hat{\boldsymbol{\sigma}}_*=\hat{\boldsymbol{\sigma}}-\rho\left[\mathbf{E}_*\cdot\mathbf{p}+\hat{\mathbf{q}}:(\boldsymbol{\nabla}\otimes\mathbf{E}_*)+\rho_m\mu'_\pi-\boldsymbol{\pi}_m\cdot\boldsymbol{\nabla}\mu'_\pi\right]\hat{\mathbf{I}},\quad(3.179)$$

$$\mathbf{F}_*=\mathbf{F}+\rho_m\boldsymbol{\nabla}\mu'_\pi-(\boldsymbol{\nabla}\otimes\boldsymbol{\nabla}\mu'_\pi)\cdot\boldsymbol{\pi}_m,\quad(3.180)$$

$$\mathbf{F}_e=\rho_e\mathbf{E}_*+\left(\mathbf{J}_{e*}+\frac{\partial\boldsymbol{\Pi}_e}{\partial t}\right)\times\mathbf{B}+\rho(\boldsymbol{\nabla}\otimes\mathbf{E}_*)\cdot\mathbf{p}+\rho\hat{\mathbf{q}}:(\boldsymbol{\nabla}\otimes\boldsymbol{\nabla}\otimes\mathbf{E}_*)^{\mathrm{T}(2,3)},\quad(3.181)$$

$$s=\frac{S}{\rho},\ \rho_m=\frac{\rho_{m\pi}}{\rho},\ \boldsymbol{\pi}_m=\frac{\boldsymbol{\Pi}_m}{\rho},\ \mathbf{p}=\frac{\mathbf{P}}{\rho},\ \hat{\mathbf{q}}=\frac{\hat{\mathbf{Q}}}{6\rho}.\quad(3.182)$$

Here, $(\boldsymbol{\nabla}\otimes\boldsymbol{\nabla}\otimes\mathbf{E}_*)^{\mathrm{T}(2,3)}$ is the third-order tensor, which is an isomer of the tensor $(\boldsymbol{\nabla}\otimes\boldsymbol{\nabla}\otimes\mathbf{E}_*)$, formed by transposition of its second and third basis vectors.

By requiring the energy balance equation (3.178) to be form-invariant under the superimposed rigid body translation, we get the balance law of linear momentum and the mass balance equation:

$$\rho\frac{d\mathbf{v}}{dt}=\boldsymbol{\nabla}\cdot\hat{\boldsymbol{\sigma}}_*+\mathbf{F}_e+\rho\mathbf{F}_*,\quad(3.183)$$

$$\frac{\partial\rho}{\partial t}+\boldsymbol{\nabla}\cdot(\rho\mathbf{v})=0.\quad(3.184)$$

It is evident from the Eqs. (3.179)–(3.181), that the electric quadrupole and mass dipole moments induce additional nonlinear body forces (i.e., ponderomotive and mass forces)

$$\mathbf{F}'_e=-\rho(\boldsymbol{\nabla}\otimes\boldsymbol{\nabla}\otimes\mathbf{E}_*):\hat{\mathbf{q}}\ \text{and}\ \mathbf{F}'=\rho_m\boldsymbol{\nabla}\mu'_\pi-\boldsymbol{\pi}_m\cdot\boldsymbol{\nabla}\otimes\boldsymbol{\nabla}\mu'_\pi$$

and coupled stresses

$$\hat{\boldsymbol{\sigma}}_* = -\rho\left[\hat{\mathbf{q}}:(\boldsymbol{\nabla}\otimes\mathbf{E}_*) + \rho_m\mu'_\pi - \boldsymbol{\pi}_m\cdot\boldsymbol{\nabla}\,\mu'_\pi\right]\hat{\mathbf{I}}$$

within the dielectric body.

Since $\hat{\boldsymbol{\sigma}}_*$ is a symmetric tensor and taking Eqs. (3.183) and (3.184) into account, the conservation law (3.178) of the internal energy can be simplified and written as follows:

$$\rho\frac{du}{dt} = \hat{\boldsymbol{\sigma}}_* : \frac{d\hat{e}}{dt} + \rho T\frac{ds}{dt} + \rho\mathbf{E}_*\cdot\frac{d\mathbf{p}}{dt}$$

$$+\rho(\boldsymbol{\nabla}\otimes\mathbf{E}_*)\cdot\frac{d\hat{\mathbf{q}}}{dt} + \rho\mu'_\pi\frac{d\rho_m}{dt} - \rho\boldsymbol{\nabla}\mu'_\pi\cdot\frac{d\boldsymbol{\pi}_m}{dt}$$

$$+\mathbf{J}_{e*}\cdot\mathbf{E}_* - \mathbf{J}_q\cdot\frac{\boldsymbol{\nabla}T}{T} - T\eta_s\,. \tag{3.185}$$

In order to take the irreversibility of local mass displacement into account, we represent the vector $\boldsymbol{\nabla}\mu'_\pi$ as the sum of its reversible $\left(\boldsymbol{\nabla}\mu'_\pi\right)^r$ and irreversible $\left(\boldsymbol{\nabla}\mu'_\pi\right)^i$ parts

$$\boldsymbol{\nabla}\mu'_\pi = \left(\boldsymbol{\nabla}\mu'_\pi\right)^r + \left(\boldsymbol{\nabla}\mu'_\pi\right)^i. \tag{3.186}$$

In view of representation (3.186), we can write the balance equation (3.185) for the internal energy as follows:

$$\rho\frac{du}{dt} = \hat{\boldsymbol{\sigma}}_* : \frac{d\hat{e}}{dt} + \rho T\frac{ds}{dt} + \rho\mathbf{E}_*\cdot\frac{d\mathbf{p}}{dt}$$

$$+\rho(\boldsymbol{\nabla}\otimes\mathbf{E}_*)\cdot\frac{d\hat{\mathbf{q}}}{dt} + \rho\mu'_\pi\frac{d\rho_m}{dt} - \rho(\boldsymbol{\nabla}\,\mu'_\pi)^r\cdot\frac{d\boldsymbol{\pi}_m}{dt}$$

$$+\mathbf{J}_{e*}\cdot\mathbf{E}_* - \mathbf{J}_q\cdot\frac{\boldsymbol{\nabla}T}{T} - \rho(\boldsymbol{\nabla}\mu'_\pi)^i\cdot\frac{d\boldsymbol{\pi}_m}{dt} - T\eta_s\,. \tag{3.187}$$

3.4.3 Gibbs Equation and Entropy Production: Constitutive Equations

By means of the Legendre transformation

$$f = u - Ts - \mathbf{E}_*\cdot\mathbf{p} - \hat{\mathbf{q}}:(\boldsymbol{\nabla}\otimes\mathbf{E}_*) + (\boldsymbol{\nabla}\mu'_\pi)^r\cdot\boldsymbol{\pi}_m\,,$$

we define the generalized Helmholtz free energy. Using this new thermodynamic function, Eq. (3.187) is transformed into

$$\rho\frac{df}{dt}=\hat{\boldsymbol{\sigma}}_*:\frac{d\hat{e}}{dt}-\rho s\frac{dT}{dt}-\rho\mathbf{p}\cdot\frac{d\mathbf{E}_*}{dt}-\rho\hat{\mathbf{q}}:\frac{d(\boldsymbol{\nabla}\otimes\mathbf{E}_*)}{dt}+\rho\mu_\pi'\frac{d\rho_m}{dt}$$

$$+\rho\boldsymbol{\pi}_m\cdot\frac{d(\boldsymbol{\nabla}\mu_\pi')^r}{dt}+\mathbf{J}_{e*}\cdot\mathbf{E}_*-\mathbf{J}_q\cdot\frac{\boldsymbol{\nabla}T}{T}-\rho(\boldsymbol{\nabla}\mu_\pi')^i\cdot\frac{d\boldsymbol{\pi}_m}{dt}-T\eta_s. \quad (3.188)$$

While inspecting the above equation, we assume that the Helmholtz free energy is a function of a strain tensor $\hat{e}$, temperature T, electric field vector $\mathbf{E}_*$, as well as the parameters $\boldsymbol{\nabla}\otimes\mathbf{E}_*$, ρ_m, and $(\boldsymbol{\nabla}\mu_\pi')^r$ related to the electric quadrupole and mass dipole moments, that is

$$f=f(\hat{e},T,\mathbf{E}_*,\boldsymbol{\nabla}\otimes\mathbf{E}_*,\rho_m,(\boldsymbol{\nabla}\mu_\pi')^r).$$

Thus, from Eq. (3.188), we get a generalized Gibbs equation:

$$df=\frac{1}{\rho}\hat{\boldsymbol{\sigma}}_*:d\hat{e}-sdT-\mathbf{p}\cdot d\mathbf{E}_*-\hat{\mathbf{q}}:d(\boldsymbol{\nabla}\otimes\mathbf{E}_*)+\mu_\pi'd\rho_m+\boldsymbol{\pi}_m\cdot d(\boldsymbol{\nabla}\mu_\pi')^r \quad (3.189)$$

and the following relation for entropy production:

$$\eta_s=\frac{1}{T}\left(-\mathbf{J}_q\cdot\frac{\boldsymbol{\nabla}T}{T}+\mathbf{J}_{e*}\cdot\mathbf{E}_*-\rho\frac{d\boldsymbol{\pi}_m}{dt}\cdot(\boldsymbol{\nabla}\mu_\pi')^i\right). \quad (3.190)$$

Since the parameters $\hat{e}$, T, $\mathbf{E}_*$, $\boldsymbol{\nabla}\otimes\mathbf{E}_*$, ρ_m, and $(\boldsymbol{\nabla}\mu_\pi')^r$ are independent, we obtain the following constitutive equations from the Gibbs relation (3.189):

$$\hat{\boldsymbol{\sigma}}_*=\rho\left.\frac{\partial f}{\partial\hat{e}}\right|_{T,\mathbf{E}_*,\boldsymbol{\nabla}\otimes\mathbf{E}_*,\rho_m,(\boldsymbol{\nabla}\mu_\pi')^r}, \quad s=-\left.\frac{\partial f}{\partial T}\right|_{\hat{e},\mathbf{E}_*,\boldsymbol{\nabla}\otimes\mathbf{E}_*,\rho_m,(\boldsymbol{\nabla}\mu_\pi')^r}, \quad (3.191)$$

$$\mathbf{p}=-\left.\frac{\partial f}{\partial\mathbf{E}_*}\right|_{\hat{e},T,\boldsymbol{\nabla}\otimes\mathbf{E}_*,\rho_m,(\boldsymbol{\nabla}\mu_\pi')^r}, \quad \hat{\mathbf{q}}=-\left.\frac{\partial f}{\partial(\boldsymbol{\nabla}\otimes\mathbf{E}_*)}\right|_{\hat{e},T,\mathbf{E}_*,\rho_m,(\boldsymbol{\nabla}\mu_\pi')^r}, \quad (3.192)$$

$$\mu_\pi'=\left.\frac{\partial f}{\partial\rho_m}\right|_{\hat{e},T,\mathbf{E}_*,\boldsymbol{\nabla}\otimes\mathbf{E}_*,(\boldsymbol{\nabla}\mu_\pi')^r}, \quad \boldsymbol{\pi}_m=\left.\frac{\partial f}{\partial(\boldsymbol{\nabla}\mu_\pi')^r}\right|_{\hat{e},T,\mathbf{E}_*,\boldsymbol{\nabla}\otimes\mathbf{E}_*,\rho_m}. \quad (3.193)$$

As it follows from Eqs. (3.191)–(3.193), due to accounting for electric quadrupoles, the electric field gradient is additionally incorporated into the space of constitutive parameters. The specific electric quadrupole $\hat{\mathbf{q}}$ is a thermodynamic conjugate of the electric field gradient $\boldsymbol{\nabla}\otimes\mathbf{E}_*$.

We should write Eqs. (3.191)–(3.193) in an explicit form. In order to obtain linear constitutive relations, we expand f into a Taylor series about $\hat{\boldsymbol{e}}=0$, $T = T_0$, $\mathbf{E}_* = 0$, $\nabla\otimes\mathbf{E}_* = 0$, $\rho_m = 0$, $\mu'_\pi = \mu'_{\pi 0}$, and $(\nabla\mu'_\pi)^r = 0$, where T_0 is a reference temperature and $\mu'_{\pi 0}$ is the modified chemical potential of an infinite medium. Denoting $I_1 = \hat{\boldsymbol{e}}:\hat{\mathbf{I}} = e$, $I_2 = \hat{\boldsymbol{e}}:\hat{\boldsymbol{e}}$, $I_{E1} = \nabla\cdot\mathbf{E}_*$, $I_{E2} = (\nabla\otimes\mathbf{E}_*):(\nabla\otimes\mathbf{E}_*)$ and keeping linear and quadratic terms only, we can write the following for isotropic materials:

$$f = f_0 - s_0\theta + \mu'_{\pi 0}\rho_m + \frac{1}{2\rho_0}\left(K - \frac{2}{3}G\right)I_1^2 + \frac{G}{\rho_0}I_2$$
$$-\frac{C_V}{2T_0}\theta^2 + \frac{d_\rho}{2}\rho_m^2 - \frac{\chi_E}{2}\mathbf{E}_*\cdot\mathbf{E}_* - \frac{\chi_m}{2}\left(\nabla\,\mu'_\pi\right)^r\cdot\left(\nabla\,\mu'_\pi\right)^r$$
$$+\frac{\chi_{q1}}{2}I_{E1}^2 - \chi_{q2}I_{E2} - \frac{K\alpha_T}{\rho_0}I_1\theta - \frac{K\alpha_{E1}}{\rho_0}I_1 I_{E1}$$
$$-\frac{K\alpha_\rho}{\rho_0}I_1\rho_m - \beta_{T\rho}\rho_m\theta + \beta_{TE}I_{E1}\theta + \beta_{E\rho}I_{E1}\rho_m$$
$$+\chi_{Em}\mathbf{E}_*\cdot(\nabla\,\mu'_\pi)^r + 2G\frac{\alpha_{E2}}{\rho_0}\hat{\boldsymbol{e}}:(\nabla\otimes\mathbf{E}_*). \qquad (3.194)$$

Here, $K, G, C_V, d_\rho, \alpha_T, \alpha_\rho, \alpha_{E1}, \alpha_{E2}, \chi_E, \chi_m, \chi_{Em}, \chi_{q1}, \chi_{q2}, \beta_{T\rho}, \beta_{TE}$, and $\beta_{E\rho}$ are material characteristics.

Formula (3.194), along with relations (3.191)–(3.193), leads to the following linear constitutive equations for isotropic dielectric materials:

$$\hat{\boldsymbol{\sigma}} = 2G\hat{\boldsymbol{e}} + 2G\alpha_{E2}\,\nabla\otimes\mathbf{E} + \left[\left(K - \frac{2}{3}G\right)e - K\alpha_T\theta - K\alpha_\rho\rho_m - K\alpha_{E1}\,\nabla\cdot\mathbf{E}\right]\hat{\mathbf{I}}, \qquad (3.195)$$

$$s = s_0 + \frac{C_V}{T_0}\theta + \frac{K\alpha_T}{\rho_0}e + \beta_{T\rho}\rho_m - \beta_{TE}\,\nabla\cdot\mathbf{E}, \qquad (3.196)$$

$$\mu'_\pi = \mu'_{\pi 0} + d_\rho\rho_m - \frac{K\alpha_\rho}{\rho_0}e - \beta_{T\rho}\theta + \beta_{E\rho}\,\nabla\cdot\mathbf{E}, \qquad (3.197)$$

$$\mathbf{p} = \chi_E\mathbf{E} - \chi_{Em}(\nabla\,\mu'_\pi)^r, \qquad (3.198)$$

$$\boldsymbol{\pi}_m = -\chi_m(\nabla \mu'_\pi)^r + \chi_{Em}\mathbf{E}, \tag{3.199}$$

$$\hat{\mathbf{q}} = 2\chi_{q2}\,\nabla \otimes \mathbf{E} - 2G\alpha_{E2}\hat{e} - \left(\chi_{q1}\,\nabla \cdot \mathbf{E} - \frac{K\alpha_{E1}}{\rho_0}e + \beta_{TE}\theta + \beta_{E\rho}\rho_m\right)\hat{\mathbf{I}}. \tag{3.200}$$

Equations (3.195)–(3.200) describe an electromechanical interaction in isotropic materials. In the framework of the developed theory, the body polarization

$$\boldsymbol{\pi}_e = \mathbf{p} - \nabla \cdot \hat{\mathbf{q}}$$

is caused not only by the electric field but also by its spatial inhomogeneity, as well as by the gradients of the strain, the temperature, and the density of induced mass. Hence, the constitutive equations (3.195)–(3.200) for isotropic materials make it possible to describe the flexoelectric and thermopolarization effects. Note that the linear classical theories of dielectrics are incapable of describing these effects.

Compared to the relations (2.76)–(2.80), the constitutive equations (3.195)–(3.200) contain new material constants α_{E1}, α_{E2}, β_{TE}, $\beta_{E\rho}$, χ_{q2}, and χ_{q1} that characterize the physical properties of a dielectric medium with electric quadrupoles.

Now, we shall specify the expressions for fluxes. We represent Eq. (3.190) for entropy production as follows: $\eta_s = \frac{1}{T}\sum_{k=1}^{3}\mathbf{j}_k \cdot \mathbf{X}_k$, where

$$\mathbf{j}_1 = \mathbf{J}_q,\quad \mathbf{j}_2 = \mathbf{J}_{e*},\quad \mathbf{j}_3 = \rho\frac{d\boldsymbol{\pi}_m}{dt},$$

$$\mathbf{X}_1 = \nabla T/T,\quad \mathbf{X}_2 = \mathbf{E}_*,\quad \mathbf{X}_3 = -(\nabla \mu'_\pi)^i$$

are the thermodynamic fluxes and forces. Let us assume that thermodynamic fluxes are functions of thermodynamic forces, that is:

$$\mathbf{j}_i = \mathbf{j}_i(\mathbf{X}_1, \mathbf{X}_2, \mathbf{X}_3, A),\ i = \overline{1,3}. \tag{3.201}$$

Note that $\mathbf{j}_i(0, 0, 0, A) = 0$, $i = \overline{1,3}$. For a linear approximation, we can write

$$\mathbf{j}_i = \sum_{j=1}^{3} L_{ij}\mathbf{X}_j,\quad i = \overline{1,3},$$

where L_{ij} $(i, j = \overline{1,3})$ are kinetic coefficients. According to the second law of thermodynamics and Onsager theorem (Gyarmati 1970), the following relations must be satisfied:

$$\mathbf{j}_1 \cdot \mathbf{X}_1 + \mathbf{j}_2 \cdot \mathbf{X}_2 + \mathbf{j}_3 \cdot \mathbf{X}_3 \geq 0, \quad L_{ii} \geq 0, \quad L_{ij} = L_{ji}.$$

Using the formula (3.186), we can exclude the irreversible $(\nabla\mu'_\pi)^i$ and reversible $(\nabla\mu'_\pi)^r$ parts of vector $\nabla\mu'_\pi$ from Eqs. (3.198), (3.199), and (3.201). As a result, we obtain the following relations for the heat flux $\mathbf{J}_q$, electric current density $\mathbf{J}_e$, mass dipole $\boldsymbol{\pi}_m$, and electric dipole $\mathbf{p}$:

$$\mathbf{J}_q = -\lambda\nabla T + \pi_t \mathbf{J}_e + (\pi_l - \eta'\pi_t)\rho_0 \frac{\partial \boldsymbol{\pi}_m}{\partial t}, \tag{3.202}$$

$$\mathbf{J}_e = \sigma_e \mathbf{E} - \eta\nabla T + \eta'\rho_0 \frac{\partial \boldsymbol{\pi}_m}{\partial t}, \tag{3.203}$$

$$\frac{\partial \boldsymbol{\pi}_m}{\partial t} + \frac{1}{\tau_\pi}\boldsymbol{\pi}_m = -\frac{\chi_m}{\tau_\pi}\left[\nabla\mu'_\pi + \pi_l \frac{\nabla T}{T_0} - \left(\eta' + \frac{\chi_{Em}}{\chi_m}\right)\mathbf{E}\right], \tag{3.204}$$

$$\mathbf{p} = \chi_E\left(1 - \frac{\chi_{Em}^2}{\chi_E \chi_m}\right)\mathbf{E} + \frac{\chi_{Em}}{\chi_m}\boldsymbol{\pi}_m. \tag{3.205}$$

Here,

$$\sigma_e = L_{22} - \frac{L_{23}^2}{L_{33}}, \quad \lambda = \frac{1}{T_0}\left[L_{11} - \frac{L_{13}^2}{L_{33}} - \frac{(L_{12}L_{33} - L_{13}L_{23})^2}{L_{33}(L_{22}L_{33} - L_{23}^2)}\right], \quad \tau_\pi = \frac{\rho_0 \chi_m}{L_{33}},$$

$$\eta = \frac{1}{T_0}\left(L_{12} - \frac{L_{23}L_{13}}{L_{33}}\right), \quad \eta' = \frac{L_{23}}{L_{33}}, \quad \pi_t = \frac{L_{12}L_{33} - L_{13}L_{23}}{L_{22}L_{33} - L_{23}^2}, \quad \pi_l = \frac{L_{13}}{L_{33}}.$$

Note that in Eqs. (3.202)–(3.205), we took into consideration that, in a linear approximation, $\mathbf{E}_* = \mathbf{E}$, $\mathbf{J}_{e*} = \mathbf{J}_e$, and $\dfrac{d}{dt} = \dfrac{\partial}{\partial t}$.

Due to consideration of the local mass displacement as a dissipative process, we obtained the rheological constitutive relations for vectors of the local mass displacement $\boldsymbol{\pi}_m$, the electric dipoles $\mathbf{p}$, the heat flux $\mathbf{J}_q$, and the electric current $\mathbf{J}_e$.

The Maxwell equations (2.20), (2.21), (2.25), (2.26); field equations (2.11), (2.41), (3.183), and (3.184); the constitutive equations (2.17), (3.170), (3.195)–(3.197), (3.200), and (3.202)–

(3.205) with the formulae (3.182) and (3.190) compose a complete set of linear equations that describe the electrothermomechanical processes in elastic dielectrics with quadrupole polarization and irreversible local mass displacement. The final form of the governing equations can be obtained by substituting the constitutive and geometric relations into the field equations.

3.4.4 Governing Equations when Neglecting the Dissipation of Local Mass Displacement

If we describe the local mass displacement as a non-dissipative process, then the linear relation for a polarization vector looks as follows:

$$\boldsymbol{\pi}_e = \chi_E \mathbf{E} - \left(\chi_{Em} - \frac{\beta_{E\rho}}{d_\rho} \right) \boldsymbol{\nabla} \tilde{\mu}'_\pi + \left(\chi_{q1} - \frac{\beta_{E\rho}^2}{d_\rho} \right) \boldsymbol{\nabla} (\boldsymbol{\nabla} \cdot \mathbf{E}) - 2\chi_{q2} \Delta \mathbf{E}$$

$$+ 2G\alpha_{E2} \boldsymbol{\nabla} \cdot \hat{\boldsymbol{e}} - \frac{K\alpha_{E1}}{\rho_0} \left(1 - \frac{\alpha_\rho \beta_{E\rho}}{d_\rho \alpha_{E1}} \right) \boldsymbol{\nabla} e + \left(\beta_{TE} + \frac{\beta_{T\rho} \beta_{E\rho}}{d_\rho} \right) \boldsymbol{\nabla} \theta .$$

We can see that the electric polarization vector $\boldsymbol{\pi}_e$ depends not only on the electric field $\mathbf{E}$ and on the gradient of modified chemical potential $\boldsymbol{\nabla} \mu'_\pi$, as it follows from Eq. (2.79), but also on the temperature gradient, the strain gradient and the spatial derivatives of an electric field.

In what follows, we shall consider an isothermal approximation. Thus, the fundamental field equations expressed in terms of the displacement vector $\mathbf{u}$, induced mass ρ_m, electric field $\mathbf{E}$, and magnetic induction $\mathbf{B}$ can be written as follows:

$$\rho_0 \frac{\partial^2 \mathbf{u}}{\partial t^2} = \left(K + \frac{1}{3} G \right) \boldsymbol{\nabla} (\boldsymbol{\nabla} \cdot \mathbf{u}) + G \Delta \mathbf{u} - K\alpha_{E1} \boldsymbol{\nabla} (\boldsymbol{\nabla} \cdot \mathbf{E}) + 2G\alpha_{E2} \Delta \mathbf{E}$$

$$- K\alpha_\rho \boldsymbol{\nabla} \rho_m + \rho_0 \mathbf{F} , \tag{3.206}$$

$$\Delta \rho_m - \lambda_\mu^2 \rho_m = \frac{K\alpha_\rho}{\rho_0 d_\rho} \Delta(\boldsymbol{\nabla} \cdot \mathbf{u}) - \frac{\beta_{E\rho}}{d_\rho} \Delta(\boldsymbol{\nabla} \cdot \mathbf{E}) + \chi_{Em} \lambda_\mu^2 \boldsymbol{\nabla} \cdot \mathbf{E} , \tag{3.207}$$

$$\boldsymbol{\nabla} \times \mathbf{E} = -\frac{\partial \mathbf{B}}{\partial t} , \tag{3.208}$$

$$\nabla \cdot \mathbf{B} = 0, \tag{3.209}$$

$$\frac{1}{\mu_0}\nabla \times \mathbf{B} = \sigma_e \mathbf{E} + \varepsilon \frac{\partial \mathbf{E}}{\partial t} + \rho_0(\chi_{q1} - \beta_{E\rho}\chi_{Em})\frac{\partial \nabla(\nabla \cdot \mathbf{E})}{\partial t}$$
$$-2\rho_0\chi_{q2}\frac{\partial \Delta \mathbf{E}}{\partial t} + \rho_0(\beta_{E\rho} - d_\rho \chi_{Em})\frac{\partial \nabla \rho_m}{\partial t} + \rho_0 G\alpha_{E2}\frac{\partial \Delta \mathbf{u}}{\partial t}$$
$$+(K\alpha_\rho \chi_{Em} + \rho_0 G\alpha_{E2} - K\alpha_{E1})\frac{\partial(\nabla\nabla \cdot \mathbf{u})}{\partial t}, \tag{3.210}$$

$$\varepsilon \nabla \cdot \mathbf{E} + \rho_0(\chi_{q1} - 2\chi_{q2} - \chi_{Em}\beta_{E\rho})\Delta(\nabla \cdot \mathbf{E})$$
$$+(2\rho_0 G\alpha_{E2} - K\alpha_{E1} + K\alpha_\rho \chi_{Em})\Delta(\nabla \cdot \mathbf{u})$$
$$+\rho_0(\beta_{E\rho} - d_\rho \chi_{Em})\Delta \rho_m = \rho_e. \tag{3.211}$$

Here, $\varepsilon = \varepsilon_0 + \rho_0 \chi_E$.

As a result of accounting for the electric quadrupoles, summands proportional to a second-order spatial derivatives of the electric field vector **E** appear in the balance of the momentum (3.206). Equations (3.210) and (3.211) additionally contain summands proportional to a third-order mixed partial derivative of the vector of electric field.

Problems

3.1 Starting with Eq. (3.8), make detailed calculations leading to the relations (3.10) and (3.14)–(3.16).

3.2 Construct an explicit form of the constitutive relations (3.22) and (3.23) for isotropic materials. Write down the corresponding Maxwell relations. Obtain a governing set of equations for a local gradient mathematical model of an elastic isotropic polarized media that takes into account the effect of the local mass displacement on the shear stresses.

3.3 Give detailed calculations leading from Eq. (3.37) to the relation (3.40).

3.4 Get a generalized Gibbs equation and an expression for the entropy production for the local gradient theory of a viscous incompressible polarized liquid that considers the irreversibility of electric polarization and local mass displacement. Based on these equations, construct the corresponding constitutive relations.

3.5 Make detailed calculations leading to the relation (3.178). Get the expressions (3.179) and (3.181).

3.6 Construct the constitutive relations of the local gradient theory of dielectric media with electric quadrupoles, taking the density of internal energy $u = u(\hat{e}, s, \mathbf{p}, \hat{\mathbf{q}}, \rho_m, \pi_m)$ as a thermodynamic potential. Make an explicit form of linear constitutive relations for centrosymmetric cubic crystals.

3.7 Make the governing equations for the linear local gradient theory of dielectric media with electric quadrupoles taking the irreversibility of local mass displacement into consideration.

3.8 Construct a local gradient theory of a viscous incompressible polarized liquid with electric quadrupoles following a step similar to Section 3.4.

Section II
APPLICATIONS

Chapter 4

Near-Surface Inhomogeneity of Electromechanical Fields

In Chapter 4, the linear relations of the local gradient theory of dielectrics are tested on some simple problems. In particular, they are used to study the effect of a free surface on the stress–strain state and polarization of elastic bodies having plane and cylindrical surfaces, as well as of a thermoelastic hollow sphere. Using the relations of the local gradient theory of elasticity, which involves the irreversibility of the local mass displacement, the formation of a near-surface inhomogeneity of the fields in an isotropic infinite layer is investigated. A relation for the surface energy of deformation and polarization for non-ferromagnetic polarized solid bodies is obtained.

The solutions to the formulated stationary boundary-value problems enabled us to theoretically describe some observed phenomena, namely,

- The near-surface inhomogeneity of electromechanical fields,
- Thermopolarization and piroelectric effects in isotropic materials,
- Nonlinear dependence of the inverse capacitance of a plane capacitor on the thickness of a dielectric layer (i.e., Mead's anomaly),
- The emergence of a bound electric charge on the free surfaces of dielectric bodies,

Local Gradient Theory for Dielectrics: Fundamentals and Applications
Olha Hrytsyna and Vasyl Kondrat

ISBN 978-981-4800-62-4 (Hardcover), 978-1-003-00686-2 (eBook)
www.jennystanford.com

- Electromagnetic emission that accompanies the process of formation of new surfaces in the body,
- Size effects of the stresses, bound electric charge, surface energy of deformation and polarization, and their dependence on the surface curvature.

These solutions are also used to substantiate the emergence of a disjoining pressure in thin solid films and lateral force in thin films of variable thickness. The classical theory of dielectrics is incapable of substantiating the mentioned phenomena.

It is also shown that for a concentrated electric line source and a point charge, the potential of electric field is not singular, which is different from the classical theory of piezoelectricity.

4.1 Surface Energy of Deformation and Polarization

The notion of surface energy of deformation and polarization was originally introduced by Mindlin (1965, 1968). He showed that in order to separate an arbitrary material continuum into two parts along a certain surface (Σ), it is necessary to surmount the energy of bonding between the particles located on this surface and in its vicinity. The bonding energy is the energy that should be spent to break the atomic bonds along the surface (Σ), retaining the strain and polarization of the body constant (Maugin 1988). This is achieved by applying an external field. If the external field is removed, then the body deforms and polarizes in the vicinity of the surface. As a result, the surface energy of deformation and polarization arises.

It was established that the surface energy of deformation and polarization is always negative and constitutes up to 30% of the total energy of the bonding (Maugin 1988). Consequently, this energy cannot be neglected.

4.1.1 Tensor-Like Representation of Parameters Related to the Local Mass Displacement

Here we use the relation of the mechanics of dielectric solids, which takes into account the tensor nature of the parameters related to the local mass displacement in order to define the surface energy of deformation and polarization. For simplicity and clarity, we accept an isothermal approximation.

Let us consider the equilibrium state of a solid that occupies a domain (V_*) of the Euclidean space bounded by a smooth surface (Σ_*) with an external normal $\mathbf{n}$. Assume that the body, which is an ideal dielectric, contacts with vacuum (domain (V_v)) and undergoes the external mechanical and electromagnetic loads.

Using constitutive equations (3.26)–(3.30), we can rewrite the representation (3.25) for the free energy f as follows:

$$f - f_0 = \frac{1}{2\rho_0}\hat{\boldsymbol{\sigma}}_* : \hat{e} + \frac{1}{2}\hat{\boldsymbol{\mu}}_0^{\prime\pi} : \hat{\boldsymbol{\rho}}^m + \frac{1}{2}\hat{\boldsymbol{\mu}}^{\prime\pi} : \hat{\boldsymbol{\rho}}^m + \frac{1}{2}\hat{\boldsymbol{\pi}}^{m(3)} \overset{(3)}{\cdot} \widehat{\mathbf{M}}^{(3)} - \frac{1}{2}\boldsymbol{\pi}_e \cdot \mathbf{E}. \tag{4.1}$$

Hence, for the specific internal energy u, where $u = f + \boldsymbol{\pi}_e \cdot \mathbf{E} - \hat{\boldsymbol{\pi}}^{m(3)} \overset{(3)}{\cdot} \widehat{\mathbf{M}}^{(3)}$, we get

$$u = \frac{1}{2\rho_0}\hat{\boldsymbol{\sigma}}_* : \hat{e} + \frac{1}{2}\hat{\boldsymbol{\mu}}_0^{\prime\pi} : \hat{\boldsymbol{\rho}}^m + \frac{1}{2}\hat{\boldsymbol{\mu}}^{\prime\pi} : \hat{\boldsymbol{\rho}}^m - \frac{1}{2}\hat{\boldsymbol{\pi}}^{m(3)} \overset{(3)}{\cdot} \widehat{\mathbf{M}}^{(3)} + \frac{1}{2}\boldsymbol{\pi}_e \cdot \mathbf{E}. \tag{4.2}$$

In the obtained expression, we take into account the strain–displacement relation (2.8), the linear equilibrium equation (3.17), the symmetric property of stress tensor $\hat{\boldsymbol{\sigma}}_*$, the identity

$$\boldsymbol{\nabla}\cdot\left(\hat{\boldsymbol{\pi}}^{m(3)} : \hat{\boldsymbol{\mu}}^{\prime\pi}\right) = \left(\boldsymbol{\nabla}\cdot\hat{\boldsymbol{\pi}}^{m(3)}\right) : \hat{\boldsymbol{\mu}}^{\prime\pi} + \hat{\boldsymbol{\pi}}^{m(3)} \overset{(3)}{\cdot} \left(\boldsymbol{\nabla}\otimes\hat{\boldsymbol{\mu}}^{\prime\pi}\right)^{\mathrm{T}(1,3)},$$

and the relation

$$\hat{\boldsymbol{\rho}}^m = -\boldsymbol{\nabla}\cdot\hat{\boldsymbol{\pi}}^{m(3)}.$$

The last relation is a consequence of Eqs. (3.11) and (3.13). After some algebra, the relation (4.2) may be written in the form

$$\rho_0 u = \frac{1}{2}\boldsymbol{\nabla}\cdot\left(\hat{\boldsymbol{\sigma}}_* \cdot \mathbf{u}\right) + \frac{1}{2}\rho_0\mathbf{F}_* \cdot \mathbf{u} - \frac{1}{2}\rho_0\hat{\boldsymbol{\mu}}_0^{\prime\pi} : \left(\boldsymbol{\nabla}\cdot\hat{\boldsymbol{\pi}}^{m(3)}\right)$$

$$-\frac{1}{2}\rho_0\boldsymbol{\nabla}\cdot\left(\hat{\boldsymbol{\pi}}^{m(3)} : \hat{\boldsymbol{\mu}}^{\prime\pi}\right) + \frac{1}{2}\boldsymbol{\Pi}_e \cdot \mathbf{E}. \tag{4.3}$$

In the stationary approximation, $\mathbf{E} = -\boldsymbol{\nabla}\varphi_e$. Thus, in view of the relation (2.24), we can rewrite the last summand in expression (4.3) as follows:

$$\frac{1}{2}\mathbf{\Pi}_e\cdot\mathbf{E}=-\frac{1}{2}\nabla\cdot(\mathbf{D}\varphi_e)+\frac{1}{2}(\nabla\cdot\mathbf{D})\varphi_e-\frac{1}{2}\varepsilon_0\mathbf{E}\cdot\mathbf{E}. \tag{4.4}$$

Taking relations (4.3), (4.4) and the Maxwell equation (2.26) into account, we obtain the following expression:

$$\mathcal{E}=\frac{1}{2}\nabla\cdot\left(\hat{\boldsymbol{\sigma}}_*\cdot\mathbf{u}\right)+\frac{1}{2}\rho_0\mathbf{F}_*\cdot\mathbf{u}-\frac{1}{2}\rho_0\hat{\boldsymbol{\mu}}_0^{\prime\pi}:\left(\nabla\cdot\hat{\boldsymbol{\pi}}^{m(3)}\right)$$

$$-\frac{1}{2}\rho_0\nabla\cdot\left(\hat{\boldsymbol{\pi}}^{m(3)}:\hat{\boldsymbol{\mu}}^{\prime\pi}\right)-\frac{1}{2}\nabla\cdot(\mathbf{D}\varphi_e), \tag{4.5}$$

where $\mathcal{E}=\rho_0 u+\frac{1}{2}\varepsilon_0\mathbf{E}\cdot\mathbf{E}$ is the total energy.

Having integrated the expression (4.5) over the domain (V') = $(V_*)\cup(V_v)$ occupied by the body (region (V_*)) with vacuum (region (V_v)) and using the divergence theorem, we arrive at the following integral equation:

$$\int_{(V')}\mathcal{E}\,dV=\frac{1}{2}\int_{(\Sigma_*)}\boldsymbol{\sigma}_{n*}\cdot\mathbf{u}\,d\Sigma+\frac{1}{2}\rho_0\int_{(V_*)}\mathbf{F}_*\cdot\mathbf{u}\,dV-\frac{1}{2}\int_{(\Sigma_*)}\mathbf{n}\cdot[\mathbf{D}]\varphi_e\,d\Sigma$$

$$-\frac{1}{2}\rho_0\int_{(\Sigma_*)}\mathbf{n}\cdot\hat{\boldsymbol{\pi}}^{m(3)}:\hat{\boldsymbol{\mu}}^{\prime\pi}\,d\Sigma-\frac{1}{2}\rho_0\hat{\boldsymbol{\mu}}_0^{\prime\pi}:\int_{(\Sigma_*)}\mathbf{n}\cdot\hat{\boldsymbol{\pi}}^{m(3)}\,d\Sigma. \tag{4.6}$$

Here, $[\mathbf{D}]=\mathbf{D}-\mathbf{D}_v$ denotes the finite jump of the electric induction over the surface (Σ_*).

Note that here we take into account the equality of the electric potentials of the body and vacuum on the surface (Σ_*).

Let us consider an equilibrium state of the solid with traction-free surfaces ($\boldsymbol{\sigma}_n=0$, $\forall\mathbf{r}\in(\Sigma_*)$) and without body forces ($\mathbf{F}=0$, $\forall\mathbf{r}\in(V_*)$). Since the body contacts with vacuum, we assume that $\forall\mathbf{r}\in(\Sigma_*)$: $\mu_\pi'=0$, $\mathbf{n}\cdot(\mathbf{D}-\mathbf{D}_v)=0$. As a result, the first four terms on the right-hand side of Eq. (4.6) vanish and, therefore,

$$\int_{(V')}\mathcal{E}\,dV=-\frac{1}{2}\rho_0\hat{\boldsymbol{\mu}}_0^{\prime\pi}:\int_{(\Sigma_*)}\mathbf{n}\cdot\hat{\boldsymbol{\pi}}^{m(3)}\,d\Sigma. \tag{4.7}$$

The right-hand side of this equation yields the surface energy of deformation and polarization U_Σ [J/m^2]. Within the framework of the local gradient theory of dielectrics, this energy is defined by the formula:

$$U_{\Sigma}=-\frac{1}{2}\rho_0\mathbf{n}\cdot\hat{\boldsymbol{\pi}}^{m(3)}:\hat{\boldsymbol{\mu}}_0^{\prime\pi}\Big|_{r\in(\Sigma_*)}=-\frac{1}{2}\mathbf{n}\cdot\widehat{\boldsymbol{\Pi}}^{m(3)}:\hat{\boldsymbol{\mu}}_0^{\prime\pi}\Big|_{r\in(\Sigma_*)}. \quad (4.8)$$

According to relation (4.8), the surface energy of deformation and polarization is determined by the modified chemical potential $\hat{\boldsymbol{\mu}}_0^{\prime\pi}$ of an infinite medium and the surface value of the projection of the tensor of the local mass displacement $\widehat{\boldsymbol{\Pi}}^{m(3)}$ onto the normal to the body surface (Σ_*).

This energy should be added to the bonding energy, per unit area, to obtain the total energy required to separate the material continuum into two parts along a surface (Σ_*).

4.1.2 Special Case

Within the framework of the local gradient theory of dielectrics developed in Chapter 2, it is assumed that the modified chemical potential and the density of the induced mass are scalar quantities. In such case, formula (4.8) takes the form (Hrytsyna 2013c):

$$U_{\Sigma}=-\frac{1}{2}\rho_0\mu'_{\pi 0}\mathbf{n}\cdot\boldsymbol{\pi}_m\Big|_{\mathbf{r}\in(\Sigma_*)}=-\frac{1}{2}\mu'_{\pi 0}\mathbf{n}\cdot\boldsymbol{\Pi}_m\Big|_{\mathbf{r}\in(\Sigma_*)}. \quad (4.9)$$

As it follows from formula (4.9), the surface energy of deformation and polarization is equal, in this partial case, to one-half the product of the material constant $\mu'_{\pi 0}$ and the boundary normal value of vector of the local mass displacement $\boldsymbol{\Pi}_m=\rho_0\boldsymbol{\pi}_m$.

Taking constitutive equations (2.78)–(2.80) into account, for isotropic materials, formula (4.9) can be written as follows:

$$U_{\Sigma}=\left[\frac{1}{2}\rho_0\mu'_{\pi 0}\left(\chi_m-\frac{\chi_{Em}\chi_{Em}}{\chi_E}\right)\left(d_\rho\mathbf{n}\cdot\nabla\rho_m-\frac{K\alpha_\rho}{\rho_0}\mathbf{n}\cdot\nabla(\nabla\cdot\mathbf{u})\right)-\mu'_{\pi 0}\frac{\chi_{Em}}{2\chi_E}\mathbf{n}\cdot\boldsymbol{\Pi}_e\right]\Bigg|_{\mathbf{r}\in(\Sigma_*)}. \quad (4.10)$$

Note that within the framework of the Mindlin gradient theory of dielectrics that includes the polarization gradient into the stored energy function, the following formula was obtained for the surface energy of deformation and polarization (Mindlin 1968):

$$U_{\Sigma}=\frac{1}{2}\mathbf{n}\cdot\hat{\mathbf{b}}^0\cdot\boldsymbol{\Pi}_e\Big|_{\mathbf{r}\in(\Sigma_*)}. \quad (4.11)$$

Here, $\hat{\mathbf{b}}^0$ is a material characteristic related to the polarization gradient $\nabla\otimes\mathbf{\Pi}_e$, which, within this theory, is included into the space of constitutive parameters.

On the other hand, within the gradient-type theory of elastic solids that incorporates the strain and its first and second gradients into constitutive parameters, the following formula for the surface energy of deformation was obtained (Mindlin 1965):

$$U_\Sigma = \frac{1}{2} b_0 \mathbf{n} \cdot \nabla (\nabla \cdot \mathbf{u}) \Big|_{\mathbf{r}\in(\Sigma_*)} . \tag{4.12}$$

Here, b_0 is a coefficient at the gradient of strain tensor in the representation of the potential energy density.

Formula (4.10) yields relations (4.11) and (4.12) as a partial case. According to expression (4.10), the surface energy of deformation is determined not only by the polarization vector and by the gradient of the spherical component of the strain tensor, as is assumed by relations (4.11) and (4.12), but also by the gradient of the induced mass density. Thus, formula (4.10) additionally takes into account the mass fluxes that accompany the near-surface polarization and deformation of the body caused by the changes in the system of electric charges in the vicinity of the newly formed surface.

4.1.3 Deformable Media with Electric Quadrupoles

Herein we use the equations for deformable dielectric media with electric quadrupoles (see Section 3.4) in order to determine the surface energy of deformation and polarization. In this case, in view of the constitutive equations (3.195)–(3.200), we modify Eq. (3.194) as follows:

$$f - f_0 = \frac{1}{2\rho_0} \hat{\boldsymbol{\sigma}}_* : \hat{e} + \frac{1}{2} \mu'_{\pi 0} \rho_m + \frac{1}{2} \mu'_\pi \rho_m$$
$$+ \frac{1}{2} \boldsymbol{\pi}_m \cdot \nabla \mu'_\pi - \frac{1}{2} \mathbf{p} \cdot \mathbf{E} - \frac{1}{2} \hat{\mathbf{q}} : (\nabla \otimes \mathbf{E}). \tag{4.13}$$

Using this formula and proceeding in a manner similar to Subsection 4.1.1, we express the perturbation of the total energy $\mathcal{E}$ as follows:

$$\mathcal{E} = \frac{1}{2} \rho_0 \mu'_{\pi 0} \rho_m + \frac{1}{2} \rho_0 \mathbf{F} \cdot \mathbf{u} + \frac{1}{2} \nabla \cdot (\hat{\boldsymbol{\sigma}}_* \cdot \mathbf{u})$$
$$- \frac{1}{2} \rho_0 \nabla \cdot (\boldsymbol{\pi}_m \tilde{\mu}'_\pi) - \frac{1}{2} \nabla \cdot (\varphi_e \mathbf{D}) + \frac{1}{12} \nabla \cdot (\mathbf{E} \cdot \hat{\mathbf{Q}}). \tag{4.14}$$

Here, $\tilde{\mu}'_\pi = \mu'_\pi - \mu'_{\pi 0}$.

Integrating both parts of Eq. (4.14) over the region $(V') = (V_*)\cup(V_v)$ occupied by the body and vacuum and using the divergence theorem, we obtain

$$\int_{(V')} \mathcal{E}\, dV = \frac{1}{2}\rho_0 \mu'_{\pi 0} \int_{(V_*)} \rho_m dV + \frac{1}{2}\rho_0 \int_{(V_*)} \mathbf{F}\cdot\mathbf{u} dV$$

$$+\frac{1}{2}\int_{(\Sigma_*)} \left(\hat{\boldsymbol{\sigma}}_* \cdot \mathbf{u} - \rho_0 \boldsymbol{\pi}_m \tilde{\mu}'_\pi - \varphi_e [\mathbf{D}] + \frac{1}{6}\mathbf{E}\cdot\hat{\mathbf{Q}} \right) \cdot \mathbf{n} d\Sigma.$$

Now we consider the solids with traction-free surfaces in the absence of external forces. Within the framework of the local gradient theory of dielectric media with electric quadrupoles, the above equation yields the formula

$$U_\Sigma = \frac{1}{2}\rho_0 \left(\mathbf{E}\cdot\hat{\mathbf{q}} - \boldsymbol{\pi}_m \tilde{\mu}'_\pi \right)\cdot\mathbf{n} \Big|_{\mathbf{r}\in\Sigma_*} \tag{4.15}$$

for the surface energy of deformation and polarization U_Σ. Hence, the specific surface energy of deformation and polarization is defined by the electric field vector $\mathbf{E}$, the quadrupole moment $\hat{\mathbf{q}}$, the local mass displacement vector $\boldsymbol{\pi}_m$, and a perturbation of the modified chemical potential $\tilde{\mu}'_\pi$.

If we neglect the influence of electric quadrupoles in the formula (4.15), we obtain the relation (4.9) as a partial case. Neglecting the influence of the local mass displacement on the electromagnetic field, from (4.15), we obtain the formula

$$U_\Sigma = \frac{1}{2}\rho_0 \mathbf{E}\cdot\hat{\mathbf{q}}\cdot\mathbf{n} \Big|_{\mathbf{r}\in\Sigma_*},$$

which corresponds to the gradient theory of elastic dielectrics that considers the electric field gradient vector as an independent constitutive variable.

4.2 Elastic Half-Space with Free Surfaces: Near-Surface Inhomogeneity of Electromechanical Fields

In this section, we study a free surface effect on the stress–strain state and the polarization of an elastic half-space of an ideal dielectric.

4.2.1 Problem Formulation

We consider an elastic polarized half-space of centrosymmetric cubic crystal occupying the region $(V) = \{(x, y, z) : x \geq 0, -\infty < y, z < +\infty\}$ in the Cartesian system (x, y, z) and the half-space $(V_v) = \{(x, y, z) : x < 0, -\infty < y, z < +\infty\}$ to be a vacuum. Assume the axes Oy and Oz to be located on the traction-free surface $x = 0$ of the half-space, which coincides with the crystallographic plane (100). The effect of temperature is neglected, while the processes of deformation, polarization, and local mass displacement are considered. We assume that the processes of polarization and local mass displacement are reversible. We also neglect the inertia of these processes.

Within the isothermal approximation, for centrosymmetric cubic crystals, in constitutive equations (2.70), (2.72)–(2.74), we should assume $\hat{\boldsymbol{\beta}} = \mathbf{0}$, $\boldsymbol{\beta}^E = 0$, $\boldsymbol{\beta}^\mu = 0$, $\beta_{T\rho} = 0$, $\gamma^E = 0$, $\gamma^\rho = 0$, $\hat{\mathbf{f}}^{(3)} = 0$, $\hat{\mathbf{g}}^{(3)} = 0$, and

$$C_{ijkl} = (C_{11} - C_{12} - 2C_{44})\delta_{ijkl} + C_{12}\delta_{ij}\delta_{kl} + C_{44}(\delta_{ik}\delta_{jl} + \delta_{il}\delta_{jk}),$$

$$\alpha_{ij}^\rho = \gamma_\rho\delta_{ij},\ \chi_{ij}^E = \chi_E\delta_{ij},\ \chi_{ij}^m = \chi_m\delta_{ij},\ \chi_{ij}^{Em} = \chi_{Em}\delta_{ij}. \quad (4.16)$$

Here, C_{11}, C_{12}, C_{44} are elastic constants, δ_{ij} is the Kronecker delta, and δ_{ijkl} equals one if $i = j = k = l$ and zero otherwise. Hence, the constitutive equations (2.70), (2.72)–(2.74) for cubic symmetry crystals correspond to

$$\sigma_{ij} = (C_{11} - C_{12} - 2C_{44})\delta_{ijkl}e_{lk} + C_{12}\delta_{ij}e_{kk} + 2C_{44}e_{ij} - \gamma_\rho\rho_m\delta_{ij}, \quad (4.17)$$

$$\mu_\pi' = \mu_{\pi 0}' + d_\rho\rho_m - \frac{1}{\rho_0}\gamma_\rho e_{kk}, \quad (4.18)$$

$$\pi_{ei} = \chi_E E_i - \chi_{Em}\nabla_i\mu_\pi', \quad (4.19)$$

$$\pi_{mi} = -\chi_m\nabla_i\mu_\pi' + \chi_{Em}E_i. \quad (4.20)$$

We take a displacement vector $\mathbf{u} = (u, 0, 0)$, electrical potential φ_e, and the modified chemical potential $\tilde{\mu}_\pi'$ as the basic functions. In this case, all the fields depend only on the coordinate x. Using the governing set of equations (2.153)–(2.155), in order to define the functions $u(x)$, $\tilde{\mu}_\pi' = \tilde{\mu}_\pi'(x)$, and $\varphi_e(x)$ within the region $x \geq 0$, we have the following equations:

$$\left(\bar{K} + \frac{4}{3}G\right)\frac{d^2u}{dx^2} - K\frac{\alpha_\rho}{d_\rho}\frac{d\tilde{\mu}_\pi'}{dx} = 0,\ \frac{d^2\tilde{\mu}_\pi'}{dx^2} - \lambda_{\mu E}^2\tilde{\mu}_\pi' = \lambda_{\mu E}^2\frac{K\alpha_\rho}{\rho_0}\frac{du}{dx}, \quad (4.21)$$

$$\frac{d^2}{dx^2}\left(\varphi_e + \kappa_E \tilde{\mu}'_\pi\right) = 0 . \qquad (4.22)$$

Here, φ_e is the electric potential, $E = -\frac{d\varphi_e}{dx}$, $\mathbf{E}$ = $(E(x), 0, 0)$, K = $(C_{11} + 2C_{12})/3$, $G = C_{44}$, and $\alpha_\rho = \gamma_\rho/K$.

To complete the above set of equations, we complement Eqs. (4.21) and (4.22) with the equation for electric potential φ_{ev} in vacuum: $\frac{d^2\varphi_{ev}}{dx^2} = 0$, the condition of the boundedness of functions $u(x)$, $\tilde{\mu}'_\pi = \tilde{\mu}'_\pi(x)$, and $\varphi_e(x)$ at $x \to +\infty$ and the electric potential in vacuum φ_{ev} at $x \to -\infty$, as well as the following boundary conditions on the surface $x = 0$:

$$\left(\bar{K} + \frac{4}{3}G\right)\frac{du}{dx} - K\frac{\alpha_\rho}{d_\rho}\tilde{\mu}'_\pi = 0 , \quad \tilde{\mu}'_\pi = -\mu'_{\pi 0} , \qquad (4.23)$$

$$\left(\frac{d\varphi_e}{dx} - \frac{d\varphi_{ev}}{dx}\right) + \rho_0 \pi_e = 0 . \qquad (4.24)$$

Here, the subscript "v" will designate the physical quantities associated with vacuum. The first relation of the set (4.23) corresponds to the absence of normal stresses on the body surface, and the second one corresponds to the equality of the absolute magnitude of the modified chemical potential μ'_π to zero (since the body is in contact with vacuum, and $\mu'_{\pi v} = 0$). The relation (4.24) is a continuity condition across the interface of the component of the electric induction perpendicular to the half-space surface.

4.2.2 Problem Solution and Its Analysis

The solution to the boundary-value problem (4.21)–(4.24) that vanishes at infinity can be presented as:

$$u(x) = \mu'_{\pi 0}\frac{K\alpha_\rho}{d_\rho \breve{\lambda}\left(\bar{K} + \frac{4}{3}G\right)} e^{-\breve{\lambda} x}, \quad \tilde{\mu}'_\pi(x) = -\mu'_{\pi 0}\, e^{-\breve{\lambda} x}, \quad x > 0, \qquad (4.25)$$

$$E(x) = \begin{cases} \kappa_E \breve{\lambda} \mu'_{\pi 0}\, e^{-\breve{\lambda} x}, & x > 0, \\ 0, & x < 0, \end{cases} \qquad (4.26)$$

where

$$\breve{\lambda} = \lambda_{\mu E}\sqrt{1 + \mathfrak{M}} . \qquad (4.27)$$

For a single nonzero component of the deformation tensor, we have

$$e_{xx}(x) = -\mu'_{\pi 0} \frac{K\alpha_\rho}{d_\rho \left(\bar{K} + \frac{4}{3}G \right)} e^{-\breve{\lambda}x}. \qquad (4.28)$$

Note that the deformation of surface

$$e_{xx}(0) = -\mu'_{\pi 0} K\alpha_\rho \Big/ \left(\bar{K} + \frac{4}{3}G \right) d_\rho$$

appears due to the break of atomic bonds across the surface. Such a deformation corresponds to the following displacement:

$$u(0) = \mu'_{\pi 0} \frac{K\alpha_\rho}{d_\rho \breve{\lambda} \left(\bar{K} + \frac{4}{3}G \right)} = \mu'_{\pi 0} \frac{K\alpha_\rho (1 + \mathfrak{M})}{d_\rho \breve{\lambda} \left(K + \frac{4}{3}G \right)}. \qquad (4.29)$$

The deformation exponentially decays with the distance x from the free surface $x = 0$. Here, $\breve{\lambda}$ is the attenuation constant.

Using formulae (4.26), (4.28), and constitutive relations (4.17)–(4.20), we obtain the following relations for nonzero components σ_{yy}, σ_{zz}, π_e, π_m of the stress tensor $\hat{\boldsymbol{\sigma}}$, polarization vector $\boldsymbol{\pi}_e$, and vector of the local mass displacement $\boldsymbol{\pi}_m$

$$\sigma_{yy}(x) = \sigma_{zz}(x) = \sigma_s e^{-\breve{\lambda}x}, \quad \pi_e(x) = -\mu'_{\pi 0} \breve{\lambda} \chi_{Em} \frac{\varepsilon_0}{\varepsilon} e^{-\breve{\lambda}x}, \qquad (4.30)$$

$$\pi_m(x) = -\mu'_{\pi 0} \breve{\lambda} \chi_m \left(1 - \kappa_E \frac{\chi_{Em}}{\chi_m} \right) e^{-\breve{\lambda}x}. \qquad (4.31)$$

Here, σ_s is the surface stress, introduced through the relation

$$\sigma_s = \frac{2GK\alpha_\rho \mu'_{\pi 0}}{d_\rho \left(\bar{K} + \frac{4}{3}G \right)} = \frac{2GK\alpha_\rho \mu'_{\pi 0} (1 + \mathfrak{M})}{d_\rho \left(K + \frac{4}{3}G \right)}. \qquad (4.32)$$

The analysis of the obtained solution shows that the distributions of the stresses σ_{yy}, σ_{zz} and the strain e_{xx} are inhomogeneous in the vicinity of the surface. The stresses σ_{yy} and σ_{zz} reach a maximum value σ_s at the surface $x = 0$. In the vicinity of the surface ($0 < x < d$), both the stress and the strain decay exponentially with the distance from the surface and tend to zero. Here, in order to define the magnitude d, we use the equation $\sigma_{\alpha\alpha}(d) = 0$, $\alpha = \{y, z\}$.

Note that $l_* = \breve{\lambda}^{-1}$ is a characteristic length of the considered problem since the stresses and the strain decrease by a factor of

e at a distance $x = \breve{\lambda}^{-1}$ from the body surface. This length may be determined by an experiment method (for example, the electron diffraction measurements (Germer et al. 1961)), by methods of discrete analysis (Benson and Yun 1967) or the theory of crystal lattice dynamics (Askar et al. 1970, 1971; Maugin 1988; Mindlin 1965, 1972a), etc. In the listed papers, it is also shown that the strain diminishes exponentially into the interior of the free surface of the deformable body, and the characteristic length is the magnitude of the order of the distance between the nearest atoms (see Table 4.1). The near-surface region is only a few atomic layers thin.

It should be noted that according to the formulae (4.30) in the cubic symmetric crystals, the electric field influences the characteristic length $\breve{\lambda}(\kappa_E, \chi_{Em})$ (see Eq. (4.27)) but does not change the distribution of stresses within the body in a qualitative way.

4.2.3 Surface Energy of Deformation and Polarization and Surface Tension

Substitution of the relations (4.31) into (4.9) yields the following formula to determine the surface energy of deformation and polarization of a half-space:

$$U_{\Sigma}^{\infty} = -\frac{\rho_0}{2l_*}\mu_{\pi 0}'^2\chi_m\left(1-\kappa_E\frac{\chi_{Em}}{\chi_m}\right). \tag{4.33}$$

If the effect of the electromagnetic field is neglected, then Eq. (4.33) yields a simple relation for the surface energy of deformation and polarization

$$U_{\Sigma}^{\infty} = -\frac{\rho_0}{2l_*}\mu_{\pi 0}'^2\chi_m. \tag{4.34}$$

The specific surface energy of a half-space is proportional to the mass density ρ_0, material constants χ_m, and the square of $\mu_{\pi 0}'$. It is also inversely proportional to the characteristic length l_*. Since the material constant χ_m is positive, the surface energy of deformation and polarization is negative. This result is consistent with the ones reported by Mindlin (1968) and Maugin (1988).

Table 4.1 Material characteristics for the alkali metal halides

Crystal	Density, ρ_0, 10^3 kg/m^3	Lattice constant, a, 10^{-10} m	Characteristic length, l_*, 10^{-10} m		Elastic coefficients, C_{ij}, 10^{10} Pa		
					C_{11}	C_{12}	C_{44}
	Ref. (Sirdeshmukh et al. 2001)	Ref. (Sirdeshmukh et al. 2001)	According to the formula $l_* \approx a/3.5$, (Benson and Yun 1967)	Ref. (Askar et al. 1971; Mindlin 1972a)	Ref. (Sirdeshmukh et al. 2001)		
Sodium iodide (NaI)	3.670	6.4728	1.85		3.025	0.88	0.74
Sodium chloride (NaCl)	2.164	5.6402	1.61	0.73	4.936	1.29	1.265
Potassium iodide (KI)	3.1257	7.0655	2.02		2.76	0.45	0.37
Potassium chloride (KCl)	1.9865	6.2931	1.8	0.93	4.078	0.69	0.633

The surface energy is defined as the work [Joule] required to create a unit of surface area [m^2] by separating the medium along the surface (Σ) into two parts. In crystalline bodies, the surface energy differs from the surface tension, which is defined as the work required to expand the existing surface to the same magnitude (Geguzin and Goncharenko 1962; Shuttleworth 1950). The distinction between the surface energy and the surface tension in crystalline bodies can be explained by the fact that atoms at a free surface and atoms in the bulk of a material have different energies of interaction. The surface tension is also defined as the tangential traction [Newton] within a near-surface layer, per unit of length [m] (Geguzin and Goncharenko 1962). The surface tension F_{ten}^{∞} [N/m] in a half-space is determined by the stresses σ_{yy} and σ_{zz}. We define the surface tension as

$$F_{\text{ten}}^{\infty} = \int_0^d \sigma_{\alpha\alpha}(x)dx,\ \alpha = \{y, z\}. \tag{4.35}$$

For the surface tension, the formulae (4.30) and (4.35) yield

$$F_{\text{ten}}^{\infty} = \int_0^d \sigma_s e^{-\breve{\lambda}x}dx = -\frac{\sigma_s}{\breve{\lambda}}e^{-\breve{\lambda}x}\Bigg|_0^d = \frac{\sigma_s}{\breve{\lambda}} = \sigma_s l_* . \tag{4.36}$$

Thus, the surface tension in the half-space is determined by the surface stress σ_s and by characteristic length l_*. Surface stress is proportional to the modified chemical potential $\mu'_{\pi 0}$ in an infinite media, to the material parameter $K\alpha_\rho$, to the shear modulus G, and inversely proportional to the parameters d_ρ and $\breve{K} \equiv K + \frac{4}{3}G - \frac{K^2\alpha_\rho^2}{\rho_0 d_\rho}$.

4.2.4 Evaluation of Surface Stress and Material Constants

If the surface tension F_{ten}^{∞} of a half-space and characteristic length l_* are known, then the relation (4.36) makes it possible to estimate the magnitude of surface stresses σ_s as the ratio of the surface tension to the material characteristic length, namely, $\sigma_s = F_{\text{ten}}^{\infty}/l_*$.

Using the data given in Table 4.2, it may be concluded that surface tension for potassium chloride is $F_{\text{ten}}^{\infty}\big|_{\text{KCl}}$ = 0.11÷0.22 N/m. Taking into account the values of characteristic lengths (Table 4.1), we obtain the following estimation for the surface stresses in potassium chloride and sodium chloride:

$$\sigma_s\big|_{\mathrm{KCl}} = (0.7 \div 1.3)\times 10^9\ \mathrm{N/m^2}\ ,\ \sigma_s\big|_{\mathrm{NaCl}} = (0.9 \div 2)\times 10^9\ \mathrm{N/m^2}\ . \quad (4.37)$$

The formulae (4.29) and (4.32) yield

$$\sigma_s = 2G\breve{\lambda}u(0)\ .$$

This relation enables us to determine the surface stresses through the displacement u at the body surface $x = 0$. In crystalline bodies, such displacements reach 1–3% of the interatomic spacing (Maugin 1988). In view of this, assuming $u(0) \approx ak_a$, we get a useful formula

$$\sigma_s = 2G\breve{\lambda}ak_a\ . \quad (4.38)$$

Here, a is the atom spacing, and k_a is a constant. Following Maugin (1988), we assume $k_a = 10^{-2} \div 3 \times 10^{-2}$. The formula (4.38) allows us to estimate the magnitude of surface stresses in solid crystalline bodies based on the known values of the shear modulus G, the atom spacing a, and the characteristic length $\breve{\lambda}$ of the material.

Table 4.2 Surface tension and surface energy of deformation and polarization for alkali metal halides

Crystal	Surface tension, $F_{\mathrm{ten}}^{\infty}$, N/m	Surface energy of deformation and polarization U_{Σ}^{∞}, 10^{-1} J/m^2
Sodium chloride (NaCl)	0.147 (Sdobnyakov et al. 2007) 0.116 (Sdobnyakov et al. 2007)	−0.59 (Mindlin 1968)
Potassium chloride (KCl)	0.117 (Sdobnyakov et al. 2007) 0.99 (Sdobnyakov et al. 2007) 0.221 (Jurov et al. 2011) 0.110 (Jurov et al. 2011) 0.192 (Jurov et al. 2011)	

Using the formula (4.38) and the data given in Tables 4.1 and 4.2, we can obtain the following estimation $\sigma_s\big|_{\mathrm{NaCl}} = 1.97k_a \times 10^{11}$ N/m^2 for surface stresses. Assuming $k_a = 10^{-2}$, we get $\sigma_s\big|_{\mathrm{NaCl}} = 1.97 \times 10^9$ N/m^2.

Taking the formulae (4.36) and (4.38) into account, we obtain a simple formula in order to calculate the surface tension in an isotropic half-space

$$F_{\mathrm{ten}}^{\infty} = 2Gak_a\ .$$

This formula can be useful for estimating the surface tension in cubic crystalline bodies with plane-parallel boundaries in the absence of the corresponding experimental data.

Now we evaluate the new material parameters. Neglecting the effect of the electric field on a characteristic length, we can write

$$l_*^{-2} = \tilde{\lambda}^2 = \lambda_\mu^2 \left(1 + \mathfrak{M}\right) = \frac{1 + \mathfrak{M}}{d_\rho \chi_m}. \tag{4.39}$$

In view of Eq. (4.32), we can rewrite the formula (4.36) as follows:

$$F_{\text{ten}}^{\infty} = \frac{2G\mu'_{\pi 0} l_* K \alpha_\rho}{\left(K + \dfrac{4}{3}G - \dfrac{K^2 \alpha_\rho^2}{\rho_0 d_\rho}\right) d_\rho}. \tag{4.40}$$

Hence, taking the relation (4.34) and the formulae

$$\frac{K^2 \alpha_\rho^2}{\rho_0 d_\rho} = \frac{\mathfrak{M}}{(1 + \mathfrak{M})}\left(K + \frac{4}{3}G\right), \quad K + \frac{4}{3}G - \frac{K^2 \alpha_\rho^2}{\rho_0 d_\rho} = \left(K + \frac{4}{3}G\right) \Big/ (1 + \mathfrak{M}),$$

into account, we can find

$$\mathfrak{M} = -\left(K + \frac{4}{3}G\right) \frac{(F_{\text{ten}}^{\infty})^2}{8G^2 l_* U_\Sigma^\infty}. \tag{4.41}$$

Using this expression and data from Tables 4.1 and 4.2, we obtain $\mathfrak{M}_{\text{NaCl}} = 8 \times 10^{-4}$. Consequently, the coupling factor between the local mass displacement and the mechanical fields (i.e., parameter $\mathfrak{M}$) is a small parameter.

Now we should estimate the magnitude of the material parameter χ_m. Based on the formula (4.34), we can write

$$\chi_m = -\frac{2 l_* U_\Sigma^\infty}{\rho_0 \, \mu_{\pi 0}'^2}. \tag{4.42}$$

If the values of the surface energy of deformation of a half-space U_Σ^∞, the characteristic length l_*, the mass density ρ_0, and the modified chemical potential $\mu'_{\pi 0}$ are known, then the formula (4.42) allows us to calculate the magnitude of the material parameter χ_m [s^2]. For sodium chloride, we get $\chi_m^{\text{NaCl}} = 3.89 \times 10^{-15} / \mu_{\pi 0}'^2$ [s^2]. Assuming $\mu'_{\pi 0} = O(10^4)$ [m^2 / s^2], we obtain $\chi_m = O(10^{-23})$ [s^2].

Making use of the formula (4.39), we can write $d_\rho = (1+\mathfrak{M})l_*^2/\chi_m$. Since $\mathfrak{M}$ is a small parameter, and taking into account that $\chi_m = O(10^{-23})$ [s^2], $l_* = O(10^{-9})$ [m], we get the following estimation: $d_\rho = O(10^5)$ [m^2/s^2].

Using the relations (4.34) and (4.40), we get

$$\alpha_\rho = \rho_0 \mu'_{\pi 0}\left(\frac{K}{4G} + \frac{1}{3}\right)\frac{F_{\text{ten}}^\infty}{KU_\Sigma^\infty}.$$

Hence, $\alpha_\rho = O(10^{-3} \div 10^{-2})$. According to the uniqueness theorem, the inequality $\alpha_\rho^2 < \rho_0 \frac{d_\rho}{K^2}\left(K - \frac{2}{3}G\right)$ should be satisfied as well. Assuming $d_\rho = O(10^5)$, for sodium chloride, we get the following restriction on the material parameter α_ρ: $\left|\alpha_\rho\right| < 0.07$.

Now we estimate the coefficient χ_{Em}, which links the polarization vector to the space gradient of the modified chemical potential μ'_π. Let us assume that the body polarization is caused by the local mass displacement only. In this case, $\mathbf{E} = 0$. Then, taking the constitutive equation (4.19) into account, we get

$$\Pi_{ei} = -\rho_0 \chi_{Em} \nabla_i \mu'_\pi \tag{4.43}$$

for an isotropic crystalline medium. The maximum displacements of atoms of a crystalline lattice are the quantities of the order of the atom spacing a, while the derivatives ∇_i are the quantities of the order of a^{-1}. In the case of an interaction of the atoms located at a distance commensurate with the size of the electron orbits of their valence electrons, we can assume that the effective electric charges are the quantities of the order of a charge of an electron $\bar{e}$. Then, the polarization caused by the local mass displacement is the quantity of the order of $\bar{e}a/a^3$. Consequently, in view of the formula (4.43), we obtain the estimation $\chi_{Em} \sim -\frac{\bar{e}}{\rho_0 a \mu'_\pi}$ for a material characteristic χ_{Em}. If the charge of the electron $\bar{e} = -1.6 \times 10^{-19}$ C, the atom spacing $a = O(10^{-10})$ [m], the mass density $\rho = O(10^3)$ [kg/m^3], and the potential $\mu'_\pi = O(10^4)$ [m^2/s^2], we obtain the following estimation: $\chi_{Em} = O(10^{-16})$ [m^2/V].

4.3 Dielectric Layer with the Free Surface: Size Effect

In this section, we study the stress–strain state of an infinite layer with traction-free surfaces in order to analyze the interaction of two plane surfaces of a layer when they are getting closer to one another.

4.3.1 Problem Formulation

Consider an elastic polarized layer with surfaces $x = \pm l$ of a centrosymmetric, cubic crystal (100) free of traction and infinite along the directions y and z as shown in Fig. 4.1. The layer is an ideal dielectric surrounded by vacuum. We consider the processes of deformation, polarization, and local mass displacement as the basic ones, while the thermal coupling is disregarded.

In this case, the vector functions $\mathbf{g} = \{\mathbf{u}, \mathbf{E}, \boldsymbol{\pi}_e, \boldsymbol{\pi}_m, \mathbf{E}_v\}$ and scalar potential $\tilde{\mu}_\pi'$ are functions of the space coordinate x and time t only such that $\mathbf{g} = (g, 0, 0)$, $g = \{u, E, \pi_e, \pi_m, E_v\}$, $g = g(x, t)$, and $\tilde{\mu}_\pi' = \tilde{\mu}_\pi'(x,t)$. Let us restrict ourselves to considering a quasi-static electric field for which $\dfrac{\partial \mathbf{A}}{\partial t} = 0$ and $\dfrac{\partial \mathbf{A}_v}{\partial t} = 0$. Therefore, in view of the representation (2.173), we assume that $\mathbf{E} = -\nabla \varphi_e$ and $\mathbf{E}_v = -\nabla \varphi_{ev}$.

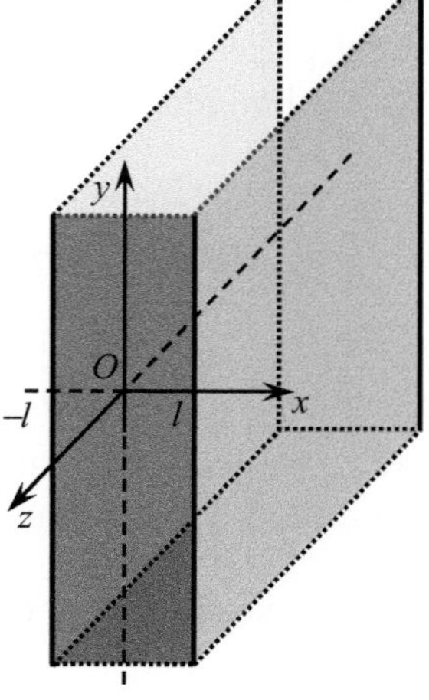

Figure 4.1 The elastic layer in rectangular coordinate system.

To study the stress–strain state of the layer, the perturbation of electromagnetic field caused by the formation of the layer surface, we use the governing set of equations (2.114), (2.116)–(2.119) for a layer (region $|x| < l$):

$$\rho_0 \frac{\partial^2 u}{\partial t^2} = \left(\bar{K} + \frac{4}{3} G \right) \frac{\partial^2 u}{\partial x^2} - K \frac{\alpha_\rho}{d_\rho} \frac{\partial \tilde{\mu}'_\pi}{\partial x}, \tag{4.44}$$

$$\frac{\partial^2 \tilde{\mu}'_\pi}{\partial x^2} - \lambda_\mu^2 \tilde{\mu}'_\pi = K \lambda_\mu^2 \frac{\alpha_\rho}{\rho_0} \frac{\partial u}{\partial x} - \frac{\chi_{Em}}{\chi_m} \frac{\partial^2 \varphi_e}{\partial x^2}, \tag{4.45}$$

$$\frac{\partial^2}{\partial x \partial t} \left(\varphi_e + \kappa_E \tilde{\mu}'_\pi \right) = 0, \quad \frac{\partial^2}{\partial x^2} \left(\varphi_e + \kappa_E \tilde{\mu}'_\pi \right) = 0, \tag{4.46}$$

and the Maxwell equation (2.217) in vacuum (region $|x| > l$):

$$\frac{\partial^2 \varphi_{ev}}{\partial x^2} - \varepsilon_0 \mu_0 \frac{\partial^2 \varphi_{ev}}{\partial t^2} = 0. \tag{4.47}$$

The electric fields E and E_v in the body and vacuum, and the electric potentials φ_e and φ_{ev} are related through the formulae: $E = -\frac{\partial \varphi_e}{\partial x}$ and $E_v = -\frac{\partial \varphi_{ev}}{\partial x}$.

To ensure the solution uniqueness, the radiation conditions (for $x \to \pm\infty$) and the boundary conditions that correspond to the traction-free surfaces $x = \pm l$, continuity condition for the electric potential, and the condition of the absolute magnitude of the modified chemical potential μ'_π being equal to zero should be considered together with Eqs. (4.44)–(4.47). The boundary and radiation conditions are as below:

$$\left(\bar{K} + \frac{4}{3} G \right) \frac{\partial u}{\partial x} - K \frac{\alpha_\rho}{a_\rho^\mu} \tilde{\mu}'_\pi = 0, \quad \tilde{\mu}'_\pi = -\mu'_{\pi 0}, \quad \varphi_e = \varphi_{ev}, \tag{4.48}$$

$$\lim_{x \to \pm\infty} \left(\frac{\partial \varphi_{ev}}{\partial x} \pm \sqrt{\varepsilon_0 \mu_0} \frac{\partial \varphi_{ev}}{\partial t} \right) = 0. \tag{4.49}$$

Relations (4.44)–(4.47) should be supplemented by the corresponding initial conditions. We consider these conditions to be zero.

Thus, to determine the displacement field and the potential μ'_π, we formulate a stationary boundary-value problem, while the problem of electrodynamics is formulated as a contact problem.

4.3.2 Problem Solution and Its Analysis

Let us neglect the inertia term in Eq. (4.44). Then, the solution to this set of equations that satisfy the boundary conditions (4.48) and (4.49) is as follows:

$$u(x) = -\mu'_{\pi 0}\frac{K\alpha_\rho}{d_\rho\breve{\lambda}\left(\bar{K}+\frac{4}{3}G\right)}\frac{\mathrm{sh}\left(\breve{\lambda}x\right)}{\mathrm{ch}\left(\breve{\lambda}l\right)}\vartheta(t), \tag{4.50}$$

$$\tilde{\mu}'_\pi(x,t) = -\mu'_{\pi 0}\frac{\mathrm{ch}\left(\breve{\lambda}x\right)}{\mathrm{ch}\left(\breve{\lambda}l\right)}\vartheta(t), \tag{4.51}$$

$$E(x,t) = -\kappa_E\breve{\lambda}\mu'_{\pi 0}\frac{\mathrm{sh}\left(\breve{\lambda}x\right)}{\mathrm{ch}\left(\breve{\lambda}l\right)}\vartheta(t),\ \varphi_e(x,t) = \kappa_E\mu'_{\pi 0}\frac{\mathrm{ch}\left(\breve{\lambda}x\right)}{\mathrm{ch}\left(\breve{\lambda}l\right)}\vartheta(t) \tag{4.52}$$

for the layer $|x| < l$;

$$\varphi_{ev}(x,t) = \begin{cases} \kappa_E\,\mu'_{\pi 0}\vartheta\left[t+\sqrt{\varepsilon_0\mu_0}\,(x+l)\right], & \forall x: x<-l \\ \kappa_E\,\mu'_{\pi 0}\,\vartheta\left[t-\sqrt{\varepsilon_0\mu_0}\,(x-l)\right], & \forall x: x>l \end{cases} \tag{4.53}$$

for the vacuum $|x| > l$. Here, $\breve{\lambda}$ and $\lambda_{\mu E}$ are defined by the formulae (4.27) and (2.111), and $\vartheta(t)$ is the Heaviside unit step function, i.e.,

$$\vartheta(t) = \begin{cases} 0 \text{ for } t \le 0, \\ 1 \text{ for } t > 0. \end{cases}$$

Then, using the relations (4.17)–(4.20), the stresses σ_{yy} and σ_{zz} as well as the components π_e and π_m of the polarization vector $\boldsymbol{\pi}_e = (\pi_e, 0, 0)$ and vector of the local mass displacement $\boldsymbol{\pi}_m = (\pi_m, 0, 0)$ are given by the formulae

$$\sigma_{\alpha\alpha}(x,t) = \sigma_s\frac{\mathrm{ch}\left(\breve{\lambda}x\right)}{\mathrm{ch}\left(\breve{\lambda}l\right)}\vartheta(t),\ \alpha = \{y, z\}, \tag{4.54}$$

$$\pi_e(x,t) = \kappa_E\breve{\lambda}\mu'_{\pi 0}\frac{\varepsilon_0}{\rho_0}\frac{\mathrm{sh}\left(\breve{\lambda}x\right)}{\mathrm{ch}\left(\breve{\lambda}l\right)}\vartheta(t), \tag{4.55}$$

$$\pi_m(x,t) = \mu'_{\pi 0}\breve{\lambda}(\chi_m - \kappa_E\chi_{Em})\frac{\mathrm{sh}\left(\breve{\lambda}x\right)}{\mathrm{ch}\left(\breve{\lambda}l\right)}\vartheta(t), \tag{4.56}$$

where σ_s is the surface stress, introduced through the formula (4.32).

At time $t > 0$, using the formula

$$\sigma_{es}(\pm l) = \rho_0 \pi_e(\pm l) = \pm \kappa_E \breve{\lambda} \varepsilon_0 \mu'_{\pi 0}\, \mathrm{th}(\breve{\lambda} l), \tag{4.57}$$

we can find the density of the bound surface charge $\sigma_{es}(\pm l)$ induced on the surfaces of layer. Since $\chi_E = \chi\varepsilon_0/\rho_0$, where χ is the coefficient of dielectric susceptibility, then $\varepsilon/\varepsilon_0 = 1 + \chi$ and

$$\sigma_{es}(\pm l) = \pm \rho_0 \breve{\lambda} \mu'_{\pi 0} \frac{\chi_{Em}}{1+\chi} \mathrm{th}\left(\breve{\lambda} l\right).$$

For thick layers, this formula yields $\sigma_{es}(\pm l) = \pm \rho_0 \breve{\lambda} \mu'_{\pi 0}\, \chi_{Em}/(1+\chi) = \pm \overset{*}{\sigma}_{es}$. If the value of the surface charge $\overset{*}{\sigma}_{es}$ can be determined by experiments, then making use of the formula $\mu'_{\pi 0}\chi_{Em} = l_*(1+\chi)\overset{*}{\sigma}_{es}/\rho_0$, one can determine the magnitude of such a material constant: $m = \mu'_{\pi 0}\chi_{Em}$.

The analysis of the relations (4.51), (4.52), (4.54), and (4.55) has shown that the inhomogeneous distributions of the mechanical stresses σ_{yy}, σ_{zz}; modified chemical potential $\tilde{\mu}'_{\pi}$; electric field E; electrical potential φ_e; and polarization π_e are proper for the near-surface regions of the layer (see Fig. 4.2, where $\varphi_* = \kappa_E \mu'_{\pi 0}$ and $\pi_{e*} = \kappa_E \breve{\lambda} \mu'_{\pi 0}\, \varepsilon_0/\rho_0$). Note that the layer thickness does not affect the surface stress value $\sigma(\pm l) = \sigma_s$, but can affect the stress distribution within the body. The thick layer regions that are far from the surface (i.e., interior regions) are stress free, but in the vicinity of the surface, these stresses rapidly increase and tend to σ_s (curves 3 and 4 in Fig. 4.2). If the layer thickness decreases so much

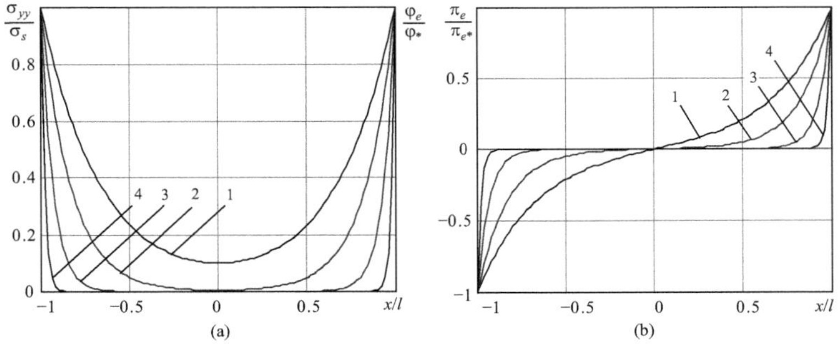

Figure 4.2 The stress σ_{yy}/σ_s, electric potential φ_e/φ_* (a) and polarization π_e/π_{e*} (b) for films of different thicknesses ($l = 3l_*$, $6l_*$, $15l_*$, $50l_*$ represented by curves 1–4, respectively).

that the surface inhomogeneities overlap, then the resulting profiles are modified. Hence, the mid region of thin layers ($l < d$) is stressed (curve 1 in Fig. 4.2). The layers characterized by overlapping of the near-surface inhomogeneities (curve 1 in Fig. 4.2) will be referred to as thin films. The stress–strain state of thin films is essentially inhomogeneous. We see that a decrease in the thickness in thin films leads to an increased stress in the film cross section $\sigma(0) = \sigma_s/\mathrm{ch}(l/l_*)$ (see Fig. 4.3). This phenomenon can be important and considerable in nano-sized structures such as nanofilms in which the surfaces are very close to each other.

We can see that on the layer surfaces, a bound charge of density (4.57) is induced, and, in vacuum, there emerge impulses of an electric field (4.53) that propagate from the layer surfaces up to the half-spaces $x < -l$ and $x > l$. Thus, the relations of the local gradient theory of dielectrics allow us to describe the electromagnetic emission accompanying the process of formation of surfaces in the body.

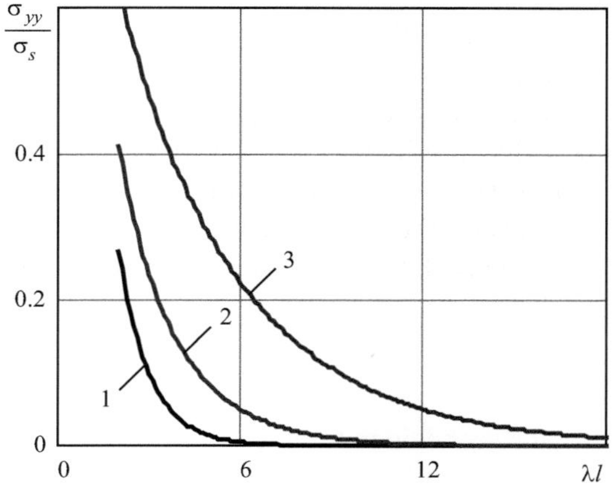

Figure 4.3 The effect of a layer thickness on the stress in cross sections $x = 0$, $x = 0.5l_*$, and $x = 0.75l_*$, that is, curves 1–3, respectively.

Note that the dependence of the electric and mechanical fields on the film thickness displays their size effect. This effect is shown on the plots of the stresses and density of bound surface charge in Figs. 4.3 and 4.4.

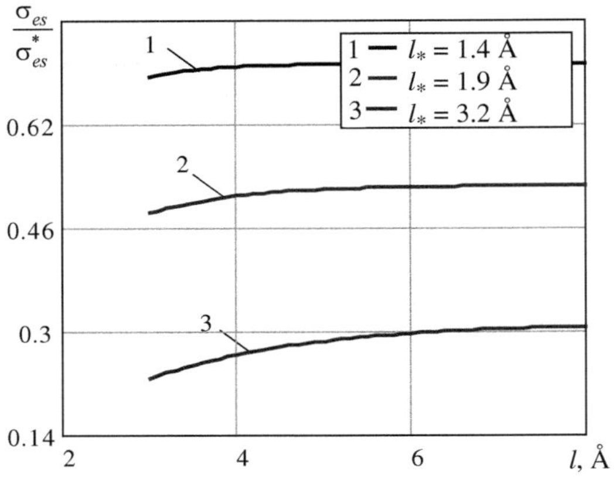

Figure 4.4 The dependence of the bound surface electric charge $\sigma_{es}(l)$ on the film thickness for different materials.

4.3.3 Size Effect of Surface Tension and Surface Energy of Deformation and Polarization

Knowing the stress distribution (4.55), we obtain the following relation in order to find the surface tension F^l_{ten} on the layer surface $x = l$:

$$F^l_{\text{ten}} = \frac{\sigma_s}{\breve{\lambda}} \text{th}\left(\breve{\lambda} l\right) = F^{\infty}_{\text{ten}} \, \text{th}\left(\breve{\lambda} l\right). \tag{4.58}$$

The surface tension in thick layers coincides with the surface tension in a half-space of the same material. However, in thin films, the absolute value of the surface tension decreases with the decrement in the thickness of the film. This is because the surfaces in thin films are so close that they begin to interact.

In view of the relation (4.58), the following scheme can be proposed to define the material scalar $\breve{\lambda}$. Knowing the ratio of the surface tension of a thin film F^l_{ten} of thickness $2l_0$ to the surface tension F^{∞}_{ten} of a thick layer for identical materials, using the formula

$$\breve{\lambda} = \frac{1}{l_0} \text{arcth}\left(\frac{F^l_{\text{ten}}}{F^{\infty}_{\text{ten}}}\right), \tag{4.59}$$

one can determine the material scalar $\breve{\lambda}$ and the characteristic length $l_* = \breve{\lambda}^{-1}$.

On the other hand, if the characteristic length l_* and surface tension of a thick layer are known, then the formula (4.59) allows us to determine the surface tension of a thin film of the same material.

The magnitudes of the surface tension F_{ten}^{∞} and F_{ten}^{l} for crystalline materials can be found in the papers (Sdobnyakov et al. 2007; Jurov et al. 2011). For sodium chloride, whose crystal structure corresponds to a simple cubic lattice, the surface tensions of a thick layer and of the layer of thickness $2l_0 = 0.564 \times 10^{-9}$ m are equal to 0.117 N/m and 0.058 N/m, respectively (Sdobnyakov et al. 2007). For potassium chloride (KCl), the surface tensions of a thick layer and of the layer of thickness $2l_0 = 0.629 \times 10^{-9}$ m are equal to 0.117 N/m and 0.058 N/m, respectively (Sdobnyakov et al. 2007). Therefore, using the formula (4.59), we get $\breve{\lambda}^{\text{NaCl}} = 2.73 \times 10^{9}\,\text{m}^{-1}$, $l_*^{\text{NaCl}} = 3.67 \times 10^{-10}\,\text{m}$, $\breve{\lambda}^{\text{KCl}} = 1.73 \times 10^{9}\,\text{m}^{-1}$, and $l_*^{\text{KCl}} = 5.78 \times 10^{-10}\,\text{m}$. It should be noted that the herein obtained values of characteristic lengths for sodium chloride and potassium chloride are one order higher than those given in Table 4.1.

Relations (4.9) and (4.56) enable us to determine the surface energy of deformation and polarization U_{Σ}^{l} within the layer

$$U_{\Sigma}^{l} = -\frac{\rho_0}{2}\breve{\lambda}\chi_m \mu_{\pi 0}^{\prime 2}\left(1-\kappa_E\frac{\chi_{Em}}{\chi_m}\right)\text{th}\left(\breve{\lambda}l\right) = U_{\Sigma}^{\infty}\,\text{th}\left(\breve{\lambda}l\right). \qquad (4.60)$$

In thick layers ($l >> l_*$), the surface energy and surface tension coincide with the surface energy U_{Σ}^{∞} and tension F_{ten}^{∞} of a half-space of the same material. In thin films, the above quantities depend on the film thickness. The absolute magnitude of the surface energy U_{Σ}^{l} and tension F_{ten}^{l} decreases when the film thickness decreases (see Fig. 4.5). Therefore, the relations (4.58) and (4.60) describe the size effect of the surface tension and the surface energy of deformation and polarization in thin dielectric films.

4.3.4 Evaluation of Additional Nonlinear Mass Force in Balance of Momentum

Let us evaluate the magnitude of the additional mass force (2.51) that arises in the equation of motion (2.54) due to the local mass displacement. To this end, we use the solutions (4.51) and (4.56) to the boundary-value problems (4.44)–(4.49). In the case of a one-dimensional problem, in order to calculate the x-component F_x' of the additional mass force $\mathbf{F}_*'$, we obtain the following expression:

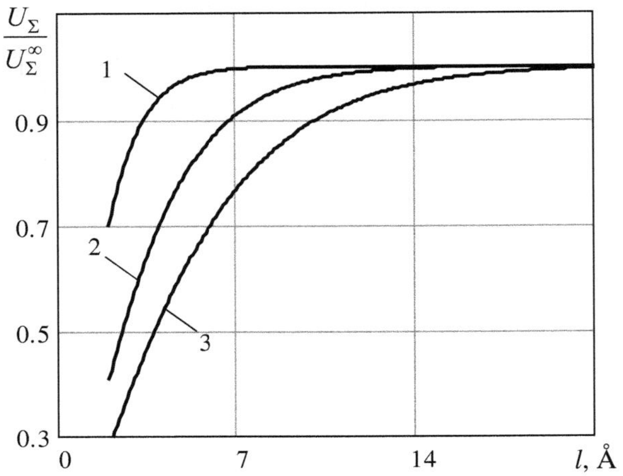

Figure 4.5 Dependences of the surface energy of deformation and polarization on the film thickness for different materials (l_* = 1.3 Å, 2.3 Å, 4.6 Å, that is, curves 1–3 respectively).

$$F_x'(x) = \rho_m \frac{d\mu_\pi'}{dx} - \pi_m \frac{d^2\mu_\pi'}{dx^2}$$

$$= 2\left(\mu_{\pi 0}'\right)^2 \breve{\lambda}^3 \chi_m \left(1 - \kappa_E \frac{\chi_{Em}}{\chi_m}\right) \frac{\mathrm{sh}\left(\breve{\lambda}x\right)\mathrm{ch}\left(\breve{\lambda}x\right)}{\mathrm{ch}^2\left(\breve{\lambda}l\right)}. \quad (4.61)$$

In view of relations (4.60), the surface value of an additional mass force can be written as:

$$\rho_0 F_x'(\pm l) = \mp 4\breve{\lambda}^2 U_\Sigma^l = \mp 4U_\Sigma^l / l_*^2 .$$

Thus, the surface value of an additional mass force is proportional to the surface energy of the deformation and polarization divided by the square of the half of the characteristic length. Since l_* is a large value, the absolute magnitude of the additional mass force is a small quantity.

The nonlinear mass force $\mathbf{F}_*'$ is localized in the near-surface areas of the layer (see Fig. 4.6, where $F_*' = 2(\mu_{\pi 0}')^2 \breve{\lambda}^3 \chi_m (1 - \kappa_E \chi_{Em}/\chi_m)$), and it is absent in the mid cross section ($F_x'(0) = 0$). The thickness of thin films affects the inhomogeneity of the force distribution therein, namely, if the thickness of a thin layer diminishes, the force distribution in the film cross section becomes more inhomogeneous.

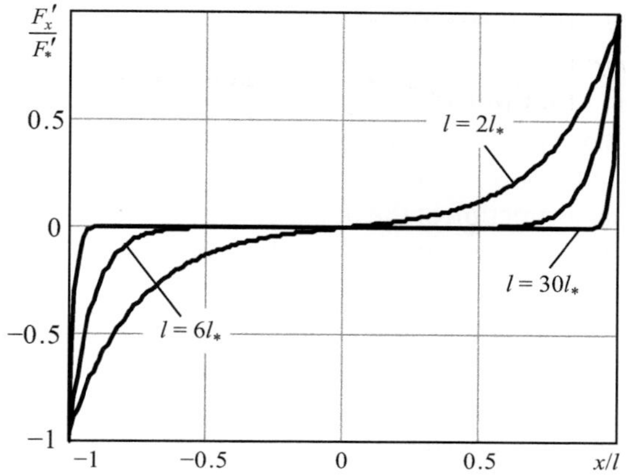

Figure 4.6 Additional mass force F'_x/F'_* distribution in layers of different thicknesses.

Using Eqs. (2.52), (4.52), and (4.55), for x-component F_{ex} of the ponderomotive force $\mathbf{F}_e = \rho(\nabla\otimes\mathbf{E}_*)\cdot\boldsymbol{\pi}_e$, we obtain

$$F_{ex}(x) = \rho_0\pi_e(x)\frac{dE}{dx} = -\varepsilon_0\left(\mu'_{\pi 0}\right)^2\kappa_E^2\breve{\lambda}^3\frac{\mathrm{sh}\left(\breve{\lambda}x\right)\mathrm{ch}\left(\breve{\lambda}x\right)}{\mathrm{ch}^2\left(\breve{\lambda}l\right)}. \quad (4.62)$$

From the formulae (4.61) and (4.62), we get the expression

$$\frac{\rho_0 F'_x}{F_{ex}} = \frac{2\varepsilon}{\varepsilon_0}\left(1 - \frac{\chi_m\varepsilon}{\rho_0\chi_{Em}^2}\right).$$

Here, $\varepsilon = \varepsilon_0(1+\chi)$. We can write the above expression as follows

$$\frac{\rho_0 F'_x}{F_{ex}} = 2(1+\chi)\left(1 - \frac{\varepsilon_0\chi_m(1+\chi)}{\rho_0\chi_{Em}^2}\right).$$

Since $\varepsilon_0 = 8.85 \times 10^{-12}$ F/m, and $\chi_m = O(10^{-23})[\mathrm{s}^2]$, $\rho_0 = O(10^3)$ [kg/m^3], $\chi_{Em} = O(10^{-16})$ [m^2/V], thus $\frac{\varepsilon_0\chi_m(1+\chi)}{\rho_0\chi_{Em}^2} \ll 1$, and, therefore, $\rho_0 F'_x \approx 2(1+\chi)F_{ex}$. Thus, in a free infinite layer of an ideal dielectric, an additional mass force that arises in the equation of motion due to the local mass displacement exceeds the ponderomotive one.

4.4 Mead's Anomaly

Within the framework of the classical theory of dielectrics, the polarization vector in a thin isotropic dielectric layer located between two metal electrodes is uniform, and the electric potential in the layer is a linear function of the thickness coordinate (Mindlin 1972a). However, the experiments provided by Mead (1961) have shown the capacitance of a thin dielectric film to be higher than the one predicted by the classical theory (see Fig. 4.7). Mead also observed some nonlinearity of the electric field distribution within the film. Here, to explain such an anomalous behavior of dielectric materials, we use a set of governing equations (2.153)–(2.155). We show that the local gradient theory of dielectrics correctly describes an anomalous dependence of the capacitance of thin dielectric films on its thickness as well as predicts a nonlinear distribution of the electric potential and polarization in thin dielectric films.

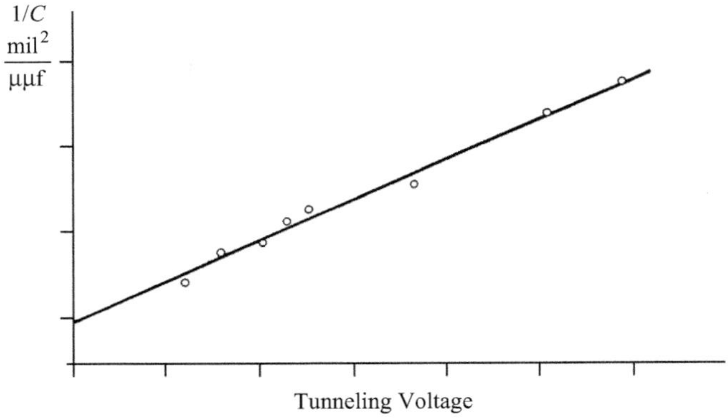

Figure 4.7 Inverse capacitance as a function of tunnel voltage for $Ta–Ta_2O_5–Au$ structures. Reprinted with permission from Mead (1961), Copyright 1961 by the American Physical Society.

4.4.1 Problem Formulation

Let us consider a stationary state of a centrosymmetric cubic crystal layer $|x| \leq l$ of an ideal dielectric with traction-free surfaces $x = \pm l$. It is supposed that the electric potential $\varphi_e = \pm V$ is kept fixed on the corresponding surfaces of the dielectric (Fig. 4.8). We restrict ourselves to an isothermal approximation and consider the deformation, polarization, and local mass displacement as the basic processes.

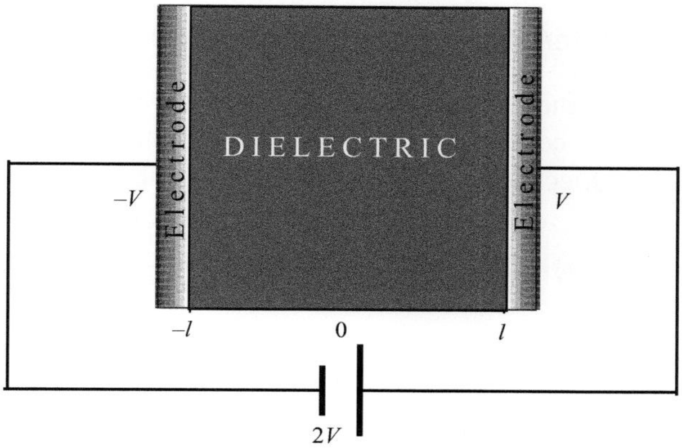

Figure 4.8 A thin plate capacitor.

Taking into account that the required functions depend only on the space coordinate *x*, we have a linear set of equations (4.21), (4.22) for the stationary approximation in order to determine the displacement vector $\mathbf{u} = (u, 0, 0)$, electric potential φ_e, and modified chemical potential $\tilde{\mu}'_\pi$.

The boundary conditions for a traction-free dielectric layer with the voltage $\pm V$ applied at $x = \pm l$ are

$$\left(\bar{K}+\frac{4}{3}G\right)\frac{du}{dx}-K\frac{\alpha_\rho}{d_\rho}\tilde{\mu}'_\pi=0, \quad \varphi_e=\pm V. \tag{4.63}$$

However, these conditions are insufficient to ensure the uniqueness of the solution to the set of equations (4.21) and (4.22). Following Mindlin (1969, 1972a), we assume that the surface polarization π_e takes up an absolute magnitude, which is smaller than the one obtained from the classical theory, namely: $\pi_e^{cl} = -\varepsilon_0\chi V/\rho_0 l$. Thus, we assume that

$$\chi_E\frac{\partial\varphi_e}{\partial x}+\chi_{Em}\frac{\partial\tilde{\mu}'_\pi}{\partial x}=k_*\frac{\varepsilon_0\chi}{\rho_0}\frac{V}{l} \quad \text{at } x=\pm l, \tag{4.64}$$

where χ is the dielectric susceptibility, and k_* $(0 \le k_* \le 1)$ is the phenomenological constant introduced earlier by Mindlin (1969).

It should be noted that the classical condition is $k^* = 1$, while $k_* = 0$ corresponds to the polarization continuity across the interface. Following Mindlin (1969), we also take $k_* = 0.1$.

4.4.2 Problem Solution and Its Analysis

Having solved the set of equations (4.21), (4.22) and having satisfied the boundary conditions (4.63), (4.64), we obtain the following relations for electric potential φ_e and specific polarization π_e:

$$\varphi_e(x) = V\frac{x}{l} + \frac{V(k_* - 1)}{1 + \chi^{-1}\breve{\lambda} l \operatorname{cth}(\breve{\lambda} l)}\left[\frac{x}{l} - \frac{\operatorname{sh}(\breve{\lambda} x)}{\operatorname{sh}(\breve{\lambda} l)}\right], \tag{4.65}$$

$$\pi_e(x) = -\chi\frac{\varepsilon_0}{\rho_0}\frac{V}{l} - \chi\frac{\varepsilon_0}{\rho_0}\frac{V}{l}(k_* - 1)\frac{1 + \chi^{-1}\breve{\lambda} l \operatorname{ch}(\breve{\lambda} x)/\operatorname{sh}(\breve{\lambda} l)}{1 + \chi^{-1}\breve{\lambda} l \operatorname{cth}(\breve{\lambda} l)}. \tag{4.66}$$

Here, $\chi_E = \chi\varepsilon_0/\rho_0$, and $\breve{\lambda}$ is presented by the formula (4.27).

The electric capacitance of the dielectric layer is defined as the ratio of the density of surface charge to the potential drop across the layer.

If the values of the electric potential φ_e and the specific polarization π_e are known, then we have the following formula to define the capacity

$$C = \frac{1}{2V}\left(\varepsilon_0\frac{d\varphi_e}{dx} - \rho_0\pi_e\right).$$

Hence, using the relation (4.65) and (4.66), for the inverse capacitance C^{-1}, we obtain

$$C^{-1} = \frac{2l\left[1 + \chi \operatorname{th}(\breve{\lambda} l)/(\breve{\lambda} l)\right]}{\varepsilon_0(1+\chi)\left[1 + k_*\chi \operatorname{th}(\breve{\lambda} l)/(\breve{\lambda} l)\right]}. \tag{4.67}$$

According to the classical theory, the inverse capacitance is proportional to the layer thickness, i.e., $C^{-1} = 2l/[\varepsilon_0(1 + \chi)]$ (Mindlin 1972a). Figure 4.9 displays the nonlinear relation between the normalized inverse capacitance and layer thickness. It can be seen that the linear part of the plot in Fig. 4.9 is well above the curve obtained from the classical theory, which is in agreement with the experimental results (see Fig. 4.7) (Mead 1961).

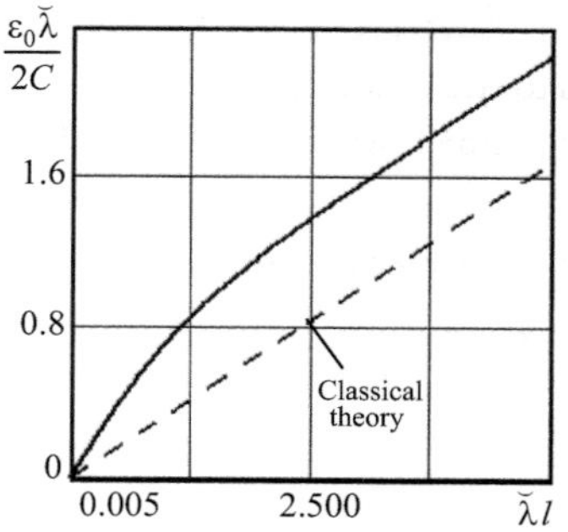

Figure 4.9 Dependence of inverse capacitance on normalized film thickness.

Figure 4.10 illustrates the distribution of polarization $\pi_e(x)/\pi_{e*}$ for $l = 5l_*$, $20l_*$, that is, curves 1 and 2, respectively, and the distribution of electric potential $\varphi_e(x)/\varphi_*$ for the same thicknesses, that is, curves 3 and 4, respectively. Here, $\pi_{e*} = -\pi_e^{cl}$ and $\varphi_* = V/l$. The solid lines are obtained using Eqs. (4.65) and (4.66), while the dashed lines correspond to the classical theory of piezoelectrics. We can see that inside the film, absolute magnitudes of polarization π_e and electric potential φ_e are smaller than those obtained within the classical theory. The film thickness affects the inhomogeneity of the polarization distribution. In particular, the region of its inhomogeneity increases with a decrease in the film thickness. Note that similar results have also been obtained by Mindlin (1969), Yang (1997), and Yang and Yang (2004), where the gradient and nonlocal polarization laws have been used.

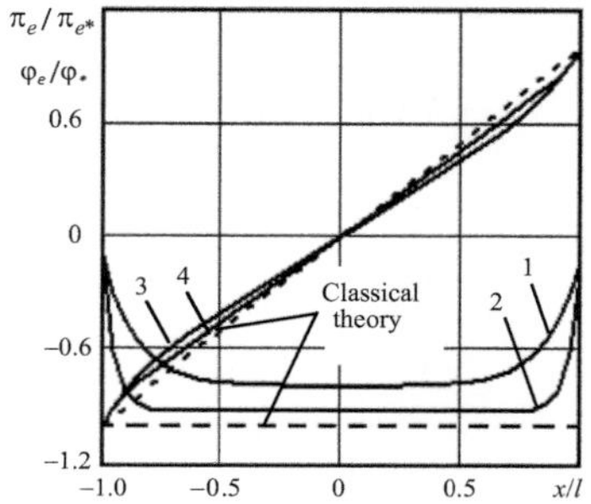

Figure 4.10 Polarization (curves 1 and 2) and electric potential (curves 3 and 4) as functions of distance x/l for $l = 5l_*$ (curves 1 and 3) and $l = 20l_*$ (curves 2 and 4).

4.5 Layer with Clamped boundaries: Disjoining Pressure

The overlap of near-surface inhomogeneities in the distribution of stresses of thin films causes the appearance of the so-called disjoining pressure. Note that such a phenomenon was obtained earlier for liquid films by Derjaguin (1980), Derjaguin and Churaev (1984).

It was Derjaguin who introduced the notion of disjoining pressure as the difference between the pressure p_1 upon the surface of a thin film and the pressure p_0 in a thick film from which a thin film is formed by means of a slow reduction in its thickness: $p_{\text{dis}}(2l) = p_1 - p_0$. It has been experimentally proved that there exists a disjoining pressure in the films having the thickness from 5 to 10 nm (Derjaguin and Churaev 1984).

Here we show that such a pressure can be present in thin solid films as well. Within the framework of the developed theory, the emergence of the disjoining pressure is associated with the changes in the structure of the near-surface regions of a thin body. To demonstrate this, we study the near-surface inhomogeneity of electromechanical fields in an infinite isotropic dielectric layer ($|x| \leq l$) with clamped boundaries. In the case of isothermal approximation, we can write the boundary conditions as follows:

$$u(\pm l) = 0, \ \tilde{\mu}'_{\pi}(\pm l) = \mu'_{\pi a}\,. \tag{4.68}$$

The solutions to Eqs. (4.21) and the boundary conditions (4.68) are as follows:

$$u(x) = \frac{B_0 K \alpha_\rho}{\left(\bar{K} + \frac{4}{3} G\right) d_\rho \breve{\lambda}} \left[\operatorname{sh}\left(\breve{\lambda} x\right) - \frac{x}{l} \operatorname{sh}\left(\breve{\lambda} l\right) \right], \tag{4.69}$$

$$\tilde{\mu}'_{\pi}(x) = B_0 \left[\operatorname{ch}\left(\breve{\lambda} x\right) + \mathfrak{M} \frac{\operatorname{sh}\left(\breve{\lambda} l\right)}{\breve{\lambda} l} \right], \tag{4.70}$$

where

$$B_0 = \frac{\mu'_{\pi a}}{\operatorname{ch}\left(\breve{\lambda} l\right) + \mathfrak{M} \operatorname{sh}\left(\breve{\lambda} l\right) / \left(\breve{\lambda} l\right)}.$$

Making use of the formulae (4.69) and (2.8), for the component e_{xx} of the deformation tensor, we get

$$e_{xx}(x) = B_0 \frac{K\alpha_\rho}{\breve{K}d_\rho}\left[\mathrm{ch}\left(\breve{\lambda}x\right) - \frac{\mathrm{sh}\left(\breve{\lambda}l\right)}{\breve{\lambda}l}\right].$$

Finally, for nonzero components σ_{xx} and $\sigma_{yy} = \sigma_{zz}$ of the stress tensor, we obtain the following formulae:

$$\sigma_{xx} = -\frac{\sigma_s'(1+\mathfrak{M})}{\breve{\lambda}l\,\mathrm{cth}(\breve{\lambda}l)+\mathfrak{M}}, \tag{4.71}$$

$$\sigma_{yy}(x) = \sigma_{zz}(x) = \frac{\sigma_s'(1+\mathfrak{M})\left[2G\breve{\lambda}l\,\mathrm{ch}\left(\breve{\lambda}x\right)+(K-2G/3)\mathrm{sh}\left(\breve{\lambda}l\right)\right]}{(K+4G/3)\left[\breve{\lambda}l\,\mathrm{ch}\left(\breve{\lambda}l\right)+\mathfrak{M}\,\mathrm{sh}\left(\breve{\lambda}l\right)\right]}. \tag{4.72}$$

Here, $\sigma_s' = K\alpha_\rho\,\mu_{\pi a}'/d_\rho$..

The distribution of normal stresses in layers of different thickness is shown in Fig. 4.11 for potassium chloride. Here we used the following material constants: $K = 1.74 \times 10^{10}$ Pa, $G = 6.24 \times 10^9$ Pa, $\mathfrak{M} = 0.5 \times 10^{-3}$. The absolute magnitude of stresses σ_{yy} and σ_{zz} reaches the highest value on the layer surface $x = \pm l$. The thickness of thin films affects the magnitude of the surface stress (see dash-dotted line in Fig. 4.12). One can see that in films with clamped boundaries, there arises a constant normal stress σ_{xx} in addition to the stresses σ_{yy} and σ_{zz}. In thick films, the stress σ_{xx} is negligibly

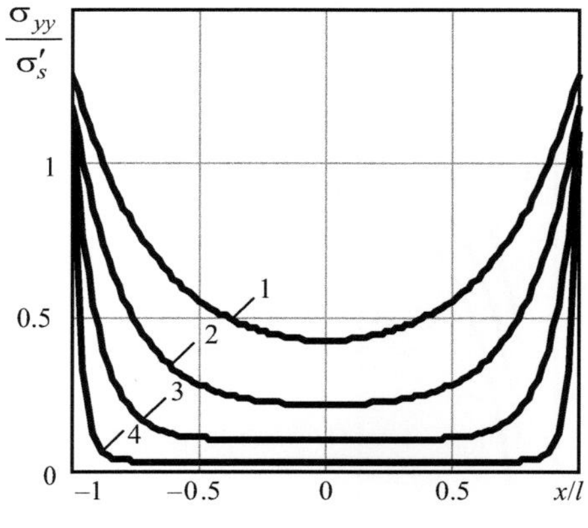

Figure 4.11 The distribution of stress σ_{yy}/σ_s' in potassium chloride layers of different thickness $l = 3l_*$, $5l_*$, $10l_*$, $30l_*$, that is, curves 1–4, respectively.

small, but a decreasing thickness of thin films leads to an increase in the absolute magnitude of this stress (see Fig. 4.13). As a result, this stress causes a disjoining pressure

$$p_{\text{dis}} = \frac{1}{2l}\int_{-l}^{l} \sigma_{xx} dx \qquad (4.73)$$

in thin solid films. Note that the characteristic lengths $l_* = 1.3$ Å and $l_* = 1.89$ Å in Fig. 4.13 correspond to crystals NaCl and KCl (Askar et al. 1971; Mindlin 1972a).

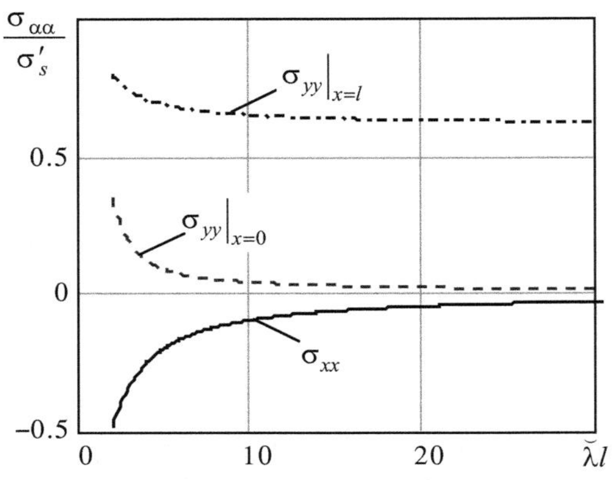

Figure 4.12 The dependence of stresses σ_{xx}, $\sigma_{yy}(l)$, and $\sigma_{yy}(0)$ on layer thickness.

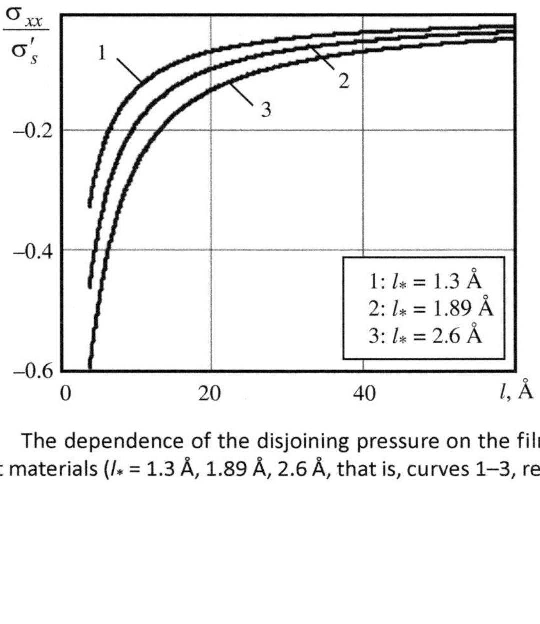

Figure 4.13 The dependence of the disjoining pressure on the film thickness for different materials ($l_* = 1.3$ Å, 1.89 Å, 2.6 Å, that is, curves 1–3, respectively).

Within the developed theory, this disjoining pressure is due to the changes in the structure of the near-surface regions of the solids. Such a pressure occurs in thin layers, in which the regions of near-surface inhomogeneity overlap. If the thickness of thin layer diminishes, the disjoining pressure becomes higher. There is no disjoining pressure in thick layers.

Since the material characteristics K and d_ρ are positive, the sign of disjoining pressure depends on the sign of material constants α_ρ and $\mu'_{\pi a}$. Quantitatively, parameter $\mu'_{\pi a}$ is determined by the material characteristics of both the layer (film) and the surrounding medium that contacts with the body. Note also that a positive disjoining pressure can prevent the reduction in the film thickness under the effect of external forces, whereas a negative pressure can reduce the thickness of the film and may lead to its destruction (i.e., rupture).

4.6 Formation of Near-Surface Inhomogeneity in an Infinite Layer

The results obtained in Sections 4.2 and 4.3 show that the near-surface inhomogeneity of the mechanical and electric fields is formed instantaneously if the process of local mass displacement is described as the reversible one. In this section, we will study the process of formation of near-surface inhomogeneity of mechanical fields. To this end, we will use the governing set of equations, which take into account the irreversibility of the local mass displacement. For simplicity, we shall consider an isothermal approximation and neglect the effect of electric field on the local mass displacement and on the process of deformation.

4.6.1 Problem Formulation

We use the dynamically coupled equations (3.120) and (3.122) to investigate the process of formation of a near-surface inhomogeneity of fields in an infinite traction-free isotropic layer. Let us assume that at time $t = 0$, the layer (region $|x| \le l$) is cut out from an infinite medium in such a way that subsequently ($t > 0$) it gets in contact with vacuum. In this case, the resulting fields ($\hat{\boldsymbol{\sigma}}$, $\mathbf{u}$, and $\tilde{\mu}'_\pi$) are functions of the spatial coordinate x and the time coordinate t. Then, in a linear approximation, the fundamental field equations expressed in terms of the displacement vector $\mathbf{u} = (u, 0, 0)$ and of the modified chemical potential $\tilde{\mu}'_\pi = \mu'_\pi - \mu'_{\pi 0}$ can be written as follows:

$$\rho_0 \frac{\partial^2 u}{\partial t^2} = \left(\bar{K} + \frac{4}{3} G \right) \frac{\partial^2 u}{\partial x^2} - K \frac{\alpha_\rho}{d_\rho} \frac{\partial \tilde{\mu}'_\pi}{\partial x}, \tag{4.74}$$

$$\mathfrak{L}_4^\mu \frac{d_\rho}{\rho_0} \frac{\partial^2 \tilde{\mu}'_\pi}{\partial x^2} + \frac{\partial \tilde{\mu}'_\pi}{\partial t} + \frac{1}{\tau_\pi} \tilde{\mu}'_\pi = -K \frac{\alpha_\rho}{\rho_0} \left(\frac{\partial^2 u}{\partial x \partial t} + \frac{1}{\tau_\pi} \frac{\partial u}{\partial x} \right). \tag{4.75}$$

Here, $$\tau_\pi = -\rho_0 \chi_m / \mathfrak{L}_4^\mu .$$

The boundary conditions that correspond to traction-free surfaces of the layer surrounded by vacuum are $\hat{\boldsymbol{\sigma}} \cdot \mathbf{n} = 0$, $\mu'_\pi = 0$ at $x = \pm l$. These conditions in terms of the displacement vector and of the disturbance of modified chemical potential $\tilde{\mu}'_\pi$ are as follows:

$$\left(\bar{K} + \frac{4}{3} G \right) \frac{\partial u}{\partial x} - \frac{K \alpha_\rho}{d_\rho} \tilde{\mu}'_\pi = 0, \quad \tilde{\mu}'_\pi = -\mu'_{\pi 0} \quad \text{at} \quad x = \pm l. \tag{4.76}$$

The solution to the boundary-value problem should satisfy the homogeneous initial conditions:

$$\tilde{\mu}'_\pi = 0, \quad u = 0, \quad \frac{\partial u}{\partial t} = 0 \quad \text{at} \quad t = 0. \tag{4.77}$$

4.6.2 Problem Solution and Its Analysis

To solve the boundary-value problem (4.74)–(4.77), we used the integral Laplace transform by the time variable (Schiff 1999). If we restrict ourselves to the quasi-static case, then for the non-vanishing components $\sigma_{yy} = \sigma_{zz}$ of the stress tensor and potential $\tilde{\mu}'_\pi$, we obtain

$$\sigma_{yy}(X,\tau) = \sigma_{zz}(X,\tau) = \sigma'_s \left\{ e^{-\breve{\lambda} l (1-X)} \operatorname{erfc}\left[\frac{\breve{\lambda} l (1-X)}{2\sqrt{\tau}} - \sqrt{\tau} \right] \right.$$

$$+ e^{\breve{\lambda} l (1-X)} \operatorname{erfc}\left[\frac{\breve{\lambda} l (1-X)}{2\sqrt{\tau}} + \sqrt{\tau} \right] + e^{-\breve{\lambda} l (1+X)} \operatorname{erfc}\left[\frac{\breve{\lambda} l (1+X)}{2\sqrt{\tau}} - \sqrt{\tau} \right]$$

$$\left. + e^{\breve{\lambda} l (1+X)} \operatorname{erfc}\left[\frac{\breve{\lambda} l (1+X)}{2\sqrt{\tau}} + \sqrt{\tau} \right] \right\}, \tag{4.78}$$

$$\tilde{\mu}'_{\pi}(X,\tau) = -\frac{1}{2}\mu'_{\pi 0}\left\{ e^{-\breve{\lambda} l(1-X)} \operatorname{erfc}\left[\frac{\breve{\lambda} l(1-X)}{2\sqrt{\tau}} - \sqrt{\tau}\right]\right.$$

$$+e^{\breve{\lambda} l(1-X)} \operatorname{erfc}\left[\frac{\breve{\lambda} l(1-X)}{2\sqrt{\tau}} + \sqrt{\tau}\right] + e^{-\breve{\lambda} l(1+X)} \operatorname{erfc}\left[\frac{\breve{\lambda} l(1+X)}{2\sqrt{\tau}} - \sqrt{\tau}\right]$$

$$\left. + e^{\breve{\lambda} l(1+X)} \operatorname{erfc}\left[\frac{\breve{\lambda} l(1+X)}{2\sqrt{\tau}} + \sqrt{\tau}\right]\right\}, \qquad (4.79)$$

where

$$\tau = \frac{t}{\tau_{\pi}}, \quad X = \frac{x}{l}, \quad \breve{\lambda} = \lambda_{\mu}\sqrt{1+\mathfrak{M}}, \quad \sigma'_s = \frac{KG\alpha_{\rho}\mu'_{\pi 0}}{d_{\rho}\left(\bar{K} + \frac{4}{3}G\right)}.$$

As far as we took the irreversibility of the local mass displacement into consideration, the constructed theory allows us, in a continuum approximation, to study the process of formation of a near-surface inhomogeneity of the stress–strain state of the solids.

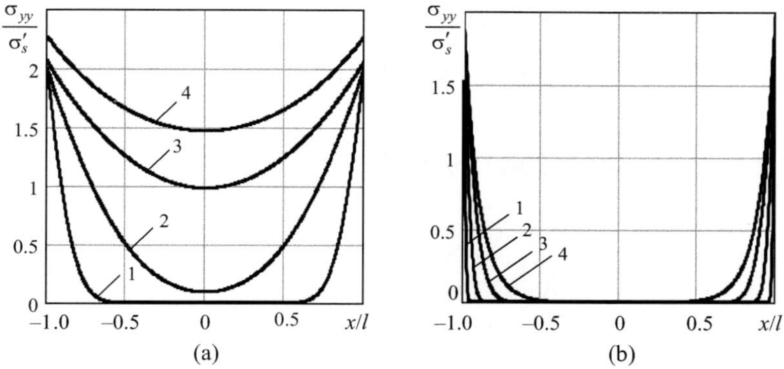

Figure 4.14 The stress σ_{yy}/σ'_s distribution for thin ($l = l_*$, a) and thick ($l = 10l_*$, b) layers at times $\tau = 10^{-2}, 10^{-1}, 5 \times 10^{-1}, 5$ (curves 1–4, respectively).

The quantitative analysis of the solution (4.78), (4.79) showed that at time $t = 0$, an inhomogeneous distribution of stresses $\sigma_{yy} = \sigma_{zz}$ and potential $\tilde{\mu}'_{\pi}$ is formed in the vicinity of the layer surfaces. Figure 4.14 displays the distributions of stresses in thin (Fig. 4.14a) and in thick (Fig. 4.14b) layers at different times. One can see that the stress distributions can be considered stationary for $\tau \geq 5$. The mid regions of thin layers are stressed. A thickness decrement of thin films causes an increase in the stress in the film cross section. An overlay of distributions of near-surface stresses in thin films causes the appearance of the disjoining pressure. Note that the disjoining pressure and an integral deformation

$$\varepsilon = \int_{-1}^{1} \varepsilon_{xx}(X)dX \tag{4.80}$$

demonstrate a similar dependence on the layer thickness.

The sign of the surface stresses as well as of the disjoining pressure (see Figs. 4.15 and 4.16) is determined by the sign of the material characteristics α_ρ and $\mu'_{\pi 0}$. Since $|\mu| > |\mu_\pi|$, and μ, K, G, and d_ρ are positive, the characteristic stress σ'_s is positive as well, provided $\alpha_\rho < 0$. In this case, the surface stress causes stretching, which worsens the stretching resistance of the layer. Figure 4.16 illustrates the dynamics of the disjoining pressure in elastic layers of different thickness.

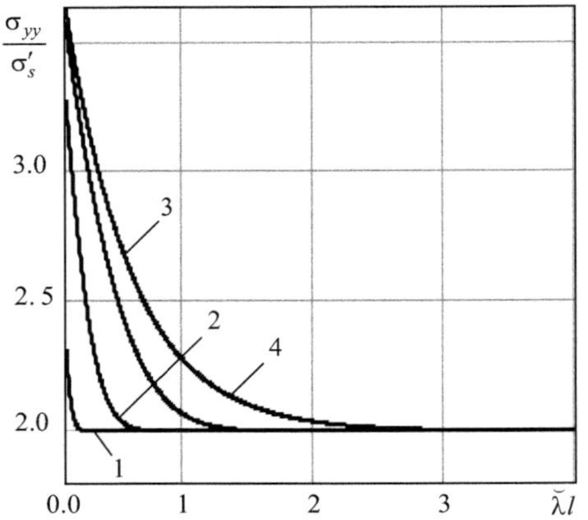

Figure 4.15 The dependence of surface stresses σ_{yy}/σ'_s on the layer thickness at $\tau = 10^{-2}$; 10^{-1}; 5×10^{-1}; 5 (curves 1–4, respectively).

4.6.3 Evaluation of the Lateral Force

We have shown that in an infinite traction-free elastic layer of constant thickness $2l$, there occur stresses $\sigma_{yy}(x, t, l)$ and $\sigma_{zz}(x, t, l)$. Now we use the obtained results to analyze the stress state in an elastic

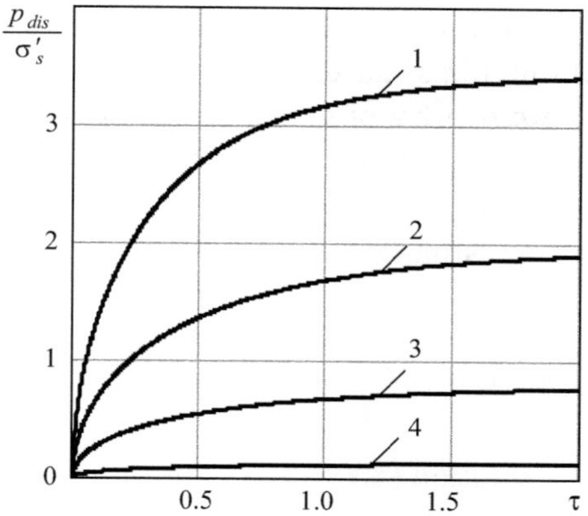

Figure 4.16 Dynamics of disjoining pressure p_{dis}/σ'_s in layers for $l = l_*$, $2l_*$, $5l_*$, $30l_*$ (curves 1–4, respectively).

layer of variable thickness. Let l be a function of the y-coordinate, i.e., $l = l(y)$. It is obvious that in arbitrary cross sections $y = y_1$ and $y = y_2$ of the layer, the stress σ_{yy} will have different values, i.e., $\sigma_{yy}(l(y_1)) \neq \sigma_{yy}(l(y_2))$. Such a difference between the stresses σ_{yy} causes the appearance of the lateral force (see Fig. 4.17). This force acts along the layer surface. Using the relation

$$\Sigma_{yy}(l) = \int_{-l}^{l} \sigma_{yy}(x)dx$$

we introduce the force Σ_{yy}, per unit length along the axis Oz. We define the linear density of the lateral force F^L by the relation

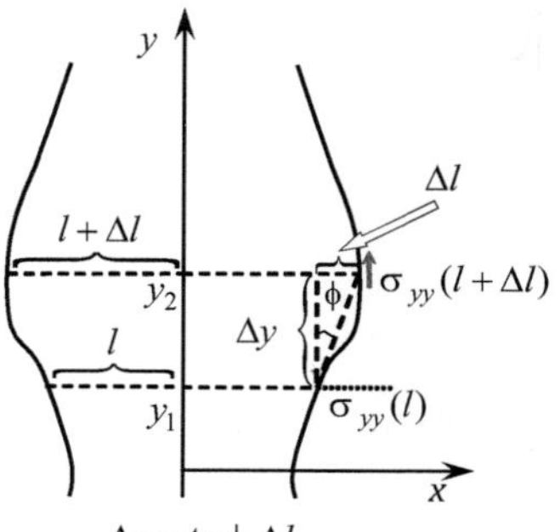

Figure 4.17 The emergence of a lateral force in a film of variable thickness.

$$F^L(l) = \frac{\Sigma_{yy}(l(y_2)) - \Sigma_{yy}(l(y_1))}{y_2 - y_1} = \frac{\Sigma_{yy}(l+\Delta l) - \Sigma_{yy}(l)}{\Delta y}$$

$$= \frac{\Sigma_{yy}(l+\Delta l) - \Sigma_{yy}(l)}{\mathrm{tg}\,\phi\,\Delta l} \to \frac{1}{\mathrm{tg}\,\phi}\frac{d\Sigma_{yy}(l)}{dl}.$$

Hence, in order to estimate the linear density of lateral force, we have a formula:

$$F^L(l) = \frac{d\Sigma_{yy}(l(y))}{dy} = \frac{1}{\mathrm{tg}\phi}\frac{d\Sigma_{yy}(l)}{dl}.$$

Here we use the solution (4.78), (4.79) to estimate the value of the lateral force $F^L_{\alpha\alpha} = \frac{\partial \sigma_{\alpha\alpha}}{\partial y}$, $\alpha = \{y, z\}$, which arises in an infinite layer of variable thickness. Figure 4.18 illustrates the dynamics of this force in thin (curves 1–3) and thick (curve 4) elastic layers. We can see that the absolute magnitude of the lateral force decreases and tends to zero when the layer thickness increases. In case $\alpha_\rho < 0$, the lateral force acts in the direction from wider to narrower parts of the layer, which means that it tends to smooth out the film thickness, thus increasing its resistance. On the contrary, if α_ρ is positive, then the lateral force will be directed from the narrower section to the wider section and, as a result, the strength of the film will deteriorate.

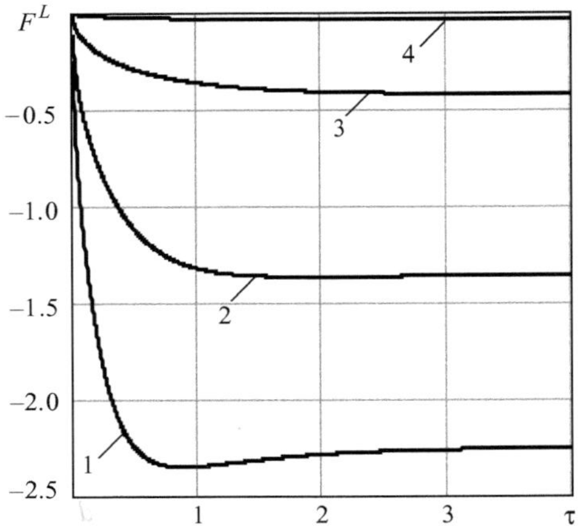

Figure 4.18 The dynamics of the lateral force F^L in the layers of different thickness (curves 1–4 correspond to $l = l_*$, $1.5l_*$, $3l_*$, $10l_*$, respectively).

Though the lateral forces as well as the disjoining pressure are negligible for thick films, their effect is quite significant for thin films. Therefore, it is important to take them into account in the case of thin films, because they can have a significant effect on the strength and mechanical stability of such films.

4.7 Solids of Cylindrical Geometry: Effect of Surface Curvature

In this section, we study the effect of surface curvature on the equilibrium stress distribution, polarization, surface energy of deformation and polarization, and the bond surface electric charge of dielectric body of cylindrical geometry.

4.7.1 Problem Formulation

Let the dielectric body be free of external loads and occupy the region (V) in a cylindrical coordinate system (r, φ, z). We assume that the body (V) is in contact with vacuum (region (V_v)). For isothermal approximation, we take the displacement vector, electric field, and the modified chemical potential $\tilde{\mu}'_\pi$ as the key functions. We also assume that the problem is centrosymmetric, and, therefore, the required functions depend on the radial coordinate only:

$$\mathbf{u} = (u_r(r), 0, 0),\ \tilde{\mu}'_\pi = \tilde{\mu}'_\pi(r), \mathbf{E} = (E_r(r), 0, 0), \mathbf{E}_v = (E_{rv}(r), 0, 0).$$

Therefore, the components of the strain tensor are determined by the relations

$$e_{rr} = \frac{du_r}{dr},\ e_{\varphi\varphi} = \frac{u_r}{r}. \tag{4.81}$$

Using the constitutive equations, we can write the following formulae for the components σ_{rr}, $\sigma_{\varphi\varphi}$, and σ_{zz} of the stress tensor, polarization vector $\boldsymbol{\pi}_e = (\pi_e(r), 0, 0)$, and vector $\boldsymbol{\pi}_m = (\pi_m(r), 0, 0)$ of the local mass displacement

$$\sigma_{rr}(r) = \left(\bar{K} + \frac{4}{3}G\right)\frac{du_r}{dr} + \left(\bar{K} - \frac{2}{3}G\right)\frac{u_r}{r} - \frac{K\alpha_\rho}{d_\rho}\tilde{\mu}'_\pi, \tag{4.82}$$

$$\sigma_{\varphi\varphi}(r) = \left(\bar{K} + \frac{4}{3}G\right)\frac{u_r}{r} + \left(\bar{K} - \frac{2}{3}G\right)\frac{du_r}{dr} - \frac{K\alpha_\rho}{d_\rho}\tilde{\mu}'_\pi, \tag{4.83}$$

$$\sigma_{zz}(r)=\left(\bar{K}-\frac{2}{3}G\right)\left(\frac{du_r}{dr}+\frac{u_r}{r}\right)-\frac{K\alpha_\rho}{d_\rho}\tilde{\mu}'_\pi, \quad (4.84)$$

$$\pi_e(r)=\chi_E E_r-\chi_{Em}\frac{d\tilde{\mu}'_\pi}{dr}, \quad (4.85)$$

$$\pi_m(r)=-\chi_m\frac{d\tilde{\mu}'_\pi}{dr}+\chi_{Em}E_r. \quad (4.86)$$

The displacement u_r, the modified chemical potential $\tilde{\mu}'_\pi$, the electric fields in the body E_r, and vacuum E_{rv} are governed by the equations

$$\frac{d}{dr}\left[\frac{1}{r}\frac{d(ru_r)}{dr}-\frac{K\alpha_\rho}{\left(\bar{K}+\frac{4}{3}G\right)d_\rho}\tilde{\mu}'_\pi\right]=0, \quad (4.87)$$

$$\frac{1}{r}\frac{d}{dr}\left(r\frac{d\tilde{\mu}'_\pi}{dr}\right)-\lambda_{\mu E}^2\tilde{\mu}'_\pi=\lambda_{\mu E}^2\frac{K\alpha_\rho}{\rho_0}\frac{1}{r}\frac{d(ru_r)}{dr}, \quad (4.88)$$

$$\frac{d}{dr}\left[r\left(E_r-\kappa_E\frac{d\tilde{\mu}'_\pi}{dr}\right)\right]=0 \quad \forall\mathbf{r}\in(V), \quad (4.89)$$

$$\frac{d(rE_{rv})}{dr}=0, \ \forall\mathbf{r}\in(V_v). \quad (4.90)$$

We can write the boundary conditions of the problem as follows:

$$\forall\mathbf{r}\in(\Sigma):\ \sigma_{rr}(r)=\sigma_{r\varphi}(r)=\sigma_{rz}(r)=0, \quad \tilde{\mu}'_\pi(r)=-\mu'_{\pi 0}, \quad (4.91)$$

$$\varepsilon_0\left[E_r(r)-E_{rv}(r)\right]+\rho_0\pi_e(r)=0. \quad (4.92)$$

Thus, the formulated problem consists in determining the functions u_r, $\tilde{\mu}'_\pi$, E_r, and E_{rv} from Eqs. (4.87)–(4.90) and boundary conditions (4.91) and (4.92). To ensure the solution's uniqueness, this set of equations should be accompanied with the condition of the boundedness of the fields at $r\to+\infty$ for an infinite medium with a cylindrical cavity and the boundedness of the fields at $r\to 0$ for an infinite cylindrical body.

4.7.2 Infinite Cylinder

In this subsection, we determine the equilibrium stress–strain state of the infinite cylindrical body (region $r\le R$), bounded by the traction-free surface $r=R$. Let the z-axis coincide with the axis of the cylinder (Fig. 4.19). Then, the region (V) corresponds to $r < R$, while the region (V_v) corresponds to $r > R$.

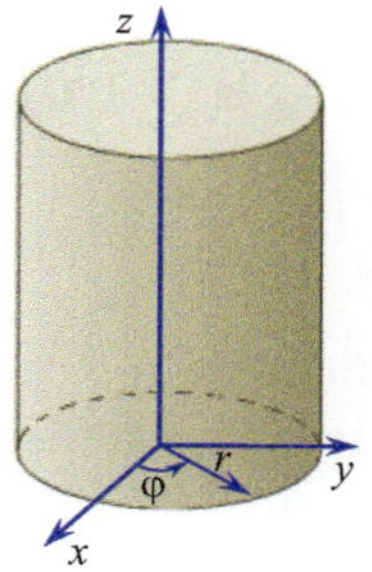

Figure 4.19 A cylindrical body.

The solution to the problems (4.87)–(4.92), which satisfies the boundedness condition at the origin ($r = 0$), looks as follows:

$$u_r(r) = \mu'_{\pi 0} \frac{K\alpha_\rho}{\breve{\lambda} d_\rho \left(\bar{K} + \frac{4}{3} G \right)} \left[\frac{1}{2} Q \breve{\lambda} r - (1 - \mathfrak{M}Q) \frac{I_1(\breve{\lambda} r)}{I_0(\breve{\lambda} R)} \right], \tag{4.93}$$

$$\tilde{\mu}'_\pi(r) = -\mu'_{\pi 0} \left[(1 - \mathfrak{M}Q) \frac{I_0(\breve{\lambda} r)}{I_0(\breve{\lambda} R)} + \mathfrak{M}Q \right], \tag{4.94}$$

$$E_r(r) = -\kappa_E \mu'_{\pi 0} \breve{\lambda} (1 - \mathfrak{M}Q) \frac{I_1(\breve{\lambda} r)}{I_0(\breve{\lambda} R)} \quad \forall r : r < R. \tag{4.95}$$

Here, $I_j(r)$ is the first-kind modified Bessel function of the order j (the Macdonald function), and

$$Q = -\frac{2GI_1(\breve{\lambda} R)}{(K + G/3)\breve{\lambda} R I_0(\breve{\lambda} R) - 2G\mathfrak{M} I_1(\breve{\lambda} R)}.$$

Making use of the formulae (4.82)–(4.86) and (4.93)–(4.95), for components of the stress tensor, polarization and local mass displacement vectors, we obtain the following relations:

$$\sigma_{rr}(r) = \sigma_s \left[\frac{Q}{2G} \left(K + \frac{1}{3} G \right) + (1 - \mathfrak{M}Q) \frac{I_1(\breve{\lambda} r)}{\breve{\lambda} r I_0(\breve{\lambda} R)} \right], \tag{4.96}$$

$$\sigma_{\varphi\varphi}(r) = \sigma_s \left[\frac{Q}{2G} \left(K + \frac{1}{3} G \right) + \frac{(1 - \mathfrak{M}Q)}{I_0(\breve{\lambda} R)} \left(I_0(\breve{\lambda} r) - \frac{I_1(\breve{\lambda} r)}{\breve{\lambda} r} \right) \right], \tag{4.97}$$

$$\sigma_{zz}(r)=\sigma_s\left[\frac{Q}{2G}\left(K-\frac{2}{3}G\right)+(1-\mathfrak{M}Q)\frac{I_0(\breve{\lambda}r)}{I_0(\breve{\lambda}R)}\right], \quad (4.98)$$

$$\pi_e(r)=\chi_{Em}\mu'_{\pi 0}\,\breve{\lambda}(1-\mathfrak{M}Q)\frac{\varepsilon_0}{\varepsilon}\frac{I_1(\breve{\lambda}r)}{I_0(\breve{\lambda}R)}, \quad (4.99)$$

$$\pi_m(r)=\mu'_{\pi 0}\,\breve{\lambda}\chi_m\left(1-\kappa_E\frac{\chi_{Em}}{\chi_m}\right)(1-\mathfrak{M}Q)\frac{I_1(\breve{\lambda}r)}{I_0(\breve{\lambda}R)}. \quad (4.100)$$

Here, the stress σ_s is defined by the relation (4.32). The formulae (4.96)–(4.98) allow us to analyze the effect of the free surface curvature $\kappa=\left(\breve{\lambda}R\right)^{-1}$ on the magnitude of the stresses. Such an effect is more essential in thin cylinders (fibers). An increase in the curvature of the surface decreases the level of the corresponding stresses (see Fig. 4.20, where $\mathfrak{M}=3\times 10^{-3}$, $K/G=2.79$).

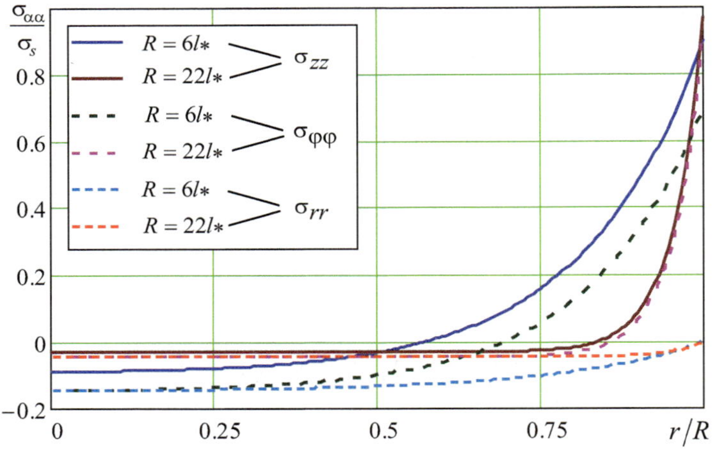

Figure 4.20 The effect of surface curvature on the stresses in fiber.

The relation (4.99) yields the following formula

$$\sigma_{es}(R)=\kappa_E\,\breve{\lambda}\varepsilon_0\mu'_{\pi 0}(1-\mathfrak{M}Q)\frac{I_1(\breve{\lambda}R)}{I_0(\breve{\lambda}R)}$$

to calculate the density of the bound surface charge. This formula can be written as follows:

$$\sigma_{es}(R)=\frac{\kappa_E\,\breve{\lambda}\varepsilon_0\mu'_{\pi 0}(K+G/3)\breve{\lambda}RI_1(\breve{\lambda}R)}{(K+G/3)\breve{\lambda}RI_0(\breve{\lambda}R)-2G\mathfrak{M}I_1(\breve{\lambda}R)}. \quad (4.101)$$

Substituting the asymptotic series $I_0(\breve{\lambda}R) \approx e^{\breve{\lambda}R}/\sqrt{2\pi\breve{\lambda}R}$, $I_1(\breve{\lambda}R) \approx e^{\breve{\lambda}R}/\sqrt{2\pi\breve{\lambda}R}$ of the Macdonald functions for large values of their argument (Korn and Korn 1968) into expression (4.101), we obtain

$$\sigma_{es}(R) \approx \kappa_E \breve{\lambda}\varepsilon_0\mu'_{\pi 0}\left(1+\frac{2G\mathfrak{M}}{(K+G/3)\breve{\lambda}R}\right).$$

An increase in the surface curvature of thin fibers reduces the density of the surface bound charge and increases the levels of absolute magnitude of the corresponding stresses. Since $\mathfrak{M}$ is a small parameter, for a sufficiently large radius R, the second term in the obtained formula is smaller than 1. In this case, it can be neglected. This means that the influence of the surface curvature on the magnitude of the bound surface charge in thick cylinders is negligible. However, the surface curvature shows important effects on thin fibers.

The density of surface energy of deformation and polarization, U_Σ^c, in a cylinder can be determined by inserting Eq. (4.100) into the relation (4.9). As a result, we get

$$U_\Sigma^c = -\frac{1}{2}\rho_0\mu_{\pi 0}'^2\breve{\lambda}\chi_m(1-\mathfrak{M}Q)\left(1-\kappa_E\frac{\chi_{Em}}{\chi_m}\right)\frac{I_1(\breve{\lambda}R)}{I_0(\breve{\lambda}R)}. \quad (4.102)$$

The surface energy of deformation and polarization in a cylindrical body also depends on the surface curvature. Taking into account the formula (4.33) for the surface energy in a body with planar boundaries, we can rewrite the relation (4.102) as follows:

$$U_\Sigma^c = U_\Sigma^\infty(1-\mathfrak{M}Q)\frac{I_1(\breve{\lambda}R)}{I_0(\breve{\lambda}R)} = U_\Sigma^\infty\frac{(K+G/3)\breve{\lambda}RI_1(\breve{\lambda}R)}{(K+G/3)\breve{\lambda}RI_0(\breve{\lambda}R)-2G\mathfrak{M}I_1(\breve{\lambda}R)}. \quad (4.103)$$

Using the asymptotic series of the Macdonald functions for large values of their argument (Korn and Korn 1968), the expression (4.103) takes the form

$$U_\Sigma^c \approx U_\Sigma^\infty\left(1-\frac{2G\mathfrak{M}}{(K+G/3)\breve{\lambda}R}\right)^{-1} \approx U_\Sigma^\infty\left(1+\frac{2G\mathfrak{M}}{(K+G/3)\breve{\lambda}R}\right). \quad (4.104)$$

Since the coupling factor between the mechanical fields and the local mass displacement is a small parameter $\mathfrak{M} \sim 10^{-5} \div 10^{-3}$, while

$\breve{\lambda} = O(10^{10})$ is a large quantity, the effect of the second term in the formula (4.104) is negligibly small. Thus, for large R, $U_\Sigma^c = U_\Sigma^\infty$.

For small arguments, $I_0(\breve{\lambda}R) \approx 1$, $I_1(\breve{\lambda}R) \approx \breve{\lambda}R/2$ (Korn and Korn 1968). In this case, the relation (4.103) can be written as follows:

$$U_\Sigma^c \approx U_\Sigma^\infty \frac{\breve{\lambda}R}{2}\left(1 - \frac{G\mathfrak{M}}{K + G/3}\right)^{-1} \approx U_\Sigma^\infty \frac{\breve{\lambda}R}{2}\left(1 + \frac{G\mathfrak{M}}{K + G/3}\right). \quad (4.105)$$

The formula (4.105) is valid only for thin fibers of nanosizes ($R < l_*$). In such fibers, the dependence of the surface energy of deformation on the radius is linear. Note that a linear dependence of the surface energy on the radius of small, elastic spherical and cylindrical bodies was also predicted by Tolman (1949).

Using the formulae (4.96)–(4.98), one can determine the hydrostatic stress $\sigma = (\sigma_{rr} + \sigma_{\varphi\varphi} + \sigma_{zz})/3$ in a cylinder

$$\sigma = \sigma_s \left[\frac{QK}{2G} + (1 - \mathfrak{M}Q)\frac{2I_0(\breve{\lambda}r)}{3I_0(\breve{\lambda}R)}\right].$$

On the surface of a cylindrical body, the hydrostatic stress is

$$\sigma\big|_{r=R} = \frac{2}{3}\sigma_s \left[1 - \frac{I_1(\breve{\lambda}R)(3K/2 - 2\mathfrak{M}G)}{(K + G/3)\breve{\lambda}RI_0(\breve{\lambda}R) - 2G\mathfrak{M}I_1(\breve{\lambda}R)}\right].$$

In view of the formula (4.103), we can rewrite this expression as

$$\sigma\big|_{r=R} = \frac{2}{3}\sigma_s \left[1 - \frac{U_\Sigma^c}{U_\Sigma^\infty}\frac{3(3K - 4\mathfrak{M}G)}{2(3K + G)\breve{\lambda}R}\right]. \quad (4.106)$$

From the relation (4.106), it can be concluded that the pressure on the cylindrical surface is lesser than in the body with a planar boundary by a quantity directly proportional to the ratio of the surface energy of deformation of the cylinder U_Σ^c and the surface energy of deformation of the half-space U_Σ^∞, and inversely proportional to the cylinder radius R and material length scale parameter $\breve{\lambda}$.

4.7.3 Infinite Medium with Cylindrical Cavity

Now we consider an infinite medium with cylindrical cavity ($r \geq R$). Then, the region (V) corresponds to $r > R$, while the region (V_v) corresponds to $r < R$. Assume that the axes of the cylindrical cavity coincide with the z-axis. In this case, the solution to the boundary-value problems (4.87)–(4.90), which satisfies the boundary conditions (4.91), (4.92) as well as the conditions of boundedness of u_r, $\tilde{\mu}'_\pi$, and E_r on the infinity ($r \to +\infty$) and boundedness of the electric field E_{rv} in vacuum at $r = 0$, is given by

$$u_r(r) = \mu'_{\pi 0} \frac{K\alpha_\rho}{\breve{\lambda} d_\rho \left(\bar{K} + \frac{4}{3} G \right)} \frac{K_1(\breve{\lambda} R)}{K_0(\breve{\lambda} R)} \left(\frac{K_1(\breve{\lambda} r)}{K_1(\breve{\lambda} R)} - \frac{R}{r} \right), \tag{4.107}$$

$$\tilde{\mu}'_\pi(r) = -\mu'_{\pi 0} \frac{K_0(\breve{\lambda} r)}{K_0(\breve{\lambda} R)}, \quad E_r(r) = \mu'_{\pi 0} \kappa_E \breve{\lambda} \frac{K_1(\breve{\lambda} r)}{K_0(\breve{\lambda} R)}, \quad \forall r : r > R, \tag{4.108}$$

$$E_{rv} = 0, \quad \forall r : r < R. \tag{4.109}$$

Here, $K_j(r)$ is the second-kind modified Bessel function of the order j (the Macdonald function).

Making use of the formulae (4.82)–(4.86), (4.107), and (4.108), we obtain the following relations to calculate the components of the stress tensor, polarization vector, and local mass displacement:

$$\sigma_{rr}(r) = \sigma_s \frac{K_1(\breve{\lambda} R)}{K_0(\breve{\lambda} R)} \frac{1}{\breve{\lambda} r} \left(\frac{R}{r} - \frac{K_1(\breve{\lambda} r)}{K_1(\breve{\lambda} R)} \right), \tag{4.110}$$

$$\sigma_{\varphi\varphi}(r) = \sigma_s \frac{K_1(\breve{\lambda} R)}{K_0(\breve{\lambda} R)} \left(\frac{K_0(\breve{\lambda} r)}{K_1(\breve{\lambda} R)} + \frac{1}{\breve{\lambda} r} \frac{K_1(\breve{\lambda} r)}{K_1(\breve{\lambda} R)} - \frac{R}{\breve{\lambda} r^2} \right), \tag{4.111}$$

$$\sigma_{zz}(r) = \sigma_s \frac{K_0(\breve{\lambda} r)}{K_0(\breve{\lambda} R)}, \tag{4.112}$$

$$\pi_e(r) = -\kappa_E \breve{\lambda} \mu'_{\pi 0} \frac{\varepsilon_0}{\rho_0} \frac{K_1(\breve{\lambda} r)}{K_0(\breve{\lambda} R)}, \tag{4.113}$$

$$\pi_m(r) = -\mu'_{\pi 0} \breve{\lambda} \chi_m \left(1 - \kappa_E \frac{\chi_{Em}}{\chi_m} \right) \frac{K_1(\breve{\lambda} r)}{K_0(\breve{\lambda} R)}. \tag{4.114}$$

It should be noted that the vector of electric induction in the body is equal to zero.

Surface values of the axial σ_{zz} and hoop $\sigma_{\varphi\varphi}$ stresses are equal to each other and do not depend on the radius of the curvature:

$$\sigma_{\varphi\varphi}(R) = \sigma_{zz}(R) \equiv \sigma_s = 2GK\alpha_\rho \mu'_{\pi 0} \Big/ d_\rho \left(\bar{K} + \frac{4}{3} G \right).$$

In view of the relations (4.9) and (4.114), we can define the surface energy of deformation and polarization U_Σ^m in an infinite medium with a cylindrical cavity:

$$U_\Sigma^m = -\frac{1}{2}\rho_0\mu_{\pi 0}^{\prime 2}\breve{\lambda}\chi_m\left(1-\kappa_E\frac{\chi_{Em}}{\chi_m}\right)\frac{K_1(\breve{\lambda}R)}{K_0(\breve{\lambda}R)}. \quad (4.115)$$

Taking into account the formula (4.33) for surface energy of an infinite body with planar boundary, we represent the expression (4.115) as follows:

$$U_\Sigma^m = U_\Sigma^\infty \frac{K_1(\breve{\lambda}R)}{K_0(\breve{\lambda}R)}. \quad (4.116)$$

For functions $K_0(\breve{\lambda}R)$ and $K_1(\breve{\lambda}R)$, we use the following asymptotic series for large arguments (Korn and Korn 1968):

$$K_0(\breve{\lambda}R) \approx \sqrt{\frac{\pi}{2\breve{\lambda}R}}e^{-\breve{\lambda}R}\left(1-\frac{1}{8\breve{\lambda}R}+\ldots\right), \quad (4.117)$$

$$K_1(\breve{\lambda}R) \approx \sqrt{\frac{\pi}{2\breve{\lambda}R}}e^{-\breve{\lambda}R}\left(1+\frac{3}{8\breve{\lambda}R}+\ldots\right).$$

Thus, in this case

$$\frac{K_1(\breve{\lambda}R)}{K_0(\breve{\lambda}R)} \approx 1+\frac{1}{2\breve{\lambda}R},$$

and for large R, we get

$$U_\Sigma^m \approx U_\Sigma^\infty\left(1+\frac{1}{2\breve{\lambda}R}\right).$$

4.7.4 Effect of Surface Curvature on Surface Energy of Deformation and Polarization

We denote the curvature of the surface by the following relation:

$$\kappa = \begin{cases} -\left(\breve{\lambda}R\right)^{-1}, & \text{medium with cylindrical cavity,} \\ 0, & \text{plane surface,} \\ \left(\breve{\lambda}R\right)^{-1}, & \text{infinite cylinder.} \end{cases}$$

Then, in view of the relations (4.103) and (4.116), the formula describing the effect of surface curvature on the density of the surface energy of deformation and polarization is given by

$$\frac{U_\Sigma(\kappa)}{U_\Sigma^\infty}=\begin{cases}\dfrac{K_1(-1/\kappa)}{K_0(-1/\kappa)}, & \kappa<0,\\ 1, & \kappa=0,\\ \dfrac{I_1(1/\kappa)}{I_0(1/\kappa)-2G\mathfrak{M}\kappa\left(K+G/3\right)^{-1}I_1(1/\kappa)}, & \kappa>0.\end{cases}$$

An increased surface curvature of a free cylinder decreases the absolute magnitude of the surface energy compared to the body with a planar boundary. In contrast, in the infinite medium with a thin cylindrical cavity, an increased curvature of the surface results in an increased surface energy (Fig. 4.21). Note that here the characteristic length $l_* = 1.3$ Å corresponds to crystal NaCl (Mindlin 1972a). Thus, the value of the surface energy of deformation and polarization for the body with a planar boundary ($\kappa = 0$) cannot be a minimum of the surface energy U_Σ as a function of surface curvature (Fig. 4.22). Such a result is also consistent with the results obtained for sphere and spherical cavity by Schwartz (1969), who used the Mindlin gradient theory of dielectrics.

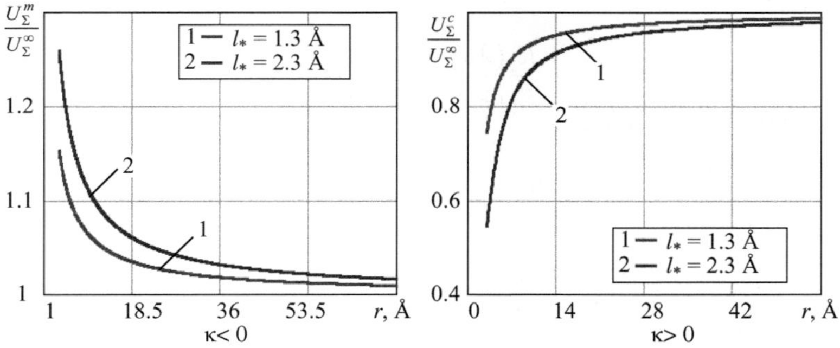

Figure 4.21 The effect of surface curvature on the surface energy of deformation and polarization of elastic bodies of cylindrical geometry.

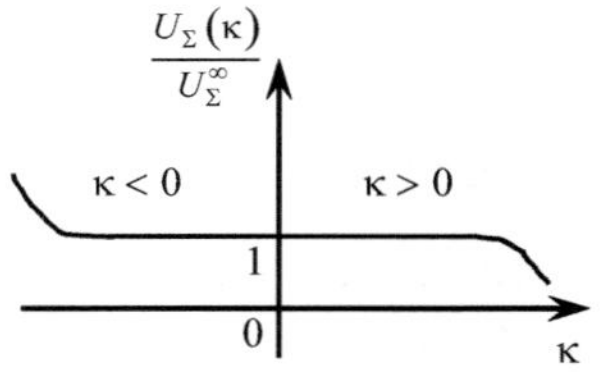

Figure 4.22 The effect of surface curvature on surface energy of deformation and polarization.

4.8 Effect of Heating on the Near-Surface Inhomogeneity of Electromechanical Fields: Piroelectric and Thermopolarization Effects

4.8.1 Problem Formulation

Let an elastic, dielectric, isotropic spherical body in a spherical coordinate system (r, φ, ϑ) occupy a region $R_1 \leq r \leq R_2$, whose boundaries $r = R_1$ and $r = R_2$ separate it from the outer vacuum (see Fig. 4.23). Assume the body to be an ideal dielectric. There are no mass forces or heat sources distributed in the body. The surfaces of the sphere are traction free and are kept at different temperatures θ_i $(i = 1, 2;\ \theta_1 \neq \theta_2)$, where θ is the temperature measured relative to the reference level T_0. We restrict ourselves to the analysis of the stationary state of the hollow sphere.

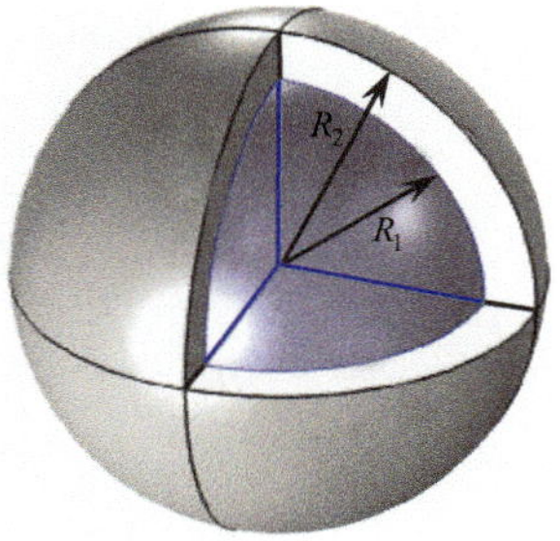

Figure 4.23 Hollow spherical body.

We take the displacement vector $\mathbf{u}$, temperature perturbation θ, specific density of induced mass ρ_m, and electric potential φ_e as the key functions. In order to study the linear response of the body to the temperature gradient, we use a linear set of governing equations (2.137)–(2.140). Since $\mathbf{E} = -\nabla\varphi_e$ for stationary approximation, we can write this set of equations in the following form:

$$\left(K+\frac{1}{3}G\right)\boldsymbol{\nabla}(\boldsymbol{\nabla}\cdot\mathbf{u})+G\Delta\mathbf{u}-K\alpha_T\boldsymbol{\nabla}\theta-K\alpha_\rho\boldsymbol{\nabla}\rho_m=0\,, \tag{4.118}$$

$$\Delta\theta=0, \tag{4.119}$$

$$\Delta\rho_m-\lambda_{\mu E}^2\,\rho_m=\frac{K\alpha_\rho}{\rho_0 d_\rho}\Delta(\boldsymbol{\nabla}\cdot\mathbf{u})\,, \tag{4.120}$$

$$\Delta\varphi_e = -\frac{\kappa_E}{\chi_m - \kappa_E \chi_{Em}} \rho_m. \tag{4.121}$$

Under the external action described above, the state of the body is characterized by a central symmetry. Hence, all required functions depend only on the radial coordinate r:

$$\mathbf{u} = (u_r(r), 0, 0), \quad \theta = \theta(r), \quad \varphi_e = \varphi_e(r), \quad \rho_m = \rho_m(r).$$

In the spherical system of coordinates, Eqs. (4.118)–(4.121) take the form

$$\frac{d}{dr}\left[\frac{1}{r^2}\frac{d}{dr}\left(r^2 u_r\right)\right] - \alpha_T' \frac{d\theta}{dr} - \alpha_\rho' \frac{d\rho_m}{dr} = 0, \tag{4.122}$$

$$\frac{d}{dr}\left(r^2 \frac{d\theta}{dr}\right) = 0, \tag{4.123}$$

$$\frac{1}{r^2}\frac{d}{dr}\left(r^2 \frac{d\rho_m}{dr}\right) - \lambda_{\mu E}^2 \rho_m = \frac{K\alpha_\rho}{\rho_0 d_\rho}\frac{1}{r^2}\frac{d}{dr}\left[r^2 \frac{d}{dr}\left(\frac{1}{r^2}\frac{d}{dr}\left(r^2 u_r\right)\right)\right], \tag{4.124}$$

$$\frac{1}{r^2}\frac{d}{dr}\left(r^2 \frac{d\varphi_e}{dr}\right) = -\frac{\kappa_E}{\chi_m - \kappa_E \chi_{Em}} \rho_m. \tag{4.125}$$

Here,

$$\alpha_T' = \frac{K\alpha_T}{K + \frac{4}{3}G}, \quad \alpha_\rho' = \frac{K\alpha_\rho}{K + \frac{4}{3}G}.$$

The set of equations (4.122)–(4.125) should be supplemented with the equation for the electric potential in vacuum (in the regions $0 \le r \le R_1$ and $r > R_2$):

$$\frac{d}{dr}\left(r^2 \frac{d\varphi_{ev}^{(i)}(r)}{dr}\right) = 0 \quad (i = 1, 2). \tag{4.126}$$

The boundary conditions on the body surfaces $r = R_1$ and $r = R_2$ are as follows:

$$\sigma_{rr}(R_i) = 0, \quad \theta(R_i) = \theta_i, \quad \tilde{\mu}_\pi'(R_i) = -\mu_{\pi 0}', \tag{4.127}$$

$$\varphi_e(R_i) = \varphi_{ev}^{(i)}(R_i), \quad D(R_i) = D_v^{(i)}(R_i), \quad i = 1, 2. \tag{4.128}$$

The conditions of boundedness of the electric potential in vacuum such as $r \to 0$ and $r \to +\infty$ take the form

$$\lim_{r \to 0} \varphi_{ev}^{(1)} < \infty, \quad \lim_{r \to +\infty} \varphi_{ev}^{(2)} = 0. \tag{4.129}$$

Here, σ_{rr} is the radial component of the stress tensor; $\varphi_{ev}^{(1)}$ and $\varphi_{ev}^{(2)}$ are, respectively, the electric potentials in the regions $0 \le r < R_1$ and $r > R_2$; $\mathbf{D} = (D(r), 0, 0)$ and $\mathbf{D}_{\mathrm{v}}^{(i)} = \left(D_{\mathrm{v}}^{(i)}(r), 0, 0\right)$ are the vectors of electric induction in the body and in vacuum.

Note that the relations (4.127) verbalize the (i) absence of mechanical traction normal to the surfaces of the body, (ii) presence of temperature perturbations θ_1 and θ_2 specified on the surfaces of the body, and (iii) equality of the potential μ_π' to zero (since the body is in contact with vacuum). Relations (4.128) are the mathematical formulation of the continuity conditions of the electric potential and normal component of the electric induction vector across the interfaces $r = R_1$ and $r = R_2$. Now, we should represent the boundary conditions (4.127) and (4.128) in terms of the required functions: u_r, θ, ρ_m, φ_e, $\varphi_{ev}^{(1)}$, and $\varphi_{ev}^{(2)}$.

In view of the strain–displacement relations (2.8) and the constitutive relation (2.76), the nonzero components of the strain tensors e_{rr}, $e_{\varphi\varphi}$, $e_{\vartheta\vartheta}$ and stresses σ_{rr}, $\sigma_{\varphi\varphi}$, $\sigma_{\vartheta\vartheta}$ satisfy the expressions

$$e_{rr} = \frac{du_r}{dr}, \quad e_{\varphi\varphi} = e_{\vartheta\vartheta} = \frac{u_r}{r}, \tag{4.130}$$

$$\sigma_{rr} = \left(K + \frac{4}{3}G\right)\frac{du_r}{dr} + 2\left(K - \frac{2}{3}G\right)\frac{u_r}{r} - K\alpha_T\theta - K\alpha_\rho\rho_m, \tag{4.131}$$

$$\sigma_{\varphi\varphi} = \sigma_{\vartheta\vartheta} = \left(K - \frac{2}{3}G\right)\frac{du_r}{dr} + 2\left(K + \frac{1}{3}G\right)\frac{u_r}{r} - K\alpha_T\theta - K\alpha_\rho\rho_m. \tag{4.132}$$

Using the constitutive equations (2.24), (2.78), (2.79) and the formulae (2.48), (4.130), the polarization vector $\boldsymbol{\pi}_e = (\pi_e(r), 0, 0)$, the potential $\mu_\pi'(r)$, and the vectors of electric induction in the body $\mathbf{D} = (D(r), 0, 0)$ and in vacuum $\mathbf{D}_{\mathrm{v}}^{(i)} = (D_{\mathrm{v}}^{(i)}(r), 0, 0)$, become

$$\pi_e = -\chi_E\frac{d\varphi_e}{dr} + \chi_{Em}\left[K\frac{\alpha_\rho}{\rho_0}\frac{d}{dr}\left(\frac{du_r}{dr} + 2\frac{u_r}{r}\right) + \beta_{T\rho}\frac{d\theta}{dr} - d_\rho\frac{d\rho_m}{dr}\right], \tag{4.133}$$

$$\mu_\pi' = \mu_{\pi 0}' + d_\rho\rho_m - \beta_{T\rho}\theta - K\frac{\alpha_\rho}{\rho_0}\left(\frac{du_r}{dr} + 2\frac{u_r}{r}\right), \tag{4.134}$$

$$D=-\varepsilon\frac{d}{dr}\left\{\varphi_e+\kappa_E\left[d_\rho\rho_m-\beta_{T\rho}\theta-K\frac{\alpha_\rho}{\rho_0}\left(\frac{du_r}{dr}+2\frac{u_r}{r}\right)\right]\right\}, \quad (4.135)$$

$$D_{\mathrm{v}}^{(i)}=-\varepsilon_0\frac{d\varphi_{ev}^{(i)}}{dr}. \quad (4.136)$$

In view of the relations (4.131), (4.134)–(4.136), the boundary and jump conditions (4.127), (4.128) can be rewritten as

$$\left.\frac{du_r}{dr}\right|_{r=R_i}+2\frac{(3K-2G)}{(3K+4G)}\frac{u_r(R_i)}{R_i}-\alpha'_\rho\rho_m(R_i)=\alpha'_T\theta_i, \quad i=1,2, \quad (4.137)$$

$$\theta(R_i)=\theta_i, \quad i=1,2, \quad (4.138)$$

$$d_\rho\rho_m(R_i)-K\frac{\alpha_\rho}{\rho_0}\left(\left.\frac{du_r}{dr}\right|_{r=R_i}+2\frac{u_r(R_i)}{R_i}\right)=-\mu'_{\pi0}+\beta_{T\rho}\theta_i, \quad (4.139)$$

$$\varphi_e(R_i)=\varphi_{ev}^{(i)}(R_i), \quad i=1,2, \quad (4.140)$$

$$\varepsilon\frac{d}{dr}\left\{\varphi_e+\kappa_E\left[d_\rho\rho_m-\beta_{T\rho}\theta-K\frac{\alpha_\rho}{\rho_0}\left(\frac{du_r}{dr}+2\frac{u_r}{r}\right)\right]\right\}\Bigg|_{r=R_i}$$

$$=\varepsilon_0\left.\frac{d\varphi_{ev}^{(i)}}{dr}\right|_{r=R_i}, \quad i=1,2. \quad (4.141)$$

Note that Eqs. (4.122) and (4.124) are coupled. However, if we replace the space derivatives of the displacement vector on the right-hand side of Eq. (4.124) by the derivatives of temperature and of the specific density of induced mass according to Eq. (4.122) and take into account the heat equation (4.123), then we get the following equation for the function ρ_m:

$$\frac{1}{r^2}\frac{d}{dr}\left(r^2\frac{d\rho_m}{dr}\right)-\lambda_\rho^2\rho_m=0, \quad (4.142)$$

where

$$\lambda_\rho^2=\lambda_{\mu E}^2\left(1-\frac{K^2\alpha_\rho^2}{\rho_0 d_\rho(K+4G/3)}\right)^{-1}.$$

Since

$$\frac{1}{r}\frac{d}{dr}\left(r^2\frac{d\rho_m}{dr}\right)=\frac{d^2(r\rho_m)}{dr^2},$$

we can rewrite Eq. (4.142) in the form

$$\frac{d^2(r\rho_m)}{dr^2} - \lambda_\rho^2 r\rho_m = 0 . \qquad (4.143)$$

4.8.2 Problem Solution and Its Analysis

The problem of determination of the stress–strain state, electric potential, and polarization of a heated hollow sphere is reduced to solving the set of equations (4.122), (4.123), (4.125), (4.126), (4.143) with the boundary and contact conditions (4.137)–(4.141) and the conditions of boundedness of the electric potential in vacuum (4.129). The components of stress tensor and polarization can be determined from the relations (4.131)–(4.133).

We propose the following scheme for the solution to the posed problem. First, we determine the temperature and the specific density of induced mass from Eqs. (4.123) and (4.143). Knowing these functions, we can find the electric potential and displacements from the balance equation (4.122) and from the equations of electrostatics (4.125) and (4.126). In this case, the problem of heat conduction (4.123) and (4.138) is not connected with the determination of the other functions, so that its solution can be given as follows:

$$\theta(r) = \theta_1 - \delta_\theta \frac{R_1 R_2}{R_2 - R_1}\left(\frac{1}{R_1} - \frac{1}{r}\right). \qquad (4.144)$$

Here, $\delta_\theta = \theta_1 - \theta_2$.

Note that Eq. (4.143) for the specific density of induced mass is not connected with the other equations of the governing set. However, the functions ρ_m, u_r, φ_e, and $\varphi_{ev}^{(i)}$ are related through the conditions (4.137), (4.139), and (4.141).

The solution to the set of equations (4.122), (4.125), (4.126), and (4.143) takes the form

$$u_r(r) = \frac{A_3}{3} r + \frac{A_4}{r^2} + \alpha'_T r\left[\frac{\theta_1}{3} + \delta_\theta \frac{R_1 R_2}{R_2 - R_1}\left(\frac{1}{2r} - \frac{1}{3R_1}\right)\right]$$

$$+ \frac{\alpha'_\rho}{\lambda_\rho r}\left\{A_1\left[\operatorname{ch}(\lambda_\rho r) - \frac{\operatorname{sh}(\lambda_\rho r)}{\lambda_\rho r}\right] + A_2\left[\operatorname{sh}(\lambda_\rho r) - \frac{\operatorname{ch}(\lambda_\rho r)}{\lambda_\rho r}\right]\right\}, \qquad (4.145)$$

$$\rho_m(r)=\frac{1}{r}\left[A_1\,\mathrm{sh}(\lambda_\rho r)+A_2\,\mathrm{ch}(\lambda_\rho r)\right], \tag{4.146}$$

$$\varphi_e(r)=\frac{\kappa_E}{r}\left\{\delta_\theta\bar{\beta}_{T\rho}\frac{R_1R_2}{R_2-R_1}-d_\rho(1-K'\alpha'_\rho)\left[A_1\,\mathrm{sh}(\lambda_\rho r)+A_2\,\mathrm{ch}(\lambda_\rho r)\right]\right\}, \tag{4.147}$$

$$\varphi_{ev}^{(1)}(r)=\varphi_*,\quad \varphi_{ev}^{(2)}(r)=0\,. \tag{4.148}$$

Here,

$$\bar{\beta}_{T\rho}=\beta_{T\rho}+\alpha'_T\frac{K\alpha_\rho}{\rho_0},\quad K'=\frac{K\alpha_\rho}{\rho_0 d_\rho},$$

and A_j $(j=\overline{1,4})$ are the constants, determined from the boundary conditions. We now represent these constants as the sum of three terms, namely, $A_j=A_j^m+A_j^\theta+A_j^\delta$ $(j=\overline{1,4})$, where

$$A_i^m=\mu_{\pi 0}a_i^m\left(K'\frac{n_1^mR_1-n_2^mR_2}{N_1R_1-N_2R_2}-\frac{1}{d_\rho}\right)\ (i=1,2),$$

$$A_i^\theta=\theta_1 a_i^m\left(K'\frac{n_1^\theta R_1-n_2^\theta R_2}{N_1R_1-N_2R_2}+\frac{\bar{\beta}_{T\rho}}{d_\rho}\right)\ (i=1,2),$$

$$A_i^\delta=\delta_\theta\left(K'a_i^m\frac{n_1^\delta R_1-n_2^\delta R_2}{N_1R_1-N_2R_2}-a_i^\delta\frac{\bar{\beta}_{T\rho}}{d_\rho}\right)\ (i=1,2),$$

$$A_3^m=\mu_{\pi 0}\frac{n_1^mR_1-n_2^mR_2}{N_1R_1-N_2R_2},\quad A_4^m=\mu_{\pi 0}\frac{R_1R_2(n_1^mN_2-n_2^mN_1)}{\lambda_\rho(N_1R_1-N_2R_2)},$$

$$A_3^\theta=\theta_1\frac{n_1^\theta R_1-n_2^\theta R_2}{N_1R_1-N_2R_2},\quad A_4^\theta=\theta_1\frac{R_1R_2\left(n_1^\theta N_2-n_2^\theta N_1\right)}{\lambda_\rho(N_1R_1-N_2R_2)},$$

$$A_3^\delta=\delta_\theta\frac{n_1^\delta R_1-n_2^\delta R_2}{N_1R_1-N_2R_2},\quad A_4^\delta=\delta_\theta\frac{R_1R_2\left(n_1^\delta N_2-n_2^\delta N_1\right)}{\lambda_\rho(N_1R_1-N_2R_2)},$$

$$C_i=\mathrm{ch}(\lambda_\rho R_i)-\frac{\mathrm{sh}(\lambda_\rho R_i)}{\lambda_\rho R_i},\quad S_i=\mathrm{sh}(\lambda_\rho R_i)-\frac{\mathrm{ch}(\lambda_\rho R_i)}{\lambda_\rho R_i},$$

$$N_i=\frac{K}{4G}\lambda_\rho R_i^2-K'\alpha'_\rho\left(a_1^mC_i+a_2^mS_i\right),$$

$$n_i^m = -\frac{\alpha'_\rho}{d_\rho}\left(a_1^m C_i + a_2^m S_i\right), \quad n_i^\theta = \bar{\beta}_{T\rho}\frac{\alpha'_\rho}{d_\rho}\left(a_1^m C_i + a_2^m S_i\right) + \frac{1}{3}\alpha'_T \lambda_\rho R_i^2,$$

$$n_i^\delta = \alpha'_T \lambda_\rho R_i^2 \frac{R_1 R_2}{\left(R_2 - R_1\right)}\left(\frac{1}{2R_i} - \frac{1}{3R_1}\right) - \bar{\beta}_{T\rho}\frac{\alpha'_\rho}{d_\rho}\left(a_1^\delta C_i + a_2^\delta S_i\right),$$

$$a_1^m = \frac{R_2 \operatorname{ch}(\lambda_\rho R_1) - R_1 \operatorname{ch}(\lambda_\rho R_2)}{\left(1 - K'\alpha'_\rho\right)\operatorname{sh}\left[\lambda_\rho\left(R_2 - R_1\right)\right]}, \quad a_2^m = \frac{R_1 \operatorname{sh}(\lambda_\rho R_2) - R_2 \operatorname{sh}(\lambda_\rho R_1)}{\left(1 - K'\alpha'_\rho\right)\operatorname{sh}\left[\lambda_\rho(R_2 - R_1)\right]},$$

$$a_1^\theta = \theta_1 a_1^m - \delta_\theta a_1^\delta, \quad a_2^\theta = \theta_1 a_2^m + \delta_\theta a_2^\delta,$$

$$a_1^\delta = \frac{R_2 \operatorname{ch}(\lambda_\rho R_1)}{\left(1 - K'\alpha'_\rho\right)\operatorname{sh}\left[\lambda_\rho(R_2 - R_1)\right]}, \quad a_2^\delta = \frac{R_2 \operatorname{sh}(\lambda_\rho R_1)}{\left(1 - K'\alpha'_\rho\right)\operatorname{sh}\left[\lambda_\rho\left(R_2 - R_1\right)\right]}.$$

Here, the superscripts "m," "θ," and "δ" refer, respectively, to the "local mass displacement," "homogeneous temperature," and "gradient of temperature" parts of the corresponding quantities.

If the displacement, electric potential, and specific density of the induced mass are found, then using the formulae (4.131) and (4.132), we obtain the following expressions for stresses:

$$\sigma_{rr}(r) = -4G\alpha'_T\left[\frac{\theta_1}{3} + \delta_\theta \frac{R_1 R_2}{R_2 - R_1}\left(\frac{1}{2r} - \frac{1}{3R_1}\right)\right] + KA_3 - 4G\frac{A_4}{r^3}$$
$$-\frac{4G\alpha'_\rho}{\lambda_\rho r^2}\left\{A_1\left[\operatorname{ch}(\lambda_\rho r) - \frac{\operatorname{sh}(\lambda_\rho r)}{\lambda_\rho r}\right] + A_2\left[\operatorname{sh}(\lambda_\rho r) - \frac{\operatorname{ch}(\lambda_\rho r)}{\lambda_\rho r}\right]\right\}, \tag{4.149}$$

$$\sigma_{\varphi\varphi}(r) = \sigma_{\vartheta\vartheta}(r) = -4G\alpha'_T\left[\frac{\theta_1}{3} + \delta_\theta \frac{R_1 R_2}{R_2 - R_1}\left(\frac{1}{4r} - \frac{1}{3R_1}\right)\right] + KA_3 + 2G\frac{A_4}{r^3}$$
$$-2G\frac{\alpha'_\rho}{r}\left\{A_1\left[\operatorname{sh}(\lambda_\rho r) - \frac{\operatorname{ch}(\lambda_\rho r)}{\lambda_\rho r} + \frac{\operatorname{sh}(\lambda_\rho r)}{(\lambda_\rho r)^2}\right]\right.$$
$$\left.+A_2\left[\operatorname{ch}(\lambda_\rho r) - \frac{\operatorname{sh}(\lambda_\rho r)}{\lambda_\rho r} + \frac{\operatorname{ch}(\lambda_\rho r)}{(\lambda_\rho r)^2}\right]\right\}. \tag{4.150}$$

The electric field and polarization are

$$E(r) = \frac{\kappa_E}{r}\left\langle \bar{\beta}_{T\rho}\frac{R_1 R_2}{R_2 - R_1}\frac{\delta_\theta}{r} + d_\rho \lambda_\rho \left(1 - K'\alpha'_\rho\right)\right.$$

$$\times\left\{A_1\left[\operatorname{ch}(\lambda_\rho r)-\frac{\operatorname{sh}(\lambda_\rho r)}{\lambda_\rho r}\right]+A_2\left[\operatorname{sh}(\lambda_\rho r)-\frac{\operatorname{ch}(\lambda_\rho r)}{\lambda_\rho r}\right]\right\}\Bigg\rangle, \quad (4.151)$$

$$\pi_e(r)=\frac{\chi_E\kappa_E-\chi_{Em}}{r}\Bigg\langle\bar{\beta}_{T\rho}\frac{R_1R_2}{R_2-R_1}\frac{\delta_\theta}{r}+d_\rho\lambda_\rho\left(1-K'\alpha'_\rho\right)$$

$$\times\left\{A_1\left[\operatorname{ch}(\lambda_\rho r)-\frac{\operatorname{sh}(\lambda_\rho r)}{\lambda_\rho r}\right]+A_2\left[\operatorname{sh}(\lambda_\rho r)-\frac{\operatorname{ch}(\lambda_\rho r)}{\lambda_\rho r}\right]\right\}\Bigg\rangle. \quad (4.152)$$

In this case, the vector of electric induction in the body is equal to zero, i.e., $D = 0$.

It is easy to see that the mechanical stresses, specific density of induced mass, electric field, and polarization are linear functions of temperature perturbations nonuniformly distributed along the radial coordinate. Note that the first terms in relations (4.149) and (4.150) are due to the coupling of the processes of deformation and heat conduction. These terms determine the temperature stresses within the framework of classical linear theories. If we set the temperature coefficient of volume dilatation α_T equal to zero, then these terms disappear. The terms proportional to the quantities A_j $\left(j=\overline{1,4}\right)$ appear as a result of taking the local mass displacements into account and its connection with the processes of deformation and heat conduction. These terms determine the inhomogeneity of the distribution of stresses due to the interaction of the indicated processes. The indicated terms are absent in the classical theory of dielectrics.

The functions $f = \{\sigma_{rr}, \sigma_{\varphi\varphi}, \sigma_{\vartheta\vartheta}, E, p\}$ can be represented as the sum of two terms, namely,

$$f = f^m(r) + f^T(r, \theta_1, \delta_\theta),$$

where the components $f^m=\left\{\sigma_{rr}^m,\sigma_{\varphi\varphi}^m,\sigma_{\vartheta\vartheta}^m,E^m,\pi_e^m\right\}$ caused by the structural changes in the near-surface regions of the body (by the local mass displacement) are independent of the temperature field. These components are given by the formulae

$$\sigma_{rr}^m(r)=KA_3^m-4G\frac{A_4^m}{r^3}-\frac{4G\alpha'_\rho}{\lambda_\rho r^2}\left\{A_1^m\left[\operatorname{ch}(\lambda_\rho r)-\frac{\operatorname{sh}(\lambda_\rho r)}{\lambda_\rho r}\right]\right.$$

$$\left.+A_2^m\left[\operatorname{sh}(\lambda_\rho r)-\frac{\operatorname{ch}(\lambda_\rho r)}{\lambda_\rho r}\right]\right\}, \quad (4.153)$$

$$\sigma_{\varphi\varphi}^{m}(r)=\sigma_{\vartheta\vartheta}^{m}(r)=KA_3^m+2G\frac{A_4^m}{r^3}$$
$$-2G\frac{\alpha_\rho'}{r}\left\{A_1^m\left[\mathrm{sh}(\lambda_\rho r)-\frac{\mathrm{ch}(\lambda_\rho r)}{\lambda_\rho r}+\frac{\mathrm{sh}(\lambda_\rho r)}{(\lambda_\rho r)^2}\right]\right.$$
$$\left.+A_2^m\left[\mathrm{ch}(\lambda_\rho r)-\frac{\mathrm{sh}(\lambda_\rho r)}{\lambda_\rho r}+\frac{\mathrm{ch}(\lambda_\rho r)}{(\lambda_\rho r)^2}\right]\right\}, \qquad (4.154)$$

$$E^m(r)=d_\rho\lambda_\rho(1-K'\alpha_\rho')\frac{\kappa_E}{r}\left\{A_1^m\left[\mathrm{ch}(\lambda_\rho r)-\frac{\mathrm{sh}(\lambda_\rho r)}{\lambda_\rho r}\right]\right.$$
$$\left.+A_2^m\left[\mathrm{sh}(\lambda_\rho r)-\frac{\mathrm{ch}(\lambda_\rho r)}{\lambda_\rho r}\right]\right\}, \qquad (4.155)$$

$$\pi_e^m(r)=d_\rho\lambda_\rho(1-K'\alpha_\rho')\frac{\chi_E\kappa_E-\chi_{Em}}{r}\left\{A_1^m\left[\mathrm{ch}(\lambda_\rho r)-\frac{\mathrm{sh}(\lambda_\rho r)}{\lambda_\rho r}\right]\right.$$
$$\left.+A_2^m\left[\mathrm{sh}(\lambda_\rho r)-\frac{\mathrm{ch}(\lambda_\rho r)}{\lambda_\rho r}\right]\right\}. \qquad (4.156)$$

Hence, the terms $f^T(r;\theta_1,\delta_\theta)=\left\{\sigma_{rr}^T,\sigma_{\varphi\varphi}^T,\sigma_{\vartheta\vartheta}^T,E^T,\pi_e^T\right\}$ are presented by the formulae

$$\sigma_{rr}^T(r)=-\frac{4}{3}G\alpha_T'\theta_1-2G\alpha_T'\delta_\theta\frac{R_1R_2}{R_2-R_1}\left(\frac{1}{r}-\frac{2}{3R_1}\right)$$
$$+K\left(A_3^\theta+A_3^\delta\right)-4G\frac{\left(A_4^\theta+A_4^\delta\right)}{r^3}$$
$$-\frac{4G\alpha_\rho'}{\lambda_\rho r^2}\left\{\left(A_1^\theta+A_1^\delta\right)\left[\mathrm{ch}\left(\lambda_\rho r\right)-\frac{\mathrm{sh}(\lambda_\rho r)}{\lambda_\rho r}\right]\right.$$
$$\left.+\left(A_2^\theta+A_2^\delta\right)\left[\mathrm{sh}(\lambda_\rho r)-\frac{\mathrm{ch}(\lambda_\rho r)}{\lambda_\rho r}\right]\right\}, \qquad (4.157)$$

$$\sigma_{\varphi\varphi}^T(r)=\sigma_{\vartheta\vartheta}^T(r)=-\frac{4}{3}G\alpha_T'\theta_1-G\alpha_T'\delta_\theta\frac{R_1R_2}{R_2-R_1}\left(\frac{1}{r}-\frac{4}{3R_1}\right)$$
$$+K\left(A_3^\theta+A_3^\delta\right)+2G\frac{\left(A_4^\theta+A_4^\delta\right)}{r^3}$$

$$-2G\frac{\alpha'_\rho}{r}\left\{\left(A_1^\theta+A_1^\delta\right)\left[\operatorname{sh}(\lambda_\rho r)-\frac{\operatorname{ch}(\lambda_\rho r)}{\lambda_\rho r}+\frac{\operatorname{sh}(\lambda_\rho r)}{(\lambda_\rho r)^2}\right]\right.$$

$$\left.+\left(A_2^\theta+A_2^\delta\right)\left[\operatorname{ch}(\lambda_\rho r)-\frac{\operatorname{sh}(\lambda_\rho r)}{\lambda_\rho r}+\frac{\operatorname{ch}(\lambda_\rho r)}{(\lambda_\rho r)^2}\right]\right\}, \quad (4.158)$$

$$E^T(r)=\frac{\kappa_E}{r}\left\langle\frac{\bar{\beta}_{T\rho}R_1R_2}{(R_2-R_1)}\frac{\delta_\theta}{r}\right.$$

$$+d_\rho\lambda_\rho\left(1-K'\alpha'_\rho\right)\left\{\left(A_1^\theta+A_1^\delta\right)\left[\operatorname{ch}(\lambda_\rho r)-\frac{\operatorname{sh}(\lambda_\rho r)}{\lambda_\rho r}\right]\right.$$

$$\left.\left.+\left(A_2^\theta+A_2^\delta\right)\left[\operatorname{sh}(\lambda_\rho r)-\frac{\operatorname{ch}(\lambda_\rho r)}{\lambda_\rho r}\right]\right\}\right\rangle, \quad (4.159)$$

$$\pi_e^T(r)=\frac{\chi_E\kappa_E-\chi_{Em}}{r}\left\langle\bar{\beta}_{T\rho}\frac{R_1R_2}{(R_2-R_1)}\frac{\delta_\theta}{r}\right.$$

$$+d_\rho\lambda_\rho\left(1-K'\alpha'_\rho\right)\left\{\left(A_1^\theta+A_1^\delta\right)\left[\operatorname{ch}(\lambda_\rho r)-\frac{\operatorname{sh}(\lambda_\rho r)}{\lambda_\rho r}\right]\right.$$

$$\left.\left.+\left(A_2^\theta+A_2^\delta\right)\left[\operatorname{sh}(\lambda_\rho r)-\frac{\operatorname{ch}(\lambda_\rho r)}{\lambda_\rho r}\right]\right\}\right\rangle. \quad (4.160)$$

The components $E^T, \pi_e^T, \sigma_{rr}^T, \sigma_{\varphi\varphi}^T$, and $\sigma_{\vartheta\vartheta}^T$ describe the additional electric field, polarization of the hollow sphere and stresses due to the influence of the temperature field. The first terms in the relations (4.157) and (4.158) along with the terms proportional to A_j^θ $(j=\overline{1,4})$ determine the radial and meridional stresses caused by the constant temperature $\theta = \theta_1$, whereas the second terms in these formulae proportional to the temperature perturbations δ_θ and the terms proportional to A_j^δ $\left(j=\overline{1,4}\right)$ are the temperature stresses induced by the gradient of surface temperature.

Figure 4.24 displays the distributions of stress $\sigma_{\varphi\varphi}/\sigma_{s*}$ along the dimensionless radial coordinate $\zeta = \lambda_\rho r$ in a thin hollow sphere with the following radii: $R_1=10\lambda_\rho^{-1}$ and $R_2=11\lambda_\rho^{-1}$. Here, $\sigma_{s*}=4G\alpha'_T\theta*$, where θ* is a characteristic temperature. The solid lines correspond to the meridional stresses in the absence of temperature gradients in the body ($\delta_\theta = 0$, curve 1) and in the presence of the temperature gradient ($\delta_\theta/\theta^* = -0.1$, curve 2). The numerical calculations are carried out for the following values of the dimensionless parameters: $K/(4G) = 0.5$; $K'\alpha'_\rho = 4\cdot10^{-3}$; $\theta_1/\theta^* = 3.2$; $\overline{\beta}_{T\rho}\alpha_\rho/(d_\rho\alpha_T) = 10^{-8}$; $d_\rho\theta^*\alpha_T/(\mu_{\pi0}\alpha_\rho) = -0.9$.

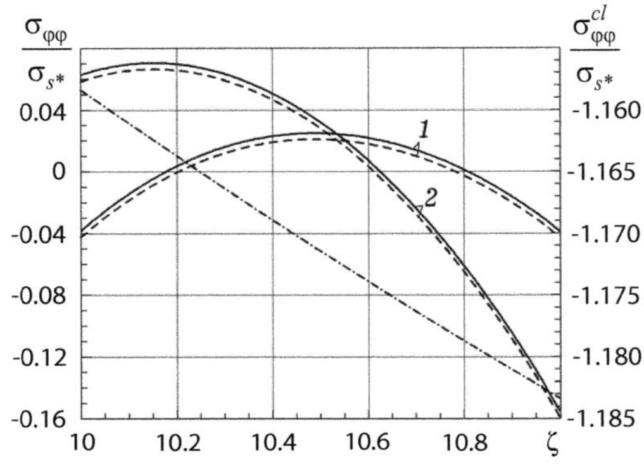

Figure 4.24 Meridional stress distribution along the dimensionless radial coordinate $\zeta = \lambda_\rho r$ for $\delta_\theta = 0$ (curve 1) and $\delta_\theta/\theta^* = -0.1$ (curve 2).

For comparison, the dashed lines show the distributions of stresses in a hollow sphere in the absence of a uniform temperature field ($\theta_1 = 0$). Thus, curve 1 is plotted in the absence of the temperature gradient ($\delta_\theta = 0$), whereas curve 2 corresponds to its presence ($\delta_\theta/\theta^* = -0.1$). The dash-dotted line corresponds to the temperature stress $\sigma^{cl}_{\varphi\varphi}/\sigma_{s*}$ computed according to the classical theory:

$$\sigma^{cl}_{\varphi\varphi}(r) = -4G\alpha'_T\left[\frac{\theta_1}{3} + \delta_\theta\frac{R_1R_2}{R_2 - R_1}\left(\frac{1}{4r} - \frac{1}{3R_1}\right)\right].$$

It is easy to see that the influence of the process of local mass displacement leads to qualitative and quantitative changes in the distribution of stresses in the body. Note that the perturbation of the temperature field θ_1 weakly affects the distribution and the level of meridional stresses in a thick hollow sphere. Moreover, unlike the uniform temperature field, the temperature gradient formed in the body causes significant changes in the stressed state of a hollow sphere.

The curves in Fig. 4.25 illustrate distributions of meridional $\sigma_{\varphi\varphi}/\sigma_{s*}$ (Fig. 4.25a) and radial σ_{rr}/σ_{s*} (Fig. 4.25b) stresses along the

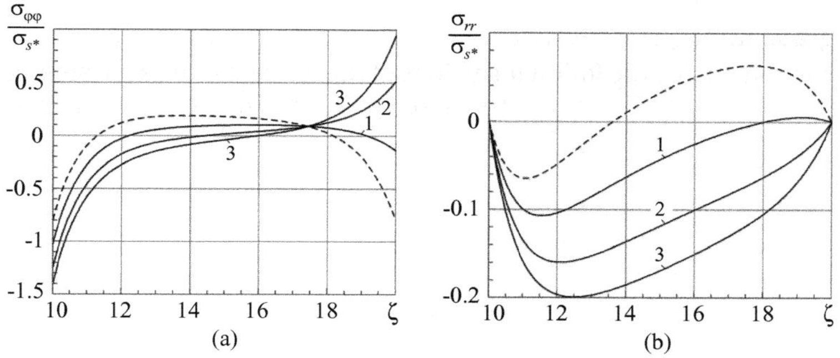

Figure 4.25 Meridional (a) and radial (b) stresses distribution along the dimensionless radial coordinate $\zeta = \lambda_\rho r$ for $\delta_\theta/\theta^* = 0.3, 0.6, 0.8$ (curves 1–3, respectively).

dimensionless radial coordinate ζ in a sphere with the inner and outer radii $R_1 = 10l_*$ and $R_2 = 20l_*$, respectively. The solid lines 1–3 correspond to $\delta_\theta/\theta^* = 0.3, 0.6, 0.8$, and $d_\rho\theta^*\alpha_T/(\mu_{\pi0}\alpha_\rho) = -0.5$. The remaining parameters are the same as in Fig. 4.24. The dashed curves illustrate the inhomogeneity of distributions of meridional and radial stresses in a thin hollow sphere in the absence of temperature perturbations and temperature gradients ($\theta_1 = 0$, $\delta_\theta = 0$). It is easy to see that for a fixed temperature of the inner surface of a sphere, the procedure of cooling of its outer surface may change the character of meridional surface stresses from compressive (curve 1 in Fig. 4.25a) into tensile stresses (curves 2 and 3 in the same figure).

From the relation (4.159), it follows that the procedure of heating of the spherical surfaces up to the temperature θ_1 results in a perturbation of the electric field E^θ:

$$E^\theta(r) = d_\rho\lambda_\rho\left(1 - K'\alpha'_\rho\right)\frac{\kappa_E}{r}\left\{A_1^\theta\left[\operatorname{ch}(\lambda_\rho r) - \frac{\operatorname{sh}(\lambda_\rho r)}{\lambda_\rho r}\right]\right.$$

$$\left. + A_2^\theta\left[\operatorname{sh}(\lambda_\rho r) - \frac{\operatorname{ch}(\lambda_\rho r)}{\lambda_\rho r}\right]\right\},$$

At the same time, the difference in temperatures $\delta_\theta = \theta_1 - \theta_2$ (i.e., temperature gradient) specified on the surfaces of the body perturbs the electric field E^δ:

$$E^{\delta}(r)=\frac{\kappa_E}{r}\left\langle\frac{\bar{\beta}_{T\rho}R_1R_2}{R_2-R_1}\frac{\delta_\theta}{r}+d_\rho\lambda_\rho\left(1-K'\alpha'_\rho\right)\left\{A_1^{\delta}\left[\operatorname{ch}(\lambda_\rho r)\right.\right.\right.$$

$$\left.\left.\left.-\frac{\operatorname{sh}(\lambda_\rho r)}{\lambda_\rho r}\right]+A_2^{\delta}\left[\operatorname{sh}(\lambda_\rho r)-\frac{\operatorname{ch}(\lambda_\rho r)}{\lambda_\rho r}\right]\right\}\right\rangle.$$

The analysis of relation (4.160) shows that the term π_e^T of the nonzero component of the polarization vector is caused by the coupling of the local mass displacement with the processes of polarization, heat conduction, and deformation. In the classical stationary linear theory of isotropic dielectrics, perturbations of the temperature field do not affect the electric field. Hence, the relations of the classical theory fail to describe the pyroelectric and thermopolarization effects. Unlike the classical theory, the relation (4.160) describes the *pyroelectric effect*, i.e., the additional polarization $\pi_e^{\theta}(r;\theta_1)$ of the body caused by the homogeneous heating (in the analyzed case, this is the procedure of heating up the body to the temperature θ_1)

$$\pi_e^{\theta}(r;\theta_1)=d_\rho\lambda_\rho\left(1-K'\alpha'_\rho\right)\frac{\chi_E\kappa_E-\chi_{Em}}{r}$$

$$\times\left\{A_1^{\theta}\left[\operatorname{ch}(\lambda_\rho r)-\frac{\operatorname{sh}(\lambda_\rho r)}{\lambda_\rho r}\right]+A_2^{\theta}\left[\operatorname{sh}(\lambda_\rho r)-\frac{\operatorname{ch}(\lambda_\rho r)}{\lambda_\rho r}\right]\right\},$$

and the *thermopolarization effect*, i.e., the polarization $\pi_e^{\delta}(r;\delta_\theta)$ of the body in the presence of the temperature gradient δ_θ

$$\pi_e^{\delta}(r;\delta_\theta)=\frac{\chi_E\kappa_E-\chi_{Em}}{r}\left\langle\bar{\beta}_{T\rho}\frac{R_1R_2}{R_2-R_1}\frac{\delta_\theta}{r}+d_\rho\lambda_\rho\left(1-K'\alpha'_\rho\right)\right.$$

$$\left.\times\left\{A_1^{\delta}\left[\operatorname{ch}(\lambda_\rho r)-\frac{\operatorname{sh}(\lambda_\rho r)}{\lambda_\rho r}\right]+A_2^{\delta}\left[\operatorname{sh}(\lambda_\rho r)-\frac{\operatorname{ch}(\lambda_\rho r)}{\lambda_\rho r}\right]\right\}\right\rangle.$$

In this case, the bound electric charge with density $\sigma_{es}(R_i)=\rho_0\pi_e(R_i)$ is formed on the surfaces of the body due to polarization:

$$\sigma_{es}(R_i)=\rho_0\frac{\chi_E\kappa_E-\chi_{Em}}{R_i}\left[\bar{\beta}_{T\rho}\frac{R_1R_2}{(R_2-R_1)}\frac{\delta_\theta}{R_i}\right.$$

$$\left.+d_\rho\lambda_\rho\left(1-K'\alpha'_\rho\right)(A_1C_i+A_2S_i)\right].$$

The pyroelectric effect results in the appearance of an additional bound electric charge proportional to the temperature perturbation on the body surfaces $r=R_1$ and $r=R_2$:

$$\sigma_{es}^{\theta}(R_i;\theta_1)=\frac{\chi_E\kappa_E-\chi_{Em}}{R_i}d_\rho\lambda_\rho\left(1-K'\alpha'_\rho\right)\left(A_1^\theta C_i+A_2^\theta S_i\right).$$

Moreover, the thermopolarization effect is responsible for the appearance of an additional bound electric charge proportional to the temperature gradient:

$$\sigma_{es}^{\delta}(R_i;\delta_\theta)=\frac{\chi_E\kappa_E-\chi_{Em}}{R_i}\left[\delta_\theta\frac{\bar{\beta}_{T\rho}R_2R_1}{(R_2-R_1)R_i}\right.$$

$$\left.+d_\rho\lambda_\rho\left(1-K'\alpha'_\rho\right)\left(A_1^\delta C_i+A_2^\delta S_i\right)\right].$$

The curves in Figs. 4.26 and 4.27 display the influence of the thermopolarization effect on the body polarization and on the bound surface charge. The curves in Fig. 4.26 reveal the influence of the wall thickness $\delta_R = \lambda_\rho(R_2 - R_1)$ on the bound surface charge $\sigma_{es}/\sigma_{es}^*$ on the outer surface of the body $r = R_2$ for $\theta_1 = 0$. The curves 1–4 correspond to the values of the temperature gradient $\delta_\theta/\theta^* =$ 0.6, 0.3, −0.3, and −0.6. The remaining parameters are the same as for the curves in Fig. 4.24. Here, $\sigma_{es}^* = \rho_0\pi_{e*}$ and $\pi_{e*} = \mu_{\pi 0}\lambda_\rho(\kappa_E\chi_E - \chi_{Em})$. It is easy to see that as a result of heating or cooling of the outer surface of the sphere, it is possible to change not only the value but also the sign of the surface charge. This effect is more pronounced for small thicknesses characterized by large temperature gradients formed in the body.

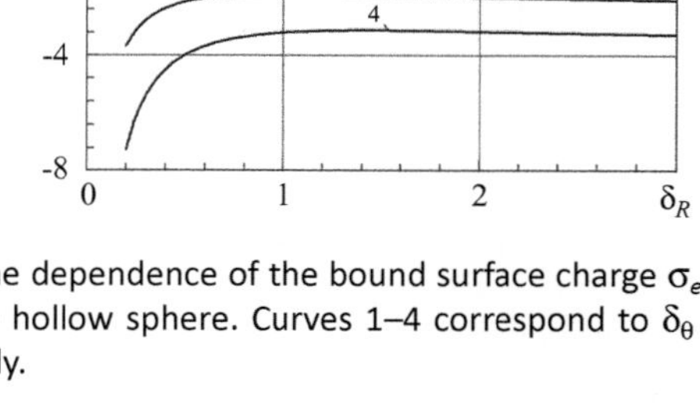

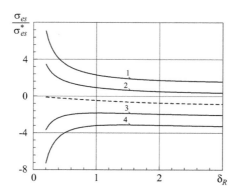

Figure 4.26 The dependence of the bound surface charge $\sigma_{es}(R_2)$ on the wall thickness of the hollow sphere. Curves 1–4 correspond to δ_θ = 0.6, 0.3, −0.3, −0.6, respectively.

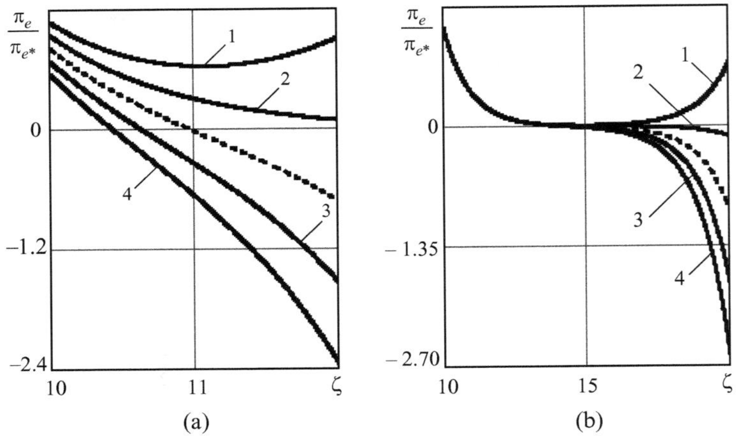

Figure 4.27 Polarization distribution along the wall thickness in a hollow sphere for $R_1 = 10\lambda_p^{-1}$, $\delta_R = 2$ [Fig. (a)], $\delta_R = 10$ [Fig. (b)], and $\delta_\theta = 0.4, 0.2, -0.2, -0.4$ (curves 1–4, respectively).

For a sufficiently large thickness of the spherical wall, in the vicinity of its surfaces, we observe a strong inhomogeneity in the distribution of polarization (Fig. 4.27b), while the regions distant from the surfaces (i.e., the middle part of the wall) remain nonpolarized. In a hollow body with a small wall thickness (Fig. 4.27a), the areas of near-surface inhomogeneities overlap, and thus the entire wall is polarized. The heating or cooling of the outer surface of a hollow spherical body with a small wall thickness leads to a change in the distribution and magnitude of the polarization in the radial direction. In a thick hollow spherical body, the influence of the temperature of the outer surface is mainly manifested in the change in the distribution of polarization near the surface. Dashed lines in Figs. 4.26 and 4.27 correspond to an isothermal approximation. We see that in the case of a fixed temperature of the inner surface of the body, by varying the temperature of the outer surface, it is possible to change not only the mechanical stresses in the body but also the distribution and magnitude of its electric polarization, as well as the magnitude and sign of the surface charge.

4.9 Electrostatic Potential of a Point Charge and a Line Source

4.9.1 Effect of Local Mass Displacement on the Potential Field of a Point Charge

In order to determine the potential field of a point electric charge, we used the relations of the local gradient theory of dielectrics represented in terms of electric φ_e, vector φ_u, and modified chemical $\tilde{\mu}'_\pi$ potentials. If the mass forces are absent, then using the formulae (2.172) and (2.173) for nonideal dielectrics we get the equations

$$\left(\bar{K}+\frac{4}{3}G\right)\Delta\varphi_u = K\frac{\alpha_\rho}{d_\rho}\tilde{\mu}'_\pi, \tag{4.161}$$

$$\Delta\tilde{\mu}'_\pi - \lambda_{\mu E}^2\,\tilde{\mu}'_\pi = \lambda_{\mu E}^2\frac{K\alpha_\rho}{\rho_0}\Delta\varphi_u + \frac{\chi_{Em}}{\varepsilon\left(\chi_m - \kappa_E\chi_{Em}\right)}\rho_e, \tag{4.162}$$

$$\Delta\varphi_e + \kappa_E\Delta\tilde{\mu}'_\pi = -\frac{\rho_e}{\varepsilon}. \tag{4.163}$$

Basing on Eqs. (4.161) and (4.162), for determining the modified chemical potential, we obtain the inhomogeneous Helmholtz equation:

$$\Delta\tilde{\mu}'_\pi - \breve{\lambda}^2\,\tilde{\mu}'_\pi = \frac{\chi_{Em}}{\varepsilon\left(\chi_m - \kappa_E\chi_{Em}\right)}\rho_e, \tag{4.164}$$

where $\quad \breve{\lambda}^2 = \lambda_{\mu E}^2\left(1+\mathfrak{M}\right).$

Applying the operator $\mathbf{L} = \Delta - \breve{\lambda}^2$ to Eq. (4.163), for an electric potential φ_e, we obtain the following fourth-order differential equation

$$\left(\Delta - \breve{\lambda}^2\right)\Delta\varphi_e = -\frac{1}{\varepsilon}\left(\frac{\chi_m}{\chi_m - \kappa_E\chi_{Em}}\Delta\rho_e - \breve{\lambda}^2\rho_e\right). \tag{4.165}$$

It should be noted that within the classical theory of dielectrics, the electric potential satisfies inhomogeneous Laplace equation

$$\Delta\varphi_e = -\frac{\rho_e}{\varepsilon}. \tag{4.166}$$

We can apply the equation of the local gradient theory of dielectrics to study the field of an electric potential of a point electric charge at the origin. Assume that Q is a constant value.

The fundamental solution to the differential equation (4.164) is given in paper (Vladimirov 1971)

$$\tilde{\mu}'_{\pi} = -\frac{\chi_{Em} Q}{4\pi\varepsilon\left(\chi_m - \kappa_E \chi_{Em}\right)} \frac{e^{-\breve{\lambda} R}}{R}, \tag{4.167}$$

where $$R = \sqrt{x^2 + y^2 + z^2}\ .$$

Taking the formula (4.167) and Eq. (4.163) into account, we get

$$\varphi_e = \frac{Q}{4\pi\varepsilon}\left(\frac{1}{R} - \frac{1}{(1-\Theta)} \frac{e^{-\breve{\lambda} R}}{R}\right), \tag{4.168}$$

where Θ and ε are given by

$$\Theta = \frac{\chi_m}{\kappa_E \chi_{Em}}, \quad \varepsilon = \varepsilon_0(1+\chi)\,.$$

Let us now estimate the magnitude of a dimensionless parameter Θ. In view of the formulae (2.100), one may write

$$\Theta = \frac{\varepsilon_0 \chi_m (1+\chi)}{\rho_0 \chi_{Em}^2}. \tag{4.169}$$

Since $\varepsilon_0 = 8.85 \cdot 10^{-12}$ F/m, and for ionic crystals $\chi_{Em} \sim 10^{-16}$ m^2/V, $\rho_0 \sim 10^3$ kg/m^2, $\chi_m \sim 10^{-23}$ s^{-1}, then using the formula (4.169), we get $\Theta << 1$. Thus, one can write

$$\varphi_e \approx \frac{Q\left(1 - e^{-\breve{\lambda} R}\right)}{4\pi\varepsilon R}.$$

Figure 4.28 displays the dependence of the normalized electric potential φ_e/φ_* on the dimensionless coordinate R/R_*. Here, $\varphi_* = Q/4\pi\varepsilon$ and curves 1 and 2 correspond to $\breve{\lambda} R_* = 1$ and $\breve{\lambda} R_* = 2$, respectively. It should be noted that within the classical theory of dielectrics, the potential of a point electric charge is given by

$$\varphi_e = \frac{Q}{4\pi\varepsilon R},$$

and therefore φ_e is a singular function at the origin. We can see that contrary to the classical theory, the local gradient theory of dielectrics allows to avoid the singularity of the electric potential φ_e of a point charge in an infinite dielectric medium. Note also that the curve with the larger value of $\breve{\lambda}$ is closer to the solution obtained within the framework of classical theory.

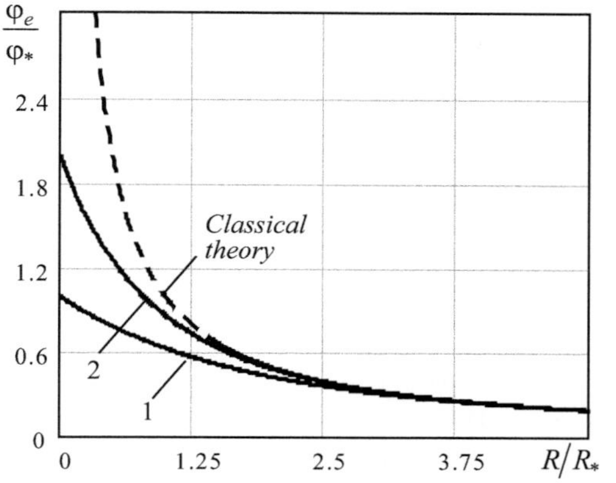

Figure 4.28 Normalized potential field φ_e/φ_* of a point charge.

4.9.2 Effect of Electric Quadrupoles on the Potential Field of a Line Source

In order to examine the electric quadrupoles effect, we now consider the static problem of the potential field of a line source Q at the origin. Within this subsection, we neglect the effect of both the deformation process and local mass displacement on the electric field. Then, using Eq. (3.216), one can obtain the following equation for the electric potential

$$\left[1-\frac{\rho_0}{\varepsilon}\left(2\chi_{q2}-\chi_{q1}\right)\Delta\right]\Delta\varphi_e=-\frac{\rho_e}{\varepsilon}. \tag{4.170}$$

Note that assuming $\chi_{q1} = 0$ and $\chi_{q2} = 0$ from the above equations, we obtain the relation (4.166) of the classical theory of dielectrics.

For line source Q, Eq. (4.170) can be rewritten as

$$\left[-\varepsilon+\rho_0\left(2\chi_{q2}-\chi_{q1}\right)\Delta\right]\Delta\varphi_e=Q\delta(x,y). \tag{4.171}$$

An analytical solution to the static problem (4.171) of the potential field of a line source looks as follows:

$$\varphi_e = -\varphi_{e*}(\ln r + K_0(\lambda_q r)). \tag{4.172}$$

Here, $r = \sqrt{x^2 + y^2}$, and λ_q and φ_{e*} are given by the relations

$$\lambda_q^2 = \frac{\varepsilon}{\rho_0\left(2\chi_{q2} - \chi_{q1}\right)}, \quad \varphi_{e*} = \frac{Q}{2\pi\varepsilon}.$$

Note that the first term, i.e., ln *r*, in the formula (4.172) is the classical solution. We can see that

$$\lim_{r\to 0} \varphi_e = \frac{Q}{4\pi\varepsilon}\ln\frac{\varepsilon}{\rho_0\left(2\chi_{q2} - \chi_{q1}\right)}$$

is a bounded function. Relation (4.172) approaches the classical solution $\varphi_e = -Q\ln r/2\pi\varepsilon$ for large r because $K_0(\lambda_q r) \to 0$ if $r \to +\infty$ (see formula (4.117)).

The normalized potential φ_e/φ_{e*} field distribution along the dimensionless coordinate r/r_* is shown in Fig. 4.29 for $\lambda_q r_* = 1.3, 1.6, 2.1$ (curves 1–3, respectively). We can see that the gradient theory of dielectrics with electric quadrupoles allowed to avoid the singularity of the potential of a linear charge that cannot be achieved within the framework of a classical theory. Note that except the origin, the gradient and the classical solutions show the same electric potential field.

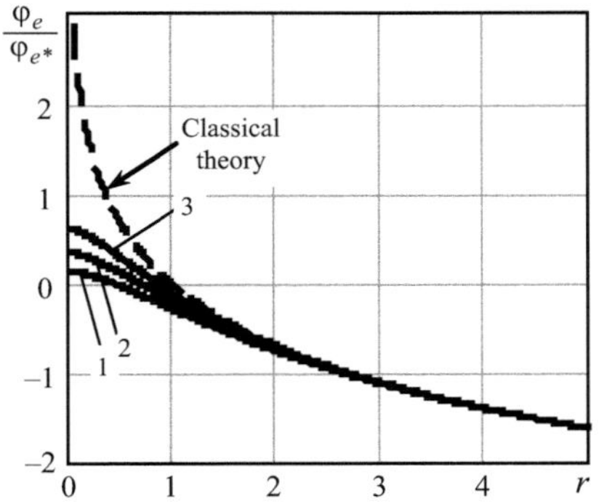

Figure 4.29 Normalized potential field φ_e/φ_{e*} of a line source.

Problems

1. The displacements in an elastic solid due to a concentrated force are known as the fundamental solutions. Write down the Eqs. (2.175), (2.176), and (2.178) for the case of static problems and isothermal approximation. Using these relations, get a fundamental solution for an unbounded isotropic elastic body under the action of concentrated mass force $\mathbf{F} = (\delta(\mathbf{r}), 0, 0)$ applied at the origin. Here, $\delta(\mathbf{r})$ is the Dirac delta function.
2. Using the expressions of differential operators in spherical coordinates (r, φ, ϑ)

$$\nabla a = \frac{\partial a}{\partial r}\mathbf{i}_r + \frac{1}{r}\frac{\partial a}{\partial \vartheta}\mathbf{i}_\vartheta + \frac{1}{r\sin\vartheta}\frac{\partial a}{\partial \varphi}\mathbf{i}_\varphi,$$

$$\nabla\cdot\mathbf{a} = \frac{1}{r^2}\frac{\partial}{\partial r}\left(r^2 a_r\right) + \frac{1}{r\sin\vartheta}\frac{\partial}{\partial \vartheta}\left(a_\vartheta \sin\vartheta\right) + \frac{1}{r\sin\vartheta}\frac{\partial a_\varphi}{\partial \varphi},$$

$$\nabla\times\mathbf{a} = \frac{1}{r\sin\vartheta}\left[\frac{\partial}{\partial \vartheta}\left(a_\varphi \sin\vartheta\right) - \frac{\partial a_\vartheta}{\partial \varphi}\right]\mathbf{i}_r$$
$$+\frac{1}{r}\left[\frac{1}{\sin\vartheta}\frac{\partial a_r}{\partial \varphi} - \frac{\partial\left(r a_\varphi\right)}{\partial r}\right]\mathbf{i}_\vartheta + \frac{1}{r}\left[\frac{\partial (r a_\vartheta)}{\partial r} - \frac{\partial a_r}{\partial \vartheta}\right]\mathbf{i}_\varphi$$

   express the set of governing equations (2.153)–(2.155) in spherical coordinates. Based on it, study the static stress–strain state of an elastic, dielectric, isotropic spherical body with the traction-free surface. Using the obtained solutions, calculate the bound surface charge and surface energy of deformation and polarization.
3. Determine the static stresses, electric potential, and polarization around a spherical cavity in an infinite isotropic media. Using the obtained solution to this boundary-value problem, calculate the surface energy of deformation and polarization. Investigate its dependence on the radius of spherical cavity.
4. Based on Eqs. (2.153)–(2.155), study the effect of stationary temperature field on the near-surface inhomogeneity of the stress–strain state, polarization, and magnitude of the bound

surface charge in an infinite centrosymmetric cubic crystal layer.

5. Get an electric potential field of a point electric charge within the local gradient theory of dielectrics with electric quadrupoles.
6. Following a step similar to Section 4.4, investigate the dependence of the capacitance of a thin circular cylindrical shell capacitor on its thickness. The inner ($r = R_1$) and outer ($r = R_2$) surfaces of the cylindrical body are traction free. A voltage V is applied across the shell thickness.
7. Following a step similar to Section 4.5, study the near-surface inhomogeneity of electromechanical fields in an infinite isotropic dielectric fiber ($|r| \leq R$) with a clamped boundary.

Chapter 5

Stationary Harmonic Wave Processes

Within the framework of isothermal approximation, the effect of local mass displacement on mechanical and electromechanical wave processes is studied in non-ferromagnetic polarized solids. The dispersion properties of the longitudinal plane elastic harmonic wave, Rayleigh wave, and non-plane surface shear wave (SH wave) are analyzed. It is shown that unlike the classical theory, these waves become dispersive in the local gradient theory of dielectrics.

The influence of the polarization inertia on the wave propagation in an isotropic infinite dielectric medium is investigated. It is shown that taking the inertia of polarization into account, the dispersion properties of a modified elastic plane longitudinal wave are manifested in two frequency ranges. In this case, there emerges a frequency range in which the character of the wave process changes. In particular, a plane longitudinal wave of the second kind may arise in a dielectric body. At the same time, there emerges a frequency range in which the wave number associated with the electromagnetic wave becomes imaginary. For the electromagnetic wave, this frequency range may be treated as the cut-off region.

It is shown that contrary to the classical theory of piezoelectricity, the local gradient theory of dielectric solids describes the direct and converse piezoelectric effect in isotropic materials and centrosymmetric cubic crystals in linear approximation.

Local Gradient Theory for Dielectrics: Fundamentals and Applications
Olha Hrytsyna and Vasyl Kondrat

ISBN 978-981-4800-62-4 (Hardcover), 978-1-003-00686-2 (eBook)
www.jennystanford.com

It is also demonstrated that for the linear theory of elasticity, which takes into account the tensor nature of parameters related to the local mass displacement, it is possible to describe the process of propagation of anti-plane surface SH waves in an isotropic half-space. These surface SH waves, however, cannot be predicted by the classical theory of elasticity.

The results presented in this chapter can be useful for a non-destructive testing and evaluation of wave propagation in dielectric materials under high-frequency loads or in the presence of large gradients of an external load.

5.1 Plane Harmonic Wave in an Infinite Medium: Dispersion of an Elastic Wave

It is well known that the classic theory of piezoelectricity does not describe the experimentally observed phenomenon of high-frequency dispersion of a longitudinal elastic wave (Axe et al. 1970). Let us show that such a dispersion in the high-frequency region may be described using the relations of local gradient electroelasticity.

5.1.1 Problem Formulation

Let a time-harmonic plane wave propagate in an isotropic medium of ideal dielectrics along the axis Ox of the Cartesian coordinate system (x, y, z). Assume that the mass forces are absent, i.e., $\mathbf{F}_* = 0$. For this problem, Eqs. (2.195), (2.196), (2.199)–(2.201) can be given as follows:

$$\frac{\partial^2 \underline{\Psi}_u}{\partial X^2} - \left(\frac{v_*}{c_2}\right)^2 \frac{\partial^2 \underline{\Psi}_u}{\partial \tau^2} = 0, \tag{5.1}$$

$$\frac{\partial^2 \underline{\varphi}_u}{\partial X^2} - \left(\frac{v_*}{c_1}\right)^2 \frac{\partial^2 \underline{\varphi}_u}{\partial \tau^2} = \frac{\mathfrak{M}}{\mathfrak{I}} \underline{\tilde{\mu}'_\pi}, \tag{5.2}$$

$$\frac{\partial^2 \underline{\tilde{\mu}'_\pi}}{\partial X^2} - \Lambda^2 \underline{\tilde{\mu}'_\pi} = \Lambda^2 \mathfrak{I} \frac{\partial^2 \underline{\varphi}_u}{\partial X^2}, \tag{5.3}$$

$$\frac{\partial^2 \underline{\varphi}_{e\mu}}{\partial X^2} - \mu_0 \varepsilon v_*^2 \frac{\partial^2 \underline{\varphi}_{e\mu}}{\partial \tau^2} = 0, \tag{5.4}$$

$$\frac{\partial^2 \underline{\mathbf{A}}}{\partial X^2} - \mu_0 \varepsilon v_*^2 \frac{\partial^2 \underline{\mathbf{A}}}{\partial \tau^2} = 0 . \tag{5.5}$$

Here,

$$\Lambda = L^* \lambda_{\mu E}, \quad \underline{\varphi}_{e\mu} = \underline{\varphi}_e + \kappa_E \underline{\tilde{\mu}}'_\pi , \tag{5.6}$$

$c_1 = \sqrt{\left(K + \frac{4}{3}G - \frac{K^2 \alpha_\rho^2}{\rho_0 d_\rho}\right) \Big/ \rho_0}$ is the velocity of propagation of a modified longitudinal elastic wave within the framework of the local gradient theory of elastic dielectrics; $c_2 = \sqrt{G/\rho_0}$ is the velocity of propagation of a transverse wave; $v_* = L^*/t^*$ and $X = x/L^*$ are the characteristic velocity and dimensionless coordinate, respectively; $\boldsymbol{\psi}_u$, $\underline{\mathbf{A}}$, φ_u, $\underline{\varphi}_{e\mu}$, and $\underline{\tilde{\mu}}'_\pi$ are the dimensionless vector and scalar potentials.

5.1.2 Problem Solution and Its Analysis

Basing on the formulae (5.2) and (5.3), we get the following equation to determine the modified longitudinal plane elastic wave:

$$\left(1 - \frac{1}{\Lambda^2} \frac{\partial^2}{\partial X^2}\right) \left[\frac{\partial^2 \underline{\varphi}_u}{\partial X^2} - \left(\frac{v_*}{c_1}\right)^2 \frac{\partial^2 \underline{\varphi}_u}{\partial \tau^2}\right] + \mathfrak{M} \frac{\partial^2 \underline{\varphi}_u}{\partial X^2} = 0 . \tag{5.7}$$

Let us represent the desired solution $f(X, \tau) = \{\varphi_u(X, \tau), \underline{\boldsymbol{\psi}}_u(X, \tau), \underline{\tilde{\mu}}'_\pi(X, \tau), \underline{\varphi}_{e\mu}(X, \tau), \underline{\mathbf{A}}(X, t)\}$ in the form of a plane wave

$$f(X, \tau) \sim e^{-i\kappa X + i\varpi\tau}, \tag{5.8}$$

where $\varpi = \omega t^*$, $\kappa = kL^*$, ω is the circular frequency of the plane wave, and k is the wave number.

Using the representation (5.8) and Eqs. (5.1), (5.4)–(5.7), we obtain the following dispersion equations:

$$\kappa^2 - \left(\frac{v_* \varpi}{c_2}\right)^2 = 0 , \tag{5.9}$$

$$\kappa^4 + \kappa^2 \left[\Lambda^2 (1 + \mathfrak{M}) - \left(\frac{v_* \varpi}{c_1}\right)^2\right] - \left(\Lambda \frac{v_* \varpi}{c_1}\right)^2 = 0 , \tag{5.10}$$

$$\kappa^2 - \mu_0 \varepsilon v_*^2 \varpi^2 = 0. \tag{5.11}$$

The solutions to the set of equations (5.9)–(5.11) for waves propagating in the positive direction of the axis *OX* are

$$\kappa_2 = \frac{v_* \varpi}{c_2}, \tag{5.12}$$

$$\kappa_{1,3} = \left\{ -\frac{1}{2}\left[\Lambda^2 (1+\mathfrak{M}) - \left(\frac{\varpi v_*}{c_1}\right)^2 \right] \right.$$

$$\left. \pm \frac{1}{2}\sqrt{\left[\Lambda^2 (1+\mathfrak{M}) - \left(\frac{\varpi v_*}{c_1}\right)^2 \right]^2 + 4\Lambda^2 \left(\frac{\varpi v_*}{c_1}\right)^2} \right\}^{1/2}, \tag{5.13}$$

$$\kappa_4 = \sqrt{\varepsilon \mu_0}\, v_* \varpi . \tag{5.14}$$

The root (5.12) corresponds to a transverse mechanical wave ($k_2 = \omega/c_2$), and the root (5.14) corresponds to an electromagnetic wave $\left(k_4 = \sqrt{\varepsilon \mu_0}\, \omega\right)$.

In view of relation (5.13) for the dimensional wave number $k_j = \kappa_j / L_*$, $j = 1, 3$, we get

$$k_{1,3} = \sqrt{-\frac{1}{2}\lambda_{\mu E}^2 (1 - \Omega^2 + \mathfrak{M}) \pm \frac{1}{2}\lambda_{\mu E}^2 (1+\Omega^2)\sqrt{1 + \mathfrak{M}\frac{2(1-\Omega^2)+\mathfrak{M}}{(1+\Omega^2)^2}}}\,. \tag{5.15}$$

Here, $\Omega = \omega/(c_1 \lambda_{\mu E})$. Using the infinitesimal character of the parameter $\mathfrak{M}$ in the formula (5.15) and restricting ourselves only to the linear components with respect to $\mathfrak{M}$, we substantially simplify the expressions for the wave numbers k_1 and k_3:

$$k_1 = \frac{\omega}{c_1}\sqrt{1 - \frac{\mathfrak{M}\,\lambda_{\mu E}^2}{\lambda_{\mu E}^2 + \omega^2/c_1^2}} = \frac{\omega}{v_1}, \tag{5.16}$$

$$k_3 = -i\lambda_{\mu E}\sqrt{1 + \frac{\mathfrak{M}\,\lambda_{\mu E}^2}{\lambda_{\mu E}^2 + \omega^2/c_1^2}} = -ik_{30}, \tag{5.17}$$

where

$$v_1 = c_1 \Big/ \sqrt{1 - \frac{\mathfrak{M}\,\lambda_{\mu E}^2}{\lambda_{\mu E}^2 + \omega^2/c_1^2}}, \quad k_{30} = \lambda_{\mu E}\sqrt{1 + \frac{\mathfrak{M}\,\lambda_{\mu E}^2}{\lambda_{\mu E}^2 + \omega^2/c_1^2}},$$

k_{30} is a real quantity, v_1 is the phase velocity of propagation of a wave that can be referred to as a modified elastic longitudinal wave.

The above relations show that the velocity of this wave depends on the frequency. Hence, this wave is dispersive. Since all processes are reversible here, the modified elastic longitudinal wave is undamped. Note that the dispersion of the modified elastic wave is due to the local mass displacement and its coupling with mechanical processes is being taken into account. Dispersion is absent when the coupling factor between these processes is equal to zero, i.e., $\mathfrak{M} = 0$.

Figure 5.1 shows the dependences of the velocity v_1 of a modified elastic longitudinal wave on the frequency Ω. We can see that for low frequencies, the value of v_1/c_1 is practically independent of ω, and for high frequencies, the velocity of the modified elastic wave decreases with frequency. For the characteristics of the material taken here, the decrement attains 6%. The parameter k_{30} has analogous frequency dependence.

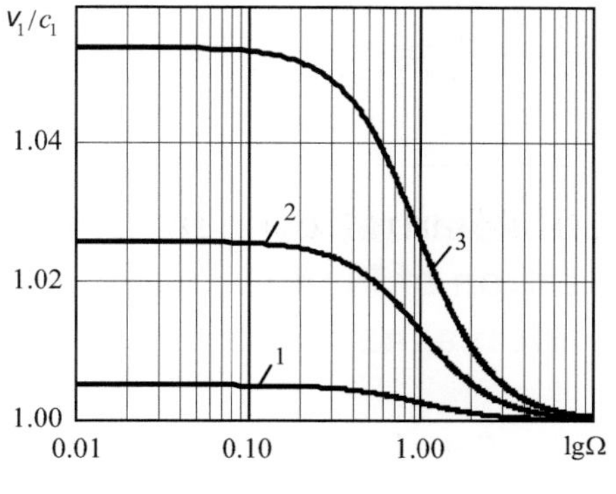

Figure 5.1 The velocity of a modified elastic longitudinal wave versus the frequency (curves 1–3 correspond to $\mathfrak{M}$ = 0.01; 0.05; 0.1).

According to solutions (5.15) of the dispersion equation (5.10), the functions $\underline{g}(X,\tau) = \left\{\underline{\varphi}_u(X,\tau), \underline{\mu}'_\pi(X,\tau)\right\}$ can be represented as the sum

$$\underline{g}(X,\tau) = \left(\underline{g}_1 e^{-i\kappa_1 X} + \underline{g}_3 e^{-i\kappa_3 X}\right) e^{i\varpi\tau}, \tag{5.18}$$

or, in a dimensional form,

$$g(x,t) = \left(g_1 e^{-i\frac{\omega}{v_1}x} + g_3 e^{-k_{30}x}\right) e^{i\omega t}. \tag{5.19}$$

The second summand in expression (5.19) does not describe the wave but just the vibrations of the spatial perturbation at the vicinity of the origin of the coordinate system (Fig. 5.2). The aforementioned vibrations rapidly decay with increasing x-coordinate. Such perturbation will be nearly absent in an infinite medium. For finite bodies, such perturbations are characteristic of near-surface regions.

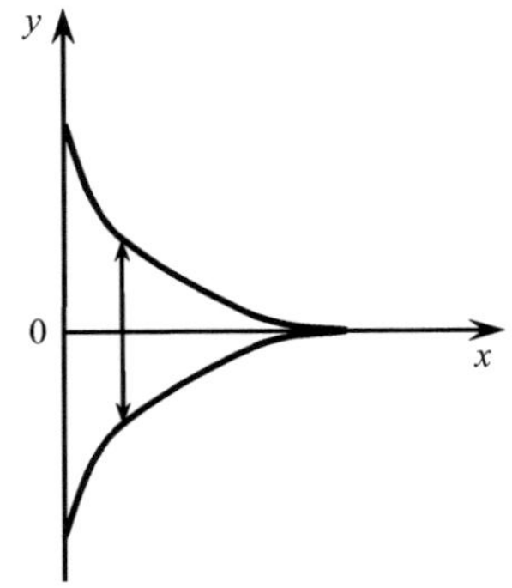

Figure 5.2 Harmonic oscillation of spatial inhomogeneity in the vicinity of the coordinate origin.

5.2 Effect of Polarization Inertia on the Propagation of Plane Waves: Dispersion of an Electromagnetic Wave

5.2.1 Governing Set of Equations: Problem Formulation

Let us study the effect of polarization inertia on the propagation and interaction of wave processes within a non-ferromagnetic isotropic dielectric. To this end, we use the governing set of equations of local gradient electromechanics, which takes the polarization inertia into account. Here we neglect the inertia of the process of local mass displacement, i.e., we assume $d_m = 0$. In this case, operator $\mathbf{L}$ is determined by the formula:

$$\mathbf{L} = 1 + \chi_E d_E \frac{\partial^2}{\partial t^2}. \tag{5.20}$$

Hence, for isothermal approximation, the set of equations (3.156), (3.157), (3.159) looks as follows:

$$G\Delta\boldsymbol{\psi}_u + \rho_0\boldsymbol{\Psi} - \rho_0\frac{\partial^2\boldsymbol{\psi}_u}{\partial t^2} = 0, \tag{5.21}$$

$$\left(K + \frac{4}{3}G - \frac{K^2\alpha_\rho^2}{\rho_0 d_\rho}\right)\Delta\varphi_u + \rho_0\Phi - \rho_0\frac{\partial^2\varphi_u}{\partial t^2} = K\frac{\alpha_\rho}{d_\rho}\tilde{\mu}'_\pi, \tag{5.22}$$

$$\left(1 + \chi_E d_E\frac{\varepsilon_0}{\varepsilon}\frac{\partial^2}{\partial t^2}\right)\left(\Delta\tilde{\mu}'_\pi - \lambda_\mu'^2\tilde{\mu}'_\pi - \lambda_\mu'^2 K\frac{\alpha_\rho}{\rho_0}\Delta\varphi_u\right) + \mathfrak{B}\Delta\tilde{\mu}'_\pi = 0. \tag{5.23}$$

Here,

$$\lambda_\mu'^2 = \frac{\chi_E}{d_\rho\left(\chi_E\chi_m - \chi_{Em}^2\right)} = \frac{\lambda_\mu^2}{1 - \chi_{Em}^2/(\chi_E\chi_m)}, \quad \mathfrak{B} = \frac{\varepsilon_0\chi_{Em}^2}{\varepsilon\left(\chi_m\chi_E - \chi_{Em}^2\right)}. \tag{5.24}$$

Using Eqs. (3.154) and (3.155) in order to determine the generalized scalar φ_{em} and vector $\mathbf{A}$ potentials, we get

$$\mathsf{L}(\Delta\varphi_{em}) - \varepsilon_0\mu_0\left(\mathsf{L} + \frac{\rho_0}{\varepsilon_0}\chi_E\right)\frac{\partial^2\varphi_{em}}{\partial t^2} = 0, \tag{5.25}$$

$$\mathsf{L}(\Delta\mathbf{A}) - \varepsilon_0\mu_0\left(\mathsf{L} + \frac{\rho_0}{\varepsilon_0}\chi_E\right)\frac{\partial^2\mathbf{A}}{\partial t^2} = 0, \tag{5.26}$$

where

$$\varphi_{em} = \varphi_e + \chi_E d_E\frac{\varepsilon_0}{\varepsilon}\frac{\partial^2\varphi_e}{\partial t^2} + \kappa_E\tilde{\mu}'_\pi. \tag{5.27}$$

We should also write the following gauge condition (see formula (3.153))

$$\mathsf{L}(\nabla\cdot\mathbf{A}) + \varepsilon\mu_0\frac{\partial\varphi_{em}}{\partial t} = 0.$$

Note that while neglecting the polarization inertia (i.e., $d_E = 0$), the modified electric potential φ_{em} coincides with the potential $\varphi_{e\mu}$ introduced in Section 2.13 by the formula (2.193).

We use the set of equations (5.21)–(5.23), (5.25), and (5.26) to study the effect of polarization inertia on the propagation of a plane harmonic wave.

Let a plane harmonic wave propagate with frequency ω in an infinite homogeneous dielectric medium in the direction of x-axis.

Assume that the solution to the set of equations (5.21)–(5.23), (5.25), and (5.26) depends only on the spatial coordinate x and time t.

In the absence of mass forces, these equations in a dimensionless form, in terms of the potentials $\underline{\boldsymbol{\psi}}_u$, $\underline{\varphi}_u$, $\underline{\mathbf{A}}$, $\underline{\varphi}_{em}$, and $\underline{\tilde{\mu}}'_{\pi}$, are as follows:

$$\frac{\partial^2 \underline{\boldsymbol{\psi}}_u}{\partial X^2} - \left(\frac{V_*}{c_2}\right)^2 \frac{\partial^2 \underline{\boldsymbol{\psi}}_u}{\partial \tau^2} = 0, \tag{5.28}$$

$$\frac{\partial^2 \underline{\varphi}_u}{\partial X^2} - \left(\frac{V_*}{c_1}\right)^2 \frac{\partial^2 \underline{\varphi}_u}{\partial \tau^2} = \frac{\mathfrak{M}}{\mathfrak{I}} \underline{\tilde{\mu}}'_{\pi}, \tag{5.29}$$

$$\frac{\partial^2 \underline{\tilde{\mu}}'_{\pi}}{\partial X^2} - \frac{\Lambda'^2}{1+\mathfrak{B}} \underline{\tilde{\mu}}'_{\pi} + \frac{\underline{\mathfrak{D}}}{1+\mathfrak{B}} \frac{\partial^4 \underline{\tilde{\mu}}'_{\pi}}{\partial X^2 \partial \tau^2} - \underline{\mathfrak{D}} \Lambda'^2 \frac{\partial^2 \underline{\tilde{\mu}}'_{\pi}}{\partial \tau^2}$$
$$= \Lambda'^2 \mathfrak{I} \frac{\partial^2}{\partial X^2} \left(\underline{\varphi}_u + \underline{\mathfrak{D}} \frac{\partial^2 \underline{\varphi}_u}{\partial \tau^2} \right), \tag{5.30}$$

$$\frac{\partial^2 \underline{\varphi}_{em}}{\partial X^2} - \varepsilon \mu_0 V_*^2 \frac{\partial^2 \underline{\varphi}_{em}}{\partial \tau^2} + \underline{\mathfrak{D}} \frac{\varepsilon}{\varepsilon_0} \frac{\partial^2}{\partial \tau^2} \left(\frac{\partial^2 \underline{\varphi}_{em}}{\partial X^2} - \varepsilon_0 \mu_0 V_*^2 \frac{\partial^2 \underline{\varphi}_{em}}{\partial \tau^2} \right) = 0, \tag{5.31}$$

$$\frac{\partial^2 \underline{\mathbf{A}}}{\partial X^2} - \varepsilon \mu_0 V_*^2 \frac{\partial^2 \underline{\mathbf{A}}}{\partial \tau^2} + \underline{\mathfrak{D}} \frac{\varepsilon}{\varepsilon_0} \frac{\partial^2}{\partial \tau^2} \left(\frac{\partial^2 \underline{\mathbf{A}}}{\partial X^2} - \varepsilon_0 \mu_0 V_*^2 \frac{\partial^2 \underline{\mathbf{A}}}{\partial \tau^2} \right) = 0. \tag{5.32}$$

Here,

$$\Lambda' = L^* \lambda'_{\mu}, \quad \underline{\mathfrak{D}} = \frac{\mathfrak{D}}{t^{*2}}, \quad \mathfrak{D} = \chi_E d_E \frac{\varepsilon_0}{\varepsilon}.$$

5.2.2 Effect of Polarization Inertia on the Propagation of Plane and Electromagnetic Waves

We represent the functions

$$\underline{f}(X,\tau) = \left\{ \underline{\varphi}_u(X,\tau),\ \underline{\boldsymbol{\psi}}_u(X,\tau),\ \underline{\mu}'_{\pi}(X,\tau),\ \underline{\varphi}_{em}(X,\tau),\ \underline{\mathbf{A}}(X,\tau) \right\}$$

according to the formula (5.8). Based on Eqs. (5.28)–(5.32), we obtain the following dispersion equations:

$$\kappa^2 - (V_* \varpi / c_2)^2 = 0, \tag{5.33}$$

$$\kappa^4 + \kappa^2 \left[g\Lambda'^2 (1+\mathfrak{M}) - \left(\frac{V_* \varpi}{c_1} \right)^2 \right] - g \left(\Lambda' \frac{V_* \varpi}{c_1} \right)^2 = 0, \quad (5.34)$$

$$\kappa^2 \left(1 - \underline{\mathfrak{D}} \varpi^2 \frac{\varepsilon}{\varepsilon_0} \right) - \varepsilon \mu_0 (V_* \varpi)^2 (1 - \underline{\mathfrak{D}} \varpi^2) = 0. \quad (5.35)$$

Here,

$$g = \frac{1-\mathfrak{D}\omega^2}{1+\mathfrak{B}-\mathfrak{D}\omega^2} = \frac{1-\underline{\mathfrak{D}}\varpi^2}{1+\mathfrak{B}-\underline{\mathfrak{D}}\varpi^2}. \quad (5.36)$$

Using Eqs. (5.33)–(5.35) for the waves propagating in the positive direction of an *x*-axis, we have the following wave numbers:

$$\kappa_2 = \frac{V_* \varpi}{c_2},$$

$$\kappa_{1,3} = \left\{ -\frac{1}{2} \left[g\Lambda'^2 (1+\mathfrak{M}) - \left(\frac{V_* \varpi}{c_1} \right)^2 \right] \right.$$

$$\left. \pm \frac{1}{2} \sqrt{ \left[g\Lambda'^2 (1+\mathfrak{M}) - \left(\frac{V_* \varpi}{c_1} \right)^2 \right]^2 + 4g\Lambda'^2 \left(\frac{V_* \varpi}{c_1} \right)^2 } \right\}^{1/2},$$

$$\kappa_4 = \sqrt{\varepsilon \mu_0} V_* \varpi \sqrt{\frac{1-\underline{\mathfrak{D}}\varpi^2}{1-\underline{\mathfrak{D}}\varpi^2 \varepsilon/\varepsilon_0}}.$$

In dimensional form, the expressions for wave numbers are as follows:

$$k_2 = \frac{\omega}{c_2}, \quad (5.37)$$

$$k_{1,3} = \left(-\frac{1}{2} \left[g\lambda_\mu'^2 (1+\mathfrak{M}) - \frac{\omega^2}{c_1^2} \right] \right.$$

$$\left. \pm \frac{1}{2} \sqrt{ \left[g\lambda_\mu'^2 (1+\mathfrak{M}) - \frac{\omega^2}{c_1^2} \right]^2 + 4g\lambda_\mu'^2 \frac{\omega^2}{c_1^2} } \right)^{1/2}, \quad (5.38)$$

$$k_4 = \sqrt{\varepsilon \mu_0} \omega \sqrt{\frac{1-\mathfrak{D}\omega^2}{1-\mathfrak{D}\omega^2 \varepsilon/\varepsilon_0}}. \quad (5.39)$$

The phase velocities v_i and the damping factors γ_i $\left(i=\overline{1,4}\right)$ of the waves are determined by formulae

$$v_i = \frac{\omega}{\mathrm{Re}(k_i)}, \quad \gamma_i = \mathrm{Im}(k_i), \ i=\overline{1,4}.$$

As it follows from the formula (5.37), the wave number k_2 corresponds to a transverse elastic wave. This wave is non-damping and non-dispersive. The local mass displacement and electromagnetic processes do not affect the transverse wave.

Due to the infinitesimal nature of the parameter $\mathfrak{M}$, in the relation (5.38), we take only linear by $\mathfrak{M}$ terms into account . Then, for the wave numbers k_1 and k_3, we obtain

$$k_1(\omega)=\frac{\omega}{c_1}\sqrt{1-\frac{\mathfrak{M}g(\omega)}{g(\omega)+\omega^2/\left(c_1\lambda'_\mu\right)^2}},$$

$$k_3(\omega)=\lambda'_\mu\sqrt{-g(\omega)\left(1+\frac{\mathfrak{M}}{g(\omega)+\omega^2/\left(c_1\lambda'_\mu\right)^2}\right)}.$$

In view of the relation (5.36), we can rewrite these formulae as follows:

$$k_1=\frac{\omega}{c_1}\sqrt{1-\frac{\mathfrak{M}}{1+\frac{\omega^2}{c_1^2\lambda'^2_\mu}\left(1+\frac{\mathfrak{B}}{1-\mathfrak{D}\omega^2}\right)}},$$

$$k_3=\lambda'_\mu\sqrt{-\frac{1-\mathfrak{D}\omega^2}{1+\mathfrak{B}-\mathfrak{D}\omega^2}\left[1+\frac{\mathfrak{M}\left(1+\mathfrak{B}-\mathfrak{D}\omega^2\right)}{1-\mathfrak{D}\omega^2+\frac{\omega^2}{c_1^2\lambda'^2_\mu}\left(1+\mathfrak{B}-\mathfrak{D}\omega^2\right)}\right]}.$$

The wave number k_1 describes the modified longitudinal elastic wave of the first order. Modification of the elastic wave is caused by the local mass displacement and its interlinking with the mechanical processes as well as due to the inertial polarization being taken into consideration. This wave is dispersive. However, contrary to the results obtained in the previous section, dispersive properties of the wave manifest themselves in two frequency ranges. For frequencies $\omega > \omega_1$, where $\omega_1 = c_1\lambda'_\mu$, one can observe a decrease in the phase velocity with the frequency increment, which agrees well with the results obtained in Section 5.1 (where the polarization inertia has not been taken into consideration). In the vicinity of frequency $\omega_2 = 1/\sqrt{\mathfrak{D}}$ at $\omega < \omega_2$, when the frequency increases, one can observe the increment of the phase velocity, which attains its maximal value at the frequency $\omega = \omega_2$, while in the region $\omega > \omega_2$, the phase velocity decreases to the reference value. The emergence of this peak of phase velocity is due to the polarization inertia being taken into account.

The wave number k_3 corresponds to the longitudinal wave of the second order. At frequencies $\omega < 1/\sqrt{\mathfrak{D}}$, the wave number does not represent the wave. It represents the inhomogeneities in the vicinity of the coordinate origin. However, contrary to the results obtained in the previous section, the polarization inertia being taken into consideration leads to the appearance of the frequency range $1/\sqrt{\mathfrak{D}} < \omega < \sqrt{(1+\mathfrak{B})/\mathfrak{D}}$, where the wave number k_3 becomes positive and describes a flat undamped wave whose phase velocity essentially depends on the frequency. For frequencies $\omega > \sqrt{(1+\mathfrak{B})/\mathfrak{D}}$, the wave number k_3 again corresponds to the stationary, harmonically oscillating inhomogeneity. Figure 5.3 presents the dependencies of the phase velocity and the damping factor γ_3 of the second-order wave in the corresponding frequency ranges on the dimensionless frequency $\Omega = \omega/\omega_2$.

The wave number k_4 corresponds to the electromagnetic wave. In case we neglect the polarization inertia ($\mathfrak{D} = 0$) from the relation (5.39) for k_4, we can get an expression $k_4 = \sqrt{\varepsilon\mu_0}\,\omega$, which agrees well with the formula (5.14) and corresponds to the undamped non-dispersive electromagnetic wave that propagates with the phase velocity $c = 1/\sqrt{\varepsilon\mu_0}$. The polarization inertia being taken into account leads to the emergence of a frequency range $\sqrt{\varepsilon_0/(\varepsilon\mathfrak{D})} < \omega < 1/\sqrt{\mathfrak{D}}$, in which the wave number k_4 becomes imaginary, which corresponds to the degeneration of the electromagnetic wave into a harmonic oscillation of spatial inhomogeneity in the vicinity of a point $x = 0$. Under the condition $\omega > 1/\sqrt{\mathfrak{D}}$, the wave number k_4, again, corresponds to the electromagnetic wave whose phase velocity at the frequency growth tends to the velocity of the electromagnetic wave propagation in vacuum (see Fig. 5.4).

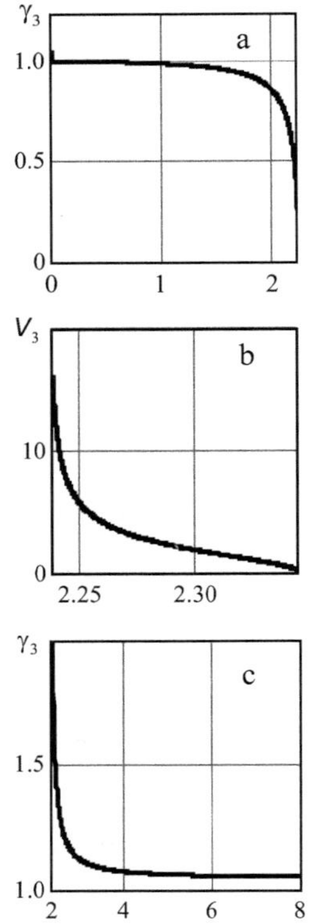

Figure 5.3 Dependence of the damping factor γ_3 (a and c) and the phase velocity V_3 (b) of the second-order longitudinal wave on the frequency.

Thus, the polarization inertia being taken into account leads not only to the changes in the character of dispersion of mechanoelectromagnetic waves but also to the emergence of a frequency range where the character of the wave process changes. At the same time, regarding the electromagnetic wave, this frequency range may be treated as the cut-off region. Surely, this result was attained due to the polarization inertia being taken into consideration. Another consequence of taking the polarization inertia into account is the existence of a frequency range in which the oscillatory process that corresponds to the number k_3 becomes a wave process.

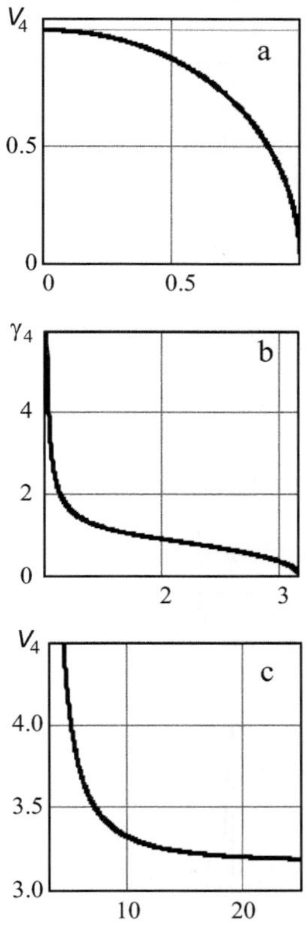

Figure 5.4 Dependence of the phase velocity V_4 (a and c) and the damping factor γ_4 (b) of the electromagnetic wave on the frequency.

5.3 Electromechanical Vibrations of Centrosymmetric Cubic Crystal Layers: Converse Piezoelectric Effect

5.3.1 Problem Formulation

We consider an infinite elastic centrosymmetric cubic crystal layer, which occupies the region $|x| \leq l$ in the Cartesian coordinate system. The surfaces $x = \pm l$ of the dielectric body are traction free and coincide with the faces of the crystals (100). These surfaces contain deposited electrode films. The potential difference that changes according to the time-harmonic voltage $\pm Ve^{i\omega t}$ is applied to these films. We consider these electrode films so thin that their mass should be neglected.

To study the vibrations of an elastic layer under the action of a harmonic electric field, we use the governing set of equations (2.146)–(2.149), constitutive equations (4.17)–(4.20), and the formula (5.40). Note that in the case of a quasi-static electric field, the electric potential is related to the vector of the electric field through the following formula:

$$\mathbf{E} = -\nabla \varphi_e. \tag{5.40}$$

5.3.2 Problem Solution and Its Analysis

For the above external loading, a steady state is considered, where the desired functions are harmonic functions of time and depend on the spatial coordinate x only. Assume that the solution to the problem $f(x,t)=\{u(x,t), \tilde{\mu}'_\pi(x,t), \varphi_e(x,t)\}$ varies in time in the harmonic manner: $f(x,t)=\underset{\sim}{f}(x)e^{i\omega t}$. Here, u is the x-component of the displacement vector. To find the functions $\underset{\sim}{f}(x)=\{\underset{\sim}{u}(x), \underset{\sim}{\tilde{\mu}}'_\pi(x), \underset{\sim}{\varphi}_e x\}$, we have a coupled set of differential equations:

$$\left(\bar{K}+\frac{4}{3}G\right)\frac{d^2\underset{\sim}{u}}{dx^2}-K\frac{\alpha_\rho}{d_\rho}\frac{d\underset{\sim}{\tilde{\mu}}'_\pi}{dx}=-\rho_0\omega^2\underset{\sim}{u}, \tag{5.41}$$

$$\frac{d^2\underset{\sim}{\tilde{\mu}}'_\pi}{dx^2}-\lambda_\mu^2\underset{\sim}{\tilde{\mu}}'_\pi=K\lambda_\mu^2\frac{\alpha_\rho}{\rho_0}\frac{d\underset{\sim}{u}}{dx}-\frac{\chi_{Em}}{\chi_m}\frac{d^2\underset{\sim}{\varphi}_e}{dx^2}, \tag{5.42}$$

$$\frac{d^2}{dx^2}\left(\underset{\sim}{\varphi}_e+\kappa_E\underset{\sim}{\tilde{\mu}}'_\pi\right)=0\cdot \tag{5.43}$$

In Eqs. (5.42) and (5.43), it is taken into account that in the one-dimensional case,

$$E=-\frac{\partial\varphi_e}{\partial x}\cdot \tag{5.44}$$

The boundary conditions of the problem that correspond to the traction-free surfaces $x = \pm l$, to which voltages are applied, can be written as follows:

$$\underset{\sim}{\sigma}_{x\alpha}\Big|_{x=\pm l} = 0, \quad \alpha = \{x, y, z\}, \quad \underset{\sim}{\varphi}_e\Big|_{x=\pm l} = \pm V. \tag{5.45}$$

Following Mindlin (1969), as the third additional boundary condition, we choose the assignment of the electric polarization proportional to its value upon the layer surfaces determined within the framework of the classical theory of dielectrics, that is,

$$\underset{\sim}{\pi}_e\Big|_{x=\pm l} = -k_* \chi \frac{V\varepsilon_0}{l\rho_0} \quad (k_* = \text{const}, 0 \le k_* \le 1). \tag{5.46}$$

Here, k_* is the phenomenological constant introduced earlier by Mindlin (1969). It should be noted that the classical condition is $k_* = 1$ while $k_* = 0$ describes the continuity of polarization across the crystal–electrode interfaces.

We should write down the boundary conditions (5.45) and (5.46) in terms of functions u, $\tilde{\mu}'_\pi$, and φ_e. Taking into account the constitutive equations (4.17)–(4.20), the strain–displacement relations (2.8), and expression (5.44), we obtain

$$\left[\left(\bar{K} + \frac{4}{3}G\right)\frac{d\underset{\sim}{u}}{dx} - K\frac{\alpha_\rho}{d_\rho}\underset{\sim}{\tilde{\mu}}'_\pi\right]\Bigg|_{x=\pm l} = 0, \tag{5.47}$$

$$\left(-\chi_E \frac{d\underset{\sim}{\varphi}_e}{dx} - \chi_{Em}\frac{d\underset{\sim}{\tilde{\mu}}'_\pi}{dx}\right)\Bigg|_{x=\pm l} = -k_* \chi \frac{V\varepsilon_0}{l\rho_0}, \tag{5.48}$$

$$\underset{\sim}{\varphi}_e\Big|_{x=\pm l} = \pm V. \tag{5.49}$$

In view of Eq. (5.43), we can rewrite Eq. (5.42) as follows:

$$\frac{d^2\underset{\sim}{\tilde{\mu}}'_\pi}{dx^2} - \lambda_{\mu E}^2 \underset{\sim}{\tilde{\mu}}'_\pi = \lambda_{\mu E}^2 \frac{K\alpha_\rho}{\rho_0}\frac{d\underset{\sim}{u}}{dx}, \tag{5.50}$$

where $\lambda_{\mu E}^2 = \lambda_\mu^2 \left(1 - \kappa_E \dfrac{\chi_{Em}}{\chi_m}\right)^{-1}$.

Thus, we have to solve Eqs. (5.41), (5.43), and (5.50) subject to the boundary conditions (5.47)–(5.49). This set of equations can be solved consecutively. To determine the displacement $\underset{\sim}{u}$ and modified chemical potential $\tilde{\mu}'_\pi$, we use the coupled equations (5.41) and (5.50). Then, we can find the electric potential $\tilde{\varphi}_e$ from Eq. (5.43) where the function $\tilde{\mu}'_\pi$ will be known.

We represent the solution to the set of differential equations (5.41) and (5.50) in the following exponential form: $e^{-i\zeta x}$. To determine the parameter ζ, we obtain a biquadratic equation

$$\zeta^4 + \left[\lambda_{\mu E}^2 (1+\mathfrak{M}) - \frac{\omega^2}{c_1^2}\right]\zeta^2 - \lambda_{\mu E}^2 \frac{\omega^2}{c_1^2} = 0\,. \tag{5.51}$$

The obtained equation coincides with Eq. (5.10).

Equation (5.51) has four roots: two real and two imaginary ones, i.e.,

$$\zeta_1 = \xi_1, \quad \zeta_2 = -\xi_1, \quad \zeta_3 = i\xi_2, \quad \zeta_4 = -i\xi_2,$$

where

$$\xi_j = \frac{\lambda_{\mu E}}{\sqrt{2}}\sqrt{\sqrt{(1+\mathfrak{M}-\Omega^2)^2 + 4\Omega^2} + (-1)^j (1+\mathfrak{M}-\Omega^2)}, \quad (j=1,2). \tag{5.52}$$

Here, $\Omega = \omega/(c_1\lambda_{\mu E})$. Root ξ_1 corresponds to a modified elastic wave induced by a harmonic electric field, and the root ξ_2 is due to the local mass displacement. The modified elastic wave is undamped since we considered all the processes to be reversible. This wave, however, is dispersive because its phase velocity $c_l = \omega/\xi_1$ is different for low and high frequencies. For $\omega << c_1\lambda_{\mu E}$ (i.e., low frequencies), to find the phase velocity of a mechanical wave, we have the following relation:

$$c_l \to c_1\sqrt{1+\mathfrak{M}} = \sqrt{\frac{K+4/3G}{\rho_0}} = c_{cl}\,.$$

Note that within the framework of classical theory of elasticity, c_{cl} denotes the velocity of propagation of a longitudinal wave in an infinite media (Nowacki 1970). For $\omega >> c_1\lambda_{\mu E}$ (i.e., high frequencies), the phase velocity is $c_l \to c_1$, where $c_1 = \sqrt{\left(K + \frac{4}{3}G - \frac{K^2\alpha_\rho^2}{\rho_0 d_\rho}\right)\Big/\rho_0}\,.$

The analysis of formula (5.52) shows that the consequence of consideration of the local mass displacement is a distinction between the group, $v_g = \dfrac{d\omega}{d\xi_1}$, and phase, $v_{ph} = \dfrac{\omega}{\xi_1}$, velocities of the modified longitudinal elastic wave.

It should be noted that within the classical theory, the elastic wave is non-dispersive, and its group and phase velocities are the same. Therefore, the linear theory of deformation of dielectrics, which takes the local mass displacement into account, in contrast to the classical theory, describes the high-frequency dispersion of an elastic wave and the converse piezoelectric effect in the bodies having a high symmetry of a crystalline lattice.

In view of the problem symmetry, a solution to the set of equations (5.41), (5.43), and (5.50) can be written as follows:

$$\begin{aligned} \underset{\sim}{u}(x) &= \alpha_1 A_1 \cos(\xi_1 x) + \alpha_2 A_2 \operatorname{ch}(\xi_2 x), \\ \underset{\sim}{\tilde{\mu}}'_{\pi}(x) &= A_1 \sin(\xi_1 x) + A_2 \operatorname{sh}(\xi_2 x), \\ \underset{\sim}{\varphi}_e(x) &= -\kappa_E \left[A_1 \sin(\xi_1 x) + A_2 \operatorname{sh}(\xi_2 x) \right] + A_3 x, \end{aligned} \tag{5.53}$$

where

$$\alpha_j(\omega) = \frac{K\alpha_\rho}{\left(\bar{K} + 4G/3\right) d_\rho} \frac{\xi_j}{\left[\omega^2/c_1^2 + (-1)^j \xi_j^2\right]} \quad (j = 1, 2).$$

The constants $A_n \left(n = \overline{1, 3}\right)$ are determined from the following algebraic set of equations:

$$\left(\xi_1 \alpha_1 + \frac{K\alpha_\rho}{\left(\bar{K} + 4G/3\right) d_\rho} \right) \sin(\xi_1 l) A_1 - \left(\xi_2 \alpha_2 - \frac{K\alpha_\rho}{\left(\bar{K} + 4G/3\right) d_\rho} \right) \operatorname{sh}(\xi_2 l) A_2 = 0, \tag{5.54}$$

$$(\kappa_E \chi_E - \chi_{Em}) \xi_1 \cos(\xi_1 l) A_1 + (\kappa_E \chi_E - \chi_{Em}) \xi_2 \operatorname{ch}(\xi_2 l) A_2 - \chi_E A_3 = -k_* \chi \frac{V \varepsilon_0}{l \rho_0}, \tag{5.55}$$

$$-\kappa_E A_1 \sin(\xi_1 l) - \kappa_E A_2 \operatorname{sh}(\xi_2 l) + A_3 l = V. \tag{5.56}$$

The resonance occurs when the determinant of this system is equal to zero. This will happen if the wave frequency, layer thickness, and material characteristics satisfy the following equality:

$$\frac{1-\left(1-\dfrac{\chi_{Em}}{\kappa_E\chi_E}\right)\xi_1 l\,\mathrm{ctg}(\xi_1 l)}{1-\left(1-\dfrac{\chi_{Em}}{\kappa_E\chi_E}\right)\xi_2 l\,\mathrm{cth}(\xi_2 l)}=\frac{1+\xi_2^2\dfrac{c_1^2}{\omega^2}}{1-\xi_1^2\dfrac{c_1^2}{\omega^2}}. \quad (5.57)$$

Since ξ_2 is proportional to the characteristic length (see formula (5.17)), this parameter acquires sufficiently large values (for example, of the order of 10^{10} m^{-1}, for crystals). Therefore, the roots of Eq. (5.57) and the roots of the equation $\sin(\xi_1 l) = 0$ are very close. Hence, we can write $\xi_1 \approx n\pi/l$, where $n = 1, 2, \ldots$. At the same time, the following relation between the frequency and wave number can be written based on Eq. (5.51):

$$\omega^2=\zeta^2c_1^2\left(1-\frac{\mathfrak{M}}{1+\zeta^2/\lambda_{\mu E}^2}\right).$$

Hence, we obtain an expression to determine the resonance frequency for a layer of finite thickness $2l$

$$\omega=n\pi\frac{c_1}{l}\left(1-\frac{\mathfrak{M}}{1+n^2\pi^2/l^2\lambda_{\mu E}^2}\right)^{1/2}, \quad n=1,2,\ldots .$$

Since $\mathfrak{M}$ is a small parameter, we can finally write

$$\omega\approx n\frac{\pi}{l}c_1=n\frac{\pi}{l}\sqrt{\left(K+\frac{4}{3}G-\frac{K^2\alpha_\rho^2}{\rho_0 d_\rho}\right)\Big/\rho_0}, \quad n=1,2,\ldots .$$

Thus, the dependence of a resonance frequency on the layer thickness is inversely proportional. For a layer of thickness 1 sm, the first resonance has a frequency of $6.28 \times 10^2\, c_1$ (s^{-1}). For sodium chloride (NaCl) crystals, this approximately corresponds to 2.9×10^6 s^{-1}.

If the determinant of the set of equations (5.54)–(5.56) is nonzero, then the solutions to Eqs. (5.41), (5.43), and (5.50) that satisfy the boundary conditions (5.47)–(5.49) are as follows:

$$\underset{\sim}{u}(x)=\alpha_1 A\left[\frac{\cos(\xi_1 x)}{\sin(\xi_1 l)}-\frac{\xi_2}{\xi_1}\frac{\mathrm{ch}(\xi_2 x)}{\mathrm{sh}(\xi_2 l)}\right], \quad (5.58)$$

$$\underset{\sim}{\tilde{\mu}'_\pi}(x)=A\left[\frac{\sin(\xi_1 x)}{\sin(\xi_1 l)}-\frac{\alpha_1\xi_2}{\alpha_2\xi_1}\frac{\mathrm{sh}(\xi_2 x)}{\mathrm{sh}(\xi_2 l)}\right], \quad (5.59)$$

$$\underset{\sim}{\varphi}_e(x) = A\kappa_E\left[\left(1-\frac{\alpha_1\xi_2}{\alpha_2\xi_1}\right)\frac{x}{l}-\frac{\sin(\xi_1 x)}{\sin(\xi_1 l)}+\frac{\alpha_1\xi_2}{\alpha_2\xi_1}\frac{\mathrm{sh}(\xi_2 x)}{\mathrm{sh}(\xi_2 l)}\right]+V\frac{x}{l}, \quad (5.60)$$

where A depends on the material characteristics, the layer thickness, and the frequency:

$$A(\omega,l) = \frac{V\chi_E(1-k_*)}{\xi_1 l(\kappa_E\chi_E-\chi_{Em})\left[\mathrm{ctg}(\xi_1 l)-\frac{\alpha_1\xi_2^2}{\alpha_2\xi_1^2}\mathrm{cth}(\xi_2 l)\right]-\kappa_E\chi_E\left(1-\frac{\alpha_1\xi_2}{\alpha_2\xi_1}\right)}.$$

Once the functions $\underset{\sim}{\varphi}_e$ and $\underset{\sim}{\tilde{\mu}}'_\pi$ are found, they can be substituted into a constitutive relation (4.19) in order to calculate the nonzero x-component of the polarization vector:

$$\underset{\sim}{\pi}_e(x) = A\xi_1(\kappa_E\chi_E-\chi_{Em})\left[\frac{\cos(\xi_1 x)}{\sin(\xi_1 l)}-\frac{\alpha_1\xi_2^2}{\alpha_2\xi_1^2}\frac{\mathrm{ch}(\xi_2 x)}{\mathrm{sh}(\xi_2 l)}\right]$$

$$-\frac{\chi_E}{l}\left[V+A\kappa_E\left(1-\frac{\alpha_1\xi_2}{\alpha_2\xi_1}\right)\right]. \quad (5.61)$$

We can see that due to the application of a differential electric potential $\pm Ve^{i\omega t}$ to the layer surfaces, there is a perturbation of mechanical harmonic vibrations $\underset{\sim}{u}(x)e^{i\omega t}$, where $\underset{\sim}{u}(x)$ is defined by the formula (5.58).

The analysis of the obtained relations testified to the fact that the effect of parameter ξ_2 in solid bodies manifests itself only in the narrow near-surface regions of the body. However, with a decreasing thickness of the body, the effect of parameter ξ_2 becomes more appreciable almost within the whole cross section of the layer, i.e., the relations (5.58)–(5.60) describe the size effect. Thus, wave numbers $\zeta_3 = i\xi_2$ and $\zeta_4 = -i\xi_2$ are related to the near-surface inhomogeneity of the fields.

5.3.3 Comparison to the Mindlin Gradient Theory of Dielectrics

Within the Mindlin gradient theory of dielectrics, electromechanical vibrations of centrosymmetric cubic crystal plates are described by the following set of field equations (Mindlin 1971):

$$C_{11}\frac{d^2u}{dx^2}+\rho_0\omega^2u+d_{11}\frac{d^2\Pi_e}{dx^2}=0\,, \tag{5.62}$$

$$d_{11}\frac{d^2u}{dx^2}+b_{11}\frac{d^2\Pi_e}{dx^2}-a_{11}\Pi_e-\frac{d\varphi_e}{dx}=0\,, \tag{5.63}$$

$$\varepsilon_0\frac{d^2\varphi_e}{dx^2}-\frac{d\Pi_e}{dx}=0\,. \tag{5.64}$$

The boundary conditions are as follows:

$$\left(C_{11}\frac{du}{dx}+d_{11}\frac{d\Pi_e}{dx}\right)\bigg|_{x=\pm l}=0\,,\ \ \Pi_e\big|_{x=\pm l}=-\frac{k_*V}{a_{11}l}\,,\ \ \varphi_e\big|_{x=\pm l}=\pm V\,. \tag{5.65}$$

Here, C_{11} is the elastic stiffness; $\varepsilon_0 a_{11}$ is the reciprocal dielectric susceptibility, i.e., $(\varepsilon_0 a_{11})^{-1} = \chi$; b_{11} and d_{11} are associated with the terms, in the energy density, involving polarization gradient, namely, $b_{11}/2$ is the coefficient of a quadratic term and d_{11} is the coefficient of a product of polarization gradient and strain tensor (Mindlin 1968).

We compare the governing set of equations (5.41)–(5.43) with the respective ones (5.62)–(5.64). To this end, we should eliminate both the modified chemical potential $\tilde{\mu}_\pi'$ from the equations (5.41)–(5.43), and the polarization Π_e from the field equations (5.62)–(5.64), taking the displacement u and electric potential φ_e as the key functions in both sets of equations.

From the relation (5.64), we get

$$\frac{d\Pi_e}{dx}=\varepsilon_0\frac{d^2\varphi_e}{dx^2}\,.$$

Substitution of this relation into Eq. (5.62) yields

$$C_{11}\frac{d^2u}{dx^2}+\rho_0\omega^2u+\varepsilon_0 d_{11}\frac{d^3\varphi_e}{dx^3}=0\,. \tag{5.66}$$

Taking into account Eq. (5.63), we can also write

$$\Pi_e=\frac{d_{11}}{a_{11}}\frac{d^2u}{dx^2}+\varepsilon_0\frac{b_{11}}{a_{11}}\frac{d^3\varphi_e}{dx^3}-\frac{1}{a_{11}}\frac{d\varphi_e}{dx}\,. \tag{5.67}$$

Substituting the relation (5.67) into Eq. (5.64), we obtain

$$\frac{d^2}{dx^2}\left[\frac{d^2\varphi_e}{dx^2}-\frac{(1+\varepsilon_0 a_{11})}{\varepsilon_0 b_{11}}\varphi_e+\frac{d_{11}}{b_{11}\varepsilon_0}\frac{du}{dx}\right]=0\,. \tag{5.68}$$

In view of the formulae (5.64) and (5.67), we can write the boundary conditions (5.65) as:

$$\left.\left(C_{11}\frac{du}{dx}+\varepsilon_0 d_{11}\frac{d^2\varphi_e}{dx^2}\right)\right|_{x=\pm l}=0, \tag{5.69}$$

$$\left.\left(\varepsilon_0 b_{11}\frac{d^3\varphi_e}{dx^3}-\frac{d\varphi_e}{dx}+d_{11}\frac{d^2u}{dx^2}\right)\right|_{x=\pm l}=-k_*\frac{V}{l}, \tag{5.70}$$

$$\varphi_e\big|_{x=\pm l}=\pm V. \tag{5.71}$$

The corresponding characteristic equation is as follows (Mindlin 1971):

$$\zeta^4+\frac{\left[C_{11}\left(1+\varepsilon_0 a_{11}\right)-\varepsilon_0 b_{11}\rho_0\omega^2\right]}{\varepsilon_0\left(b_{11}C_{11}-d_{11}^2\right)}\zeta^2-\frac{\rho_0\left(1+\varepsilon_0 a_{11}\right)}{\varepsilon_0\left(b_{11}C_{11}-d_{11}^2\right)}\omega^2=0.$$

Let us continue with the set of equations (5.41)–(5.43). Basing on Eqs. (5.41) and (5.42), we get

$$\tilde{\underset{\sim}{\mu}}'_\pi=-K\frac{\alpha_\rho}{\rho_0}\frac{d\underset{\sim}{u}}{dx}-\frac{1}{\kappa_E\lambda_\mu^2}\left(1-\frac{\kappa_E\chi_{Em}}{\chi_m}\right)\frac{d^2\underset{\sim}{\varphi}_e}{dx^2}. \tag{5.72}$$

Substituting formula (5.72) into Eqs. (5.41) and (5.43), we obtain a governing set of equations of local gradient electromechanics in terms of displacement and electric potential, namely:

$$\left(K+\frac{4}{3}G\right)\frac{d^2\underset{\sim}{u}}{dx^2}+\rho_0\omega^2\underset{\sim}{u}+K\alpha_\rho\left(\frac{\chi_m}{\kappa_E}-\chi_{Em}\right)\frac{d^3\underset{\sim}{\varphi}_e}{dx^3}=0, \tag{5.73}$$

$$\frac{d^2}{dx^2}\left(\frac{d^2\underset{\sim}{\varphi}_e}{dx^2}-\lambda_{\mu E}^2\underset{\sim}{\varphi}_e+\kappa_E\lambda_{\mu E}^2\frac{K\alpha_\rho}{\rho_0}\frac{d\underset{\sim}{u}}{dx}\right)=0. \tag{5.74}$$

The boundary conditions (5.47)–(5.49) will be written as follows:

$$\left.\left[\left(K+\frac{4}{3}G\right)\frac{d\underset{\sim}{u}}{dx}+\frac{K\alpha_\rho}{\kappa_E d_\rho\lambda_{\mu E}^2}\frac{d^2\underset{\sim}{\varphi}_e}{dx^2}\right]\right|_{x=\pm l}=0, \tag{5.75}$$

$$\left.\left(\frac{\chi_{Em}}{\kappa_E\lambda_{\mu E}^2}\frac{d^3\underset{\sim}{\varphi}_e}{dx^3}-\chi_E\frac{d\underset{\sim}{\varphi}_e}{dx}+\chi_{Em}\frac{K\alpha_\rho}{\rho_0}\frac{d^2\underset{\sim}{u}}{dx^2}\right)\right|_{x=\pm l}=-k_*\chi\frac{V\varepsilon_0}{l\rho_0}, \tag{5.76}$$

$$\underset{\sim}{\varphi}_e\Big|_{x=\pm l} = \pm V\,. \tag{5.77}$$

Equations (5.66), (5.68), (5.73), and (5.74) describe the modified longitudinal elastic wave within the framework of the Mindlin gradient theory of dielectrics and the theory of polarized solids that considers the local mass displacement. In comparison with the classical theory of electroelasticity (Nowacki 1983), these equations contain spatial derivatives of higher orders. In particular, Eq. (5.73) contains a derivative of the third order from the electric potential, which proves that the mechanical and electrical fields in centrosymmetric crystals are coupled. Equations (5.73), (5.74) and boundary conditions (5.75)–(5.77) are identical to their counterparts (5.66), (5.68), (5.69)–(5.71) in Mindlin's gradient theory of dielectrics. Comparing Eqs. (5.73), (5.74) to (5.66), (5.68), we can conclude that these sets of equations differ only by coefficients:

$$\lambda_{\mu E}^2 \Leftrightarrow \frac{1+\varepsilon_0 a_{11}}{\varepsilon_0 b_{11}},\ \mathfrak{M} \Leftrightarrow \frac{d_{11}^2}{b_{11}C_{11}-d_{11}^2},\ c_1^2 \Leftrightarrow \frac{b_{11}C_{11}-d_{11}^2}{\rho_0 b_{11}},$$

$$K\alpha_\rho \chi_{Em} \Leftrightarrow \frac{\varepsilon d_{11}}{1+\varepsilon_0 a_{11}},\ \frac{K^2\alpha_\rho^2}{\rho_0 d_\rho} \Leftrightarrow \frac{d_{11}^2}{b_{11}},\ \frac{\chi_m}{\chi_{Em}^2} \Leftrightarrow \frac{\rho_0}{\varepsilon}\left[1+\frac{\varepsilon_0}{\varepsilon}(1+\varepsilon_0 a_{11})\right],$$

$$d_\rho \chi_{Em}^2 \Leftrightarrow \frac{\varepsilon^2 b_{11}}{\rho_0(1+\varepsilon_0 a_{11})^2},\ d_\rho \chi_m \Leftrightarrow \frac{b_{11}}{1+\varepsilon_0 a_{11}}\left(\varepsilon_0+\frac{\varepsilon}{1+\varepsilon_0 a_{11}}\right).$$

5.4 Rayleigh Waves in a Piezoelectric Half-Space: Direct Piezoelectric Effect

Surface acoustic waves, including Rayleigh waves, are widely used in ultrasonic methods of non-destructive testing of mechanical properties of materials (Erofeev 2003; Jagnoux and Vincent 1989). This is due to the fact that the velocity of propagation of such waves, their damping and dispersion properties are closely related to the mechanical characteristics of materials. Within the classical theory of elasticity, the Rayleigh surface waves in homogeneous elastic solids show no dispersion, that is, their phase velocity does not depend on the frequency, including the hypersonic range (Nowacki

1970). However, the velocity of a Rayleigh wave in practice depends on the wavelength and, therefore, on the frequency. It is known that some investigators observed the Rayleigh wave dispersion in thin films and multi-layered structures.

Here, we use the local gradient theory of dielectrics to investigate the dispersion properties of Rayleigh waves and to study a slow electromagnetic wave caused by a direct piezoelectric effect.

5.4.1 Problem Formulation

We consider an isotropic homogeneous elastic half-space, which is an ideal dielectric. Let the half-space be in contact with vacuum. On the traction-free surface, a monochromatic harmonic surface wave propagates with an unknown phase velocity $c_R = \omega/k$, which is parallel to the body surface, where its amplitude attains the maximum value. Here, symbol k denotes the unknown wave number, and ω is the frequency. This is a surface wave, and, therefore, its intensity rapidly decreases with the distance from the surface.

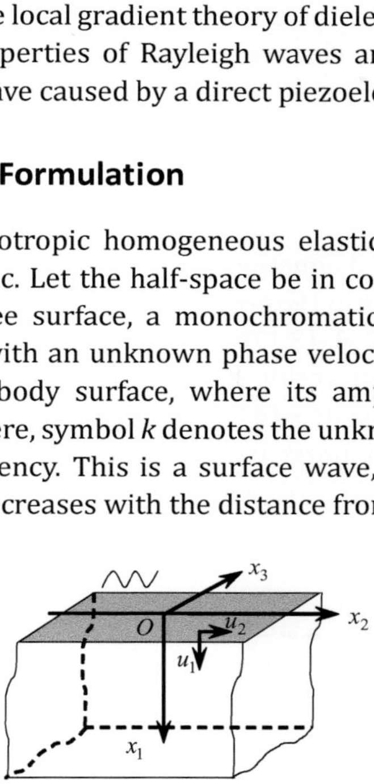

Figure 5.5 An elastic half-space in a plane strain state.

Assume that in a Cartesian coordinate system (x_1, x_2, x_3), the body occupies the region $x_1 \geq 0$, $-\infty < x_i < +\infty$ $(i = 2, 3)$. Let us direct the axis x_2 along the surface in the direction of wave propagation and the axis x_1 be directed vertically downward (see Fig. 5.5). We assume that the body forces vanish, and the external load that caused this wave does not depend on the coordinate x_3.

For such a loading, all quantities related to the medium depend on the Cartesian coordinates x_1 and x_2 and the time t. Hence, a plane strain state is realized in the body, and consequently the displacement vector has only two components $\mathbf{u} = (u_1, u_2, 0)$, and the components of the strain tensor e_{13}, e_{23}, e_{33} are equal to zero. Here, $u_i = (x_1, x_2, t)$, $i = 1, 2$. Since the wave propagates in the dielectric body, a slow electromagnetic wave caused by the piezoelectric effect is additionally induced in the body. In this case, the vectors of the electric field and the specific polarization have the following components: $\mathbf{E} = (E_1, E_2, 0)$ and $\boldsymbol{\pi}_e = (\pi_{e1}, \pi_{e2}, 0)$. As far as the characteristic length of the perturbation of the electric field in vacuum exceeds the characteristic length of the crystal, we assume that the vectors $\mathbf{E}$ and $\boldsymbol{\pi}_e$ are subjected to the Maxwell equations, written in the quasielectrostatics approximation (Maugin 1988), that is

$$\nabla \cdot \mathbf{D} = 0, \quad \nabla \times \mathbf{E} = 0, \quad x_1 > 0. \tag{5.78}$$

The linearized set of equations, describing the motion of the points of a body in the case of the propagation of the Rayleigh surface wave, includes Eqs. (2.175), (2.176) for the scalar and vector potentials of the displacement vector, Eq. (2.178) for the potential $\tilde{\mu}'_\pi$, and equation of electrostatics (5.78). For isothermal approximation, this set of equations can be written as follows:

$$\frac{\partial^2 \psi_u}{\partial x_1^2} + \frac{\partial^2 \psi_u}{\partial x_2^2} - \frac{1}{c_2^2}\frac{\partial^2 \psi_u}{\partial t^2} = 0, \tag{5.79}$$

$$\frac{\partial^2 \varphi_u}{\partial x_1^2} + \frac{\partial^2 \varphi_u}{\partial x_2^2} - \frac{1}{c_1^2}\frac{\partial^2 \varphi_u}{\partial t^2} = \frac{K\alpha_\rho}{\left(\bar{K} + 4G/3\right)d_\rho}\tilde{\mu}'_\pi, \tag{5.80}$$

$$\frac{\partial^2 \tilde{\mu}'_\pi}{\partial x_1^2} + \frac{\partial^2 \tilde{\mu}'_\pi}{\partial x_2^2} - \lambda_{\mu E}^2 \tilde{\mu}'_\pi = \lambda_{\mu E}^2 \frac{K\alpha_\rho}{\rho_0}\left(\frac{\partial^2 \varphi_u}{\partial x_1^2} + \frac{\partial^2 \varphi_u}{\partial x_2^2}\right), \tag{5.81}$$

$$\frac{\partial^2 \varphi_e}{\partial x_1^2} + \frac{\partial^2 \varphi_e}{\partial x_2^2} = -\kappa_E\left(\frac{\partial^2 \tilde{\mu}'_\pi}{\partial x_1^2} + \frac{\partial^2 \tilde{\mu}'_\pi}{\partial x_2^2}\right), \quad x_1 > 0. \tag{5.82}$$

According to the relations (2.172), components u_1 and u_2 of the displacement vector are determined through the potentials φ_u and ψ_u using the following formulae:

$$u_1 = \frac{\partial \varphi_u}{\partial x_1} + \frac{\partial \psi_u}{\partial x_2}, \quad u_2 = \frac{\partial \varphi_u}{\partial x_2} - \frac{\partial \psi_u}{\partial x_1}. \tag{5.83}$$

In order to determine the electric potential in vacuum, we have the equation

$$\frac{\partial^2 \varphi_{ev}}{\partial x_1^2} + \frac{\partial^2 \varphi_{ev}}{\partial x_2^2} = 0, \quad x_1 < 0. \tag{5.84}$$

Here, $\mathbf{E}_v = -\nabla \varphi_{ev}$. The boundary conditions are

$$\sigma_{1j}(0, x_2, t) = 0, \ j = \overline{1,3}, \tag{5.85}$$

$$\tilde{\mu}'_\pi(0, x_2, t) = 0. \tag{5.86}$$

We should present the boundary condition (5.85) in terms of φ_u, ψ_u, and $\tilde{\mu}'_\pi$. To this end, we substitute the formulae (2.8) and (5.83) into the constitutive equation (2.82). As a result, for non-vanishing components σ_{11}, σ_{12}, and σ_{22} of the stress tensor, we obtain

$$\sigma_{11} = \left(\bar{K} + \frac{4}{3}G\right)\frac{\partial^2 \varphi_u}{\partial x_1^2} + \left(\bar{K} - \frac{2}{3}G\right)\frac{\partial^2 \varphi_u}{\partial x_2^2} + 2G\frac{\partial^2 \psi_u}{\partial x_1 \partial x_2} - \frac{K\alpha_\rho}{d_\rho}\tilde{\mu}'_\pi, \tag{5.87}$$

$$\sigma_{12} = G\left(2\frac{\partial^2 \varphi_u}{\partial x_1 \partial x_2} + \frac{\partial^2 \psi_u}{\partial x_2^2} - \frac{\partial^2 \psi_u}{\partial x_1^2}\right), \tag{5.88}$$

$$\sigma_{22} = \left(\bar{K} + \frac{4}{3}G\right)\frac{\partial^2 \varphi_u}{\partial x_2^2} + \left(\bar{K} - \frac{2}{3}G\right)\frac{\partial^2 \varphi_u}{\partial x_1^2} - 2G\frac{\partial^2 \psi_u}{\partial x_1 \partial x_2} - \frac{K\alpha_\rho}{d_\rho}\tilde{\mu}'_\pi. \tag{5.89}$$

Next, employing the relations (5.87) and (5.88), the boundary conditions (5.85) are written in terms of the scalar and vector potentials of the displacement (i.e., φ_u and ψ_u) and modified chemical potential $\tilde{\mu}'_\pi$ as follows:

$$\left[\left(\bar{K} + \frac{4}{3}G\right)\frac{\partial^2 \varphi_u}{\partial x_1^2} + \left(\bar{K} - \frac{2}{3}G\right)\frac{\partial^2 \varphi_u}{\partial x_2^2} + 2G\frac{\partial^2 \psi_u}{\partial x_1 \partial x_2} - \frac{K\alpha_\rho}{d_\rho}\tilde{\mu}'_\pi\right]\Bigg|_{x_1=0} = 0, \tag{5.90}$$

$$\left(2\frac{\partial^2 \varphi_u}{\partial x_1 \partial x_2} + \frac{\partial^2 \psi_u}{\partial x_2^2} - \frac{\partial^2 \psi_u}{\partial x_1^2}\right)\Bigg|_{x_1=0} = 0. \tag{5.91}$$

These conditions are complemented by continuity conditions of the electrical potential and the normal component of the electric field induction vector across the interface $x_1 = 0$:

$$\varphi_e\big|_{x_1=0} = \varphi_{ev}\big|_{x_1=0}, \tag{5.92}$$

$$\varepsilon\left(\frac{\partial \varphi_e}{\partial x_1} + \kappa_E\frac{\partial \tilde{\mu}'_\pi}{\partial x_1}\right)\Bigg|_{x_1=0} = \varepsilon_0\frac{\partial \varphi_{ev}}{\partial x_1}\Bigg|_{x_1=0}. \tag{5.93}$$

Thus, the boundary-value problem is reduced to solving a set of Eqs. (5.79)–(5.82), (5.84) together with the radiation conditions as well as the boundary and jump conditions (5.86), (5.90)–(5.93).

5.4.2 Problem Solution and Its Analysis

We can see that the function ψ_u is determined from Eq. (5.79), which is not connected with the rest of equations of the governing set, while the potentials φ_u and $\tilde{\mu}'_\pi$ are the solutions to the coupled equations (5.80) and (5.81). However, functions ψ_u, φ_u, and $\tilde{\mu}'_\pi$ are interlinked by boundary conditions (5.86), (5.90), and (5.91). The electrical potential in the body and in vacuum is determined based on Eqs. (5.82), (5.84) and conditions of continuity (5.92), (5.93), where function $\tilde{\mu}'_\pi$ is already known. Thus, the formulated problem is solved in two steps.

The solution $f=\{\psi_u,\varphi_u,\tilde{\mu}'_\pi\}$ to the set of equations (5.79)–(5.81) is sought in the following form:

$$f(x_1,x_2,t)=f^*(x_1)e^{-i(\omega t-kx_2)}.$$

Here, $f^*(x_1)=\{\psi^*_u(x_1),\varphi^*_u(x_1),\tilde{\mu}'^*_\pi(x_1)\}$, and k is an unknown constant, which we determine from homogeneous boundary conditions (5.86), (5.90), and (5.91). Basing on Eqs. (5.79)–(5.81), in order to find the functions $f^*(x_1)$, we obtain the following set of ordinary differential equations:

$$\frac{d^2\psi^*_u}{dx_1^2}-v_2^2\psi^*_u=0, \tag{5.94}$$

$$\frac{d^2\varphi^*_u}{dx_1^2}-v_1^2\varphi^*_u=\frac{K\alpha_\rho}{(\bar{K}+4G/3)d_\rho}\tilde{\mu}'^*_\pi, \tag{5.95}$$

$$\frac{d^2\tilde{\mu}'^*_\pi}{dx_1^2}-\left(k^2+\lambda_\mu^2\right)\tilde{\mu}'^*_\pi=\lambda_\mu^2\frac{K\alpha_\rho}{\rho_0}\left(\frac{d^2\varphi^*_u}{dx_1^2}-k^2\varphi^*_u\right). \tag{5.96}$$

Here,

$$v_j^2=k^2-\omega^2/c_j^2,\ j=1,2. \tag{5.97}$$

From the general solutions to Eq. (5.94), we choose the one that corresponds to the reduction in the wave amplitude at the extension of the coordinate x_1. Thus, we can write

$$\overset{*}{\psi}_u(x_1) = a_3 e^{-v_2 x_1}, \tag{5.98}$$

where a_3 is an unknown constant, and $v_2 > 0$.

The solutions to Eqs. (5.95) and (5.96) are of the form

$$\overset{*}{\varphi}_u(x_1) = a e^{\beta x_1}, \quad \tilde{\mu}_\pi'^{*} = \tilde{a} e^{\beta x_1}. \tag{5.99}$$

Here, a and $\tilde{a}$ are unknown constants.

Based on Eqs. (5.95), (5.96), and (5.99), we get the following relation satisfied by β:

$$\beta^4 - \left[2k^2 - \frac{\omega^2}{c_1^2} + \lambda_\mu^2 (1+\mathfrak{M})\right]\beta^2 + \left(k^2 - \frac{\omega^2}{c_1^2}\right)\left(k^2 + \lambda_\mu^2\right) + \mathfrak{M} k^2 \lambda_\mu^2 = 0. \tag{5.100}$$

The roots of this biquadrate characteristic equation are

$$\beta_{1,2}^2 = \frac{1}{2}\left\langle 2k^2 - \frac{\omega^2}{c_1^2} + \lambda_\mu^2 (1+\mathfrak{M}) \right.$$

$$\left. \mp \sqrt{\left[2k^2 - \frac{\omega^2}{c_1^2} + \lambda_\mu^2 (1+\mathfrak{M})\right]^2 - 4\left[k^2\left(k^2 - \frac{\omega^2}{c_1^2}\right) + \lambda_\mu^2\left(k^2(1+\mathfrak{M}) - \frac{\omega^2}{c_1^2}\right)\right]}\right\rangle. \tag{5.101}$$

After some simplifications, relation (5.101) can be rewritten as follows:

$$2\beta_{1,2}^2 = 2k^2 - \frac{\omega^2}{c_1^2} + \lambda_\mu^2 (1+\mathfrak{M}) \mp \left[\frac{\omega^2}{c_1^2} + \lambda_\mu^2 (1+\mathfrak{M})\right]\sqrt{1-2\delta'}, \tag{5.102}$$

where

$$\delta' = \frac{2\mathfrak{M}\lambda_\mu^2 \omega^2 / c_1^2}{\left[\omega^2/c_1^2 + \lambda_\mu^2 (1+\mathfrak{M})\right]^2} \equiv \frac{2\mathfrak{M}\Omega'^2 \bar{\eta}}{\left(1+\mathfrak{M}+\Omega'^2\bar{\eta}\right)^2},$$

$$\bar{\eta} = c_2^2 / c_1^2 < 1, \ \Omega' = \omega/(c_2 \lambda_\mu).$$

Since $\mathfrak{M}$ is a small parameter, as well as $\bar{\eta}<1$ and $\Omega' < 1$ (for crystalline bodies $\lambda_\mu \sim 10^{10}$ m^{-1}), δ' is a small quantity, and thus we can write the formulae (5.102) as follows:

$$\beta_1^2 \approx k^2 - \frac{\omega^2}{c_1^2}(1-\delta') + \lambda_\mu^2(1+\mathfrak{M})\delta' , \tag{5.103}$$

$$\beta_2^2 \approx k^2 + \lambda_\mu^2\left(1 + \frac{\mathfrak{M}}{1+\mathfrak{M}+\Omega'^2\bar{\eta}}\right). \tag{5.104}$$

Note that in the absence of gradient effects (i.e., $\mathfrak{M} = 0$ and $\delta' = 0$), from Eq. (5.103), we obtain a known formula $\beta_1^2 = k^2 - \dfrac{\omega^2}{c_1^2}$ that is consistent with the results of the classical theory of elasticity (Nowacki 1970).

From expressions (5.102), we should choose those roots that correspond to the decreasing wave amplitude with depth. Thus, we get

$$\overset{*}{\varphi}_u(x_1) = a_1 e^{-\beta_1 x_1} + a_2 e^{-\beta_2 x_1}, \quad \tilde{\mu}_\pi'^{*} = \tilde{a}_1 e^{-\beta_1 x_1} + \tilde{a}_2 e^{-\beta_2 x_1}. \tag{5.105}$$

Here, $\beta_j > 0$ ($j = 1, 2$); a_j and $\tilde{a}_j$ ($j = 1, 2$) are the unknown constants.

By substituting the formulae (5.105) into Eqs. (5.95) and (5.96), we find

$$\tilde{a}_j = a_j \frac{d_\rho}{K\alpha_\rho}\left(\bar{K} + \frac{4}{3}G\right)\left(\beta_j^2 - k^2 + \frac{\omega^2}{c_1^2}\right), \quad j = 1, 2.$$

Finally, we present the potentials ψ_u, φ_u, and $\tilde{\mu}_\pi'$ in the following form:

$$\psi_u(x_1, x_2, t) = a_3 e^{-\nu_2 x_1} e^{-i(\omega t - k x_2)}, \tag{5.106}$$

$$\varphi_u(x_1, x_2, t) = \left(a_1 e^{-\beta_1 x_1} + a_2 e^{-\beta_2 x_1}\right) e^{-i(\omega t - k x_2)}, \tag{5.107}$$

$$\tilde{\mu}_\pi'(x_1, x_2, t) = \frac{d_\rho}{K\alpha_\rho}\left(\bar{K} + \frac{4}{3}G\right)\Big[a_1\left(\beta_1^2 - \nu_1^2\right)e^{-\beta_1 x_1} + a_2\left(\beta_2^2 - \nu_1^2\right)e^{-\beta_2 x_1}\Big] e^{-i(\omega t - k x_2)}. \tag{5.108}$$

The amplitudes of longitudinal and transverse vibrations decrease rapidly with depth according to the exponential law with different coefficients of decrease. The local mass displacement influences the rate of decrease in the longitudinal wave amplitude depending on the depth, which is proved by the dependence of parameters β_1 and β_2 on the quantities λ_μ and $\mathfrak{M}$, connected with the local mass displacement. It may seem that the process of local mass displacement did not influence the transverse wave (according to the formula (5.106), potential ψ_u is the same as in the classical theory of elasticity; the expression for ν_2 has not changed as well) (Nowacki 1970). In the boundless space, a transverse wave is not influenced by the local mass displacement. However, in a half-space, transverse and longitudinal waves, as well as the potential $\tilde{\mu}'_\pi$, are interlinked by boundary conditions (5.90), (5.91), and consequently the transverse wave is also indirectly influenced by the local mass displacement due to parameter k. According to the formula (5.107), the amplitude of the potential φ_u is determined by two addends. The first one, being proportional to $e^{-\beta_1 x_1}$, agrees with the results of the theory of elasticity (Nowacki 1970) if we assume $\mathfrak{M} = 0$ (in such an approximation, $\beta_1 = \nu_1$). The presence of the second addend, which is proportional to $e^{-\beta_2 x_1}$, is due to the process of local mass displacement being taken into account. This addend is not considered within the classical theory.

For frequencies $\omega << c_2 \lambda_\mu$ (for sodium chloride, this corresponds to $\omega << 10^9$ Hz), we have the following relations:

$$\beta_1^2 \approx k^2\left(1-\frac{\eta\bar{\eta}}{1+\mathfrak{M}}\right)=k^2\left[1-\eta\bar{\eta}\left(1-\frac{\mathfrak{M}}{1+\mathfrak{M}}\right)\right], \qquad (5.109)$$

$$\beta_2^2 \approx \frac{k^2}{\Omega'^2}\left[\eta(1+\mathfrak{M})+\Omega'^2\left(1-\eta\bar{\eta}\frac{\mathfrak{M}}{1+\mathfrak{M}}\right)\right] \qquad (5.110)$$

to determine β_1^2 and β_2^2. Here, $\eta = c_R^2 / c_2^2$, $c_R = \omega/k$.

Let us evaluate the distances $x_1^{(1)}$ and $x_1^{(2)}$ from the surface, at which the exponents $e^{-\beta_1 x_1}$ and $e^{-\beta_2 x_1}$ decay e times. Thus, we get

$$x_1^{(1)} = \frac{1}{\beta_1} \approx \frac{1}{k}\sqrt{\frac{1+\mathfrak{M}}{1+\mathfrak{M}-\eta\bar{\eta}}}, \qquad x_1^{(2)} = \frac{1}{\beta_2} \approx \frac{\Omega'}{k\sqrt{\eta(1+\mathfrak{M})}}.$$

Hence, $\dfrac{x_1^{(2)}}{x_1^{(1)}} \approx \dfrac{\omega}{\lambda_\mu}\sqrt{\dfrac{1}{c_R^2} - \dfrac{1}{c_1^2}}$, as well as $\eta < (1+\mathfrak{M})/\bar{\eta}$ and consequently $c_R < \sqrt{1+\mathfrak{M}}\, c_1$. The phase velocity of Rayleigh wave is lower than the phase velocity of longitudinal wave. Here, λ_μ is a large quantity (for crystals, it is of the order of 10^9–10^{10} m^{-1}). For low frequencies, $\Omega' << 1$ and $\beta_1 << \beta_2$, that is why $x_1^{(2)} \ll x_1^{(1)}$. Thus, $e^{-\beta_2 x_1}$ drops much faster than $e^{-\beta_1 x_1}$, i.e., the effect of the addend, which is proportional to $e^{-\beta_2 x_1}$, is ponderable only in a narrow near-surface region $0 \le x_1 \ll x_1^{(2)}$. It may be concluded that the frequency increase leads to an increased distance $x_1^{(2)}$.

For frequencies $\omega >> c_2\lambda_\mu$, we obtain

$$\beta_1^2 \approx k^2\left(1 - \eta\bar{\eta} + \frac{\eta\mathfrak{M}}{\Omega'^2}\right), \quad \beta_2^2 \approx k^2\left(1 + \frac{\eta}{\Omega'^2}\right) \tag{5.111}$$

from the formulae (5.101). For NaCl crystals, these frequencies are $\omega >> 10^9$ Hz (i.e., hypersonic waves). At such frequencies, the wavelength of an ultrasound is commensurable with intermolecular distances. In this case, β_1 and β_2 also become commensurable quantities; therefore, for this frequency range, the influence of the local mass displacement on the propagation of Rayleigh waves cannot be neglected.

Substituting the formulae (5.106)–(5.108) into boundary conditions (5.86), (5.90), and (5.91), we obtain the following linear homogeneous set of algebraic equations to determine the unknown amplitudes a_j, $j = \overline{1,3}$,

$$\left[2Gk^2 - \left(\bar{K} + \frac{4}{3}G\right)\frac{\omega^2}{c_1^2}\right]a_1 + \left[2Gk^2 - \left(\bar{K} + \frac{4}{3}G\right)\frac{\omega^2}{c_1^2}\right]a_2 - 2ikv_2Ga_3 = 0, \tag{5.112}$$

$$2ik\beta_1 a_1 + 2ik\beta_2 a_2 + \left(k^2 + v_2^2\right)a_3 = 0, \tag{5.113}$$

$$\left(\beta_1^2 - v_1^2\right)a_1 + \left(\beta_2^2 - v_1^2\right)a_2 = 0. \tag{5.114}$$

This set of equations has a nontrivial solution if and only if the determinant of the matrix is equal to zero:

$$4k^2\nu_2\left(\beta_1\beta_2+k^2-\frac{\omega^2}{c_1^2}\right)=\left(2k^2-\frac{\omega^2}{c_2^2}\right)^2(\beta_1+\beta_2). \qquad (5.115)$$

Note that in order to find the wave number k, we obtained here a more complicated equation comparing to the ones in the linear classic theory of elasticity (Nowacki 1970):

$$4k^2\nu_2\nu_1=\left(2k^2-\frac{\omega^2}{c_2^2}\right)^2,$$

where $\nu_1=k^2-\omega^2/c_{cl}^2$, and $c_{cl}^2=(K+4/3G)/\rho_0$.

If the right-hand and the left-hand parts of the relation (5.115) are twice squared, taking into consideration that

$$\beta_1^2\beta_2^2=k^4\left[1-\bar{\eta}\eta+\frac{\eta}{\Omega'^2}(1+\mathfrak{M}-\bar{\eta}\eta)\right],$$

$$\beta_1^2+\beta_2^2=k^2\left[2-\bar{\eta}\eta+\frac{\eta}{\Omega'^2}(1+\mathfrak{M})\right],$$

we obtain the following algebraic equation to find the unknown parameter η:

$$\left\{\left(1-\frac{1}{2}\eta\right)^4\left[2-\bar{\eta}\eta+\frac{\eta}{\Omega'^2}(1+\mathfrak{M})\right]\right.$$

$$\left.-(1-\eta)\left[1-\bar{\eta}\eta+\frac{\eta}{\Omega'^2}(1+\mathfrak{M}-\bar{\eta}\eta)+(1-\bar{\eta}\eta)^2\right]\right\}^2$$

$$=4\left[1-\bar{\eta}\eta+\frac{\eta}{\Omega'^2}(1+\mathfrak{M}-\bar{\eta}\eta)\right]\left[(1-\eta)(1-\bar{\eta}\eta)-\left(1-\frac{1}{2}\eta\right)^4\right]^2.$$

(5.116)

Equation (5.116) is a dispersion relation for the motion of Rayleigh surface waves in an elastic homogeneous isotropic half-space. This equation is quite complicated. In case we know the values of the parameters $\bar{\eta}$, Ω', and $\mathfrak{M}$, the roots of this equation may be found using numerical methods. From the found real roots, we should choose only those satisfying the conditions $\nu_2>0$ and $\beta_j>0, j=1,2$. We can see that Eq. (5.116) contains a dimensionless parameter

Ω', which is proportional to the frequency ω. Thus, we conclude that if we take the local mass displacement into consideration, the velocity of the Rayleigh wave will depend on the frequency, which means that this wave is dispersive. It should be noted that if we let the parameters $\mathfrak{M}$ and Ω' in the relation (5.116) to be tending to zero, we can get a dispersive equation from the classical theory of elasticity.

Let us consider a partial case of an incompressible medium for which $c_1^2 = \infty$, and hence $\bar{\eta} = 0$. Assuming in the dispersion equation (5.116) $\bar{\eta} = 0$, we obtain

$$(1+\mathfrak{M})^2 \Omega'^{-4} \eta^3 (\eta^3 - 8\eta^2 + 24\eta - 16) = 0. \qquad (5.117)$$

Equation (5.117) agrees with the results presented by Nowacki (1970), who studied the propagation of the Rayleigh waves based on the relations of the classical theory of elasticity. Thus, in an incompressible medium, a longitudinal elastic wave is non-dispersive. One could expect this result since it was shown in Subsection 2.10.4 that the local mass displacement is connected with the processes of compression and stretching and does not depend on the shear deformation. That is why the effect of the local mass displacement on the mechanical fields in an incompressible medium is absent altogether.

The set of equations (5.112)–(5.114) yields

$$a_3 = -2ik \frac{\left(\beta_1\beta_2 + \nu_1^2\right)(\beta_2 - \beta_1)}{\left(\beta_2^2 - \nu_1^2\right)\left(k^2 + \nu_2^2\right)} a_1, \quad a_2 = -\frac{\beta_1^2 - \nu_1^2}{\beta_2^2 - \nu_1^2} a_1. \qquad (5.118)$$

Taking the formulae (5.106)–(5.108) and (5.118) into account, we obtain the following expressions to find the functions ψ_u, φ_u, and $\tilde{\mu}'_\pi$:

$$\psi_u(x_1, x_2, t) = -2ika_1 \frac{\left(\beta_1\beta_2 + \nu_1^2\right)(\beta_2 - \beta_1)}{\left(\beta_2^2 - \nu_1^2\right)\left(k^2 + \nu_2^2\right)} e^{-\nu_2 x_1} e^{-i(\omega t - kx_2)}, \qquad (5.119)$$

$$\varphi_u(x_1, x_2, t) = a_1 \left(e^{-\beta_1 x_1} - \frac{\beta_1^2 - \nu_1^2}{\beta_2^2 - \nu_1^2} e^{-\beta_2 x_1} \right) e^{-i(\omega t - kx_2)}, \qquad (5.120)$$

$$\tilde{\mu}'_\pi(x_1, x_2, t) = \frac{a_1 d_\rho}{K\alpha_\rho} \left(\bar{K} + \frac{4}{3} G \right) \left(\beta_1^2 - \nu_1^2\right) \left(e^{-\beta_1 x_1} - e^{-\beta_2 x_1}\right) e^{-i(\omega t - kx_2)}.$$

(5.121)

Formulae (5.121) makes it possible to determine the potentials with the accuracy up to a constant a_1. This constant may be determined having specified the reason for the emergence of a surface wave.

Using the relations (5.83), (5.119), and (5.120), the components of a displacement vector can be written as

$$u_1 = a_1\left[2k^2\frac{\left(\beta_1\beta_2+v_1^2\right)\left(\beta_2-\beta_1\right)}{\left(\beta_2^2-v_1^2\right)\left(k^2+v_2^2\right)}e^{-v_2x_1}-\beta_1e^{-\beta_1x_1}\right.$$

$$\left.+\beta_2\frac{\beta_1^2-v_1^2}{\beta_2^2-v_1^2}e^{-\beta_2x_1}\right]\cos\left[\omega\left(t-\frac{x_2}{c}\right)\right],$$

$$u_2 = ka_1\left[e^{-\beta_1x_1}-2v_2\frac{\left(\beta_1\beta_2+v_1^2\right)\left(\beta_2-\beta_1\right)}{\left(\beta_2^2-v_1^2\right)\left(k^2+v_2^2\right)}e^{-v_2x_1}\right.$$

$$\left.-\frac{\beta_1^2-v_1^2}{\beta_2^2-v_1^2}e^{-\beta_2x_1}\right]\sin\left[\omega\left(t-\frac{x_2}{c}\right)\right].$$

The propagation of a surface mechanical wave in a solid dielectric body causes perturbation of the electric field. In case we know the function $\tilde{\mu}'_\pi$, we can determine the perturbation of the electric field due to the piezoelectric effect.

Using Eqs. (5.82), (5.84), (5.92), and (5.93), we get the following formulae in order to define the components of the electric field and polarization vector

$$E_i(x_1,x_2,t)=\kappa_E\frac{\partial\tilde{\mu}'_\pi}{\partial x_i},\quad \pi_{ei}(x_1,x_2,t)=-\chi_{Em}\frac{\varepsilon_0}{\varepsilon}\frac{\partial\tilde{\mu}'_\pi}{\partial x_i},\quad i=1,2.$$

Here from, in view of the formula (5.121), we have

$$E_1(x_1,x_2,t)=-\kappa_E\frac{a_1d_\rho}{K\alpha_\rho}\left(\bar{K}+\frac{4}{3}G\right)$$
$$\times\left(\beta_1^2-v_1^2\right)\left(\beta_1e^{-\beta_1x_1}-\beta_2e^{-\beta_2x_1}\right)e^{-i(\omega t-kx_2)}, \qquad (5.122)$$

$$E_2(x_1,x_2,t)=ik\kappa_E\frac{a_1d_\rho}{K\alpha_\rho}\left(\bar{K}+\frac{4}{3}G\right)$$
$$\times\left(\beta_1^2-v_1^2\right)\left(e^{-\beta_1x_1}-e^{-\beta_2x_1}\right)e^{-i(\omega t-kx_2)}, \qquad (5.123)$$

$$\rho_0\pi_{e1}(x_1,x_2,t)=\kappa_E\varepsilon_0\frac{a_1d_\rho}{K\alpha_\rho}\left(\bar{K}+\frac{4}{3}G\right)$$
$$\times\left(\beta_1^2-v_1^2\right)\left(\beta_1e^{-\beta_1x_1}-\beta_2e^{-\beta_2x_1}\right)e^{-i(\omega t-kx_2)},\ (5.124)$$

$$\rho_0\pi_{e2}(x_1,x_2,t)=-ik\kappa_E\varepsilon_0\frac{a_1d_\rho}{K\alpha_\rho}\left(\bar{K}+\frac{4}{3}G\right)$$
$$\times\left(\beta_1^2-v_1^2\right)\left(e^{-\beta_1x_1}-e^{-\beta_2x_1}\right)e^{-i(\omega t-kx_2)}. \qquad (5.125)$$

We can see that $\rho_0\pi_{ei}(x_1, x_2, t) = -\varepsilon_0 E_i(x_1, x_2, t)$; hence, the vector of the electric induction is equal to zero not only upon the surface but also in the whole solid body $\mathbf{D} = 0$.

On the body surface $x_1 = 0$, we have

$$E_2\big|_{x_1=0}=0,\ \pi_{e2}\big|_{x_1=0}=0,$$

while

$$E_1\big|_{x_1=0}=-\kappa_E\frac{a_1d_\rho}{K\alpha_\rho}\left(\bar{K}+\frac{4}{3}G\right)(\beta_1-\beta_2)\left(\beta_1^2-v_1^2\right)e^{-i(\omega t-kx_2)},$$

$$\pi_{e1}\big|_{x_1=0}=\chi_{Em}\frac{a_1d_\rho}{K\alpha_\rho}\frac{\varepsilon_0}{\varepsilon}\left(\bar{K}+\frac{4}{3}G\right)(\beta_1-\beta_2)\left(\beta_1^2-v_1^2\right)e^{-i(\omega t-kx_2)}.$$

Thus, the surface Rayleigh wave that propagates in a non-ferromagnetic isotropic half-space of an ideal dielectric induces a slow surface electric wave subjected to the relations (5.122), (5.123) as well as causes polarization of the surface and near-surface regions of the solid body (see formulae (5.124) and (5.125)). The amplitudes of the quantities defined by the formulae (5.122)–(5.125) are proportional to the difference $\beta_1^2-v_1^2$. Considering the formulae (5.97), (5.109), and (5.111), we get

$$\text{for } \omega << c_2\lambda_\mu:\quad \beta_1^2-v_1^2\approx k^2\frac{\mathfrak{M}\eta\bar{\eta}}{1+\mathfrak{M}}\approx\mathfrak{M}\frac{\omega^2}{c_1^2}\ll\mathfrak{M}\lambda_\mu^2\frac{c_2^2}{c_1^2},$$

$$\text{for } \omega >> c_2\lambda_\mu:\quad \beta_1^2-v_1^2\approx k^2\frac{\mathfrak{M}\eta}{\Omega'^2}\approx\mathfrak{M}\lambda_\mu^2.$$

Hence, we may conclude that in the high-frequency range, when the frequency increases, the amplitudes of the above-mentioned waves (the electric field and polarization) increase as well. The amplitudes of the electric field and polarization are proportional to the parameters $\mathfrak{M}$ and κ_E (i.e., coupling factors between the local mass displacement and deformation, as well as the local mass displacement and electromagnetic field). The higher are the parameters $\mathfrak{M}$ and κ_E, the more significant is the piezoelectric effect.

5.5 Surface SH Waves

Within the framework of a linear classical theory of elastic media, the propagation of anti-plane horizontally polarized surface SH waves is described by the Helmholtz equation with zero boundary conditions and the boundedness condition of mechanical fields at infinity. However, such a problem involves only a zero solution. Therefore, the classical theory of elasticity is not capable of predicting these types of surface waves in a homogeneous and isotropic half-space despite the laboratory confirmation of the existence of these waves (for example, this type of waves has been observed by Kraut (1971) and Bullen and Bolt (1985)). Vardoulakis and Georgiadis (1997) showed that propagation of SH surface waves in a homogeneous half-space can be described by the generalized theory of gradient elasticity with surface energy.

In this section, we prove that the local gradient theory of isotropic elastic materials is also capable of predicting SH surface waves in an isotropic and homogeneous half-space. To this end, we used Eqs. (3.11), (3.13), (3.17), and (3.31)–(3.33) of the generalized theory of elastic continua that consider parameters related to the local mass displacement as the tensor quantities.

5.5.1 Problem Formulation

Let a horizontally polarized surface wave propagate in a homogeneous elastic isotropic half-space. Assume that with respect to the Cartesian coordinate system $\{x_1, x_2, x_3\}$, the half-space occupies the region $x_2 \geq 0$, $-\infty < x_1 < +\infty$ (Fig. 5.6). We characterize this wave by the displacement field $\mathbf{u} = (0, 0, u_3)$, where $u_3 = u_3(x_1, x_2)$. Assume that the displacement u_3 exponentially decays with the distance from the body surface $x_2 = 0$. Since the surface of the solid body is devoid of the external action, the boundary conditions of the problem can be written down as follows:

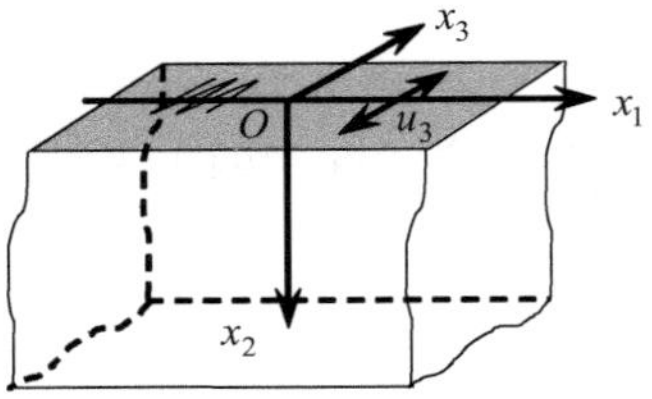

Figure 5.6 An elastic half-space under the anti-plane shear state.

$$\mathbf{n}\cdot\hat{\boldsymbol{\sigma}}\Big|_{x_2=0}=0,\ \mathbf{n}\cdot\hat{\boldsymbol{\pi}}^{m(3)}\Big|_{x_2=0}=0. \tag{5.126}$$

In the above case of external action, the strain tensor has two nonzero components only, that is

$$e_{13}=\frac{1}{2}\frac{\partial u_3}{\partial x_1},\quad e_{23}=\frac{1}{2}\frac{\partial u_3}{\partial x_2}. \tag{5.127}$$

Based on the constitutive equations (3.31) and (3.32), for components σ_{13}, σ_{23}, $\tilde{\mu}_{13}^{\prime\pi}$, and $\tilde{\mu}_{23}^{\prime\pi}$ of tensors $\hat{\boldsymbol{\sigma}}_*$ and $\hat{\tilde{\mu}}^{\prime\pi}$, we get the following relations:

$$\sigma_{13}=G\frac{\partial u_3}{\partial x_1}-2G_{\rho e}\rho_{13}^{m}=G\left(1-\bar{\bar{\gamma}}\right)\frac{\partial u_3}{\partial x_1}-\frac{G_{\rho e}}{G_\rho}\tilde{\mu}_{13}^{\prime\pi}, \tag{5.128}$$

$$\sigma_{23}=G\frac{\partial u_3}{\partial x_2}-2G_{\rho e}\rho_{23}=G\left(1-\bar{\bar{\gamma}}\right)\frac{\partial u_3}{\partial x_2}-\frac{G_{\rho e}}{G_\rho}\tilde{\mu}_{23}^{\prime\pi}, \tag{5.129}$$

$$\tilde{\mu}_{13}^{\prime\pi}=2G_\rho\rho_{13}^{m}-\frac{1}{\rho_0}G_{\rho e}\frac{\partial u_3}{\partial x_1}, \tag{5.130}$$

$$\tilde{\mu}_{23}^{\prime\pi}=2G_\rho\rho_{23}^{m}-\frac{1}{\rho_0}G_{\rho e}\frac{\partial u_3}{\partial x_2}. \tag{5.131}$$

Here, $$\bar{\bar{\gamma}}=\frac{G_{\rho e}^2}{\rho_0 G G_\rho}.$$

The constitutive relations for the components π_{ijk} of the tensor $\hat{\boldsymbol{\pi}}^{m(3)}$ of the local mass displacement look as follows:

$$\pi_{113}=\pi_{131}=-\left(\varsigma_4+\varsigma_5+\varsigma_6\right)\nabla_1\tilde{\mu}_{13}^{\prime\pi}-\varsigma_4\nabla_2\tilde{\mu}_{23}^{\prime\pi}, \tag{5.132}$$

$$\pi_{132}=\pi_{123}=-\varsigma_5\nabla_1\tilde{\mu}_{23}^{\prime\pi}-\varsigma_6\nabla_2\tilde{\mu}_{13}^{\prime\pi}, \tag{5.133}$$

$$\pi_{213} = \pi_{231} = -\varsigma_5 \nabla_2 \tilde{\mu}_{13}^{\prime\pi} - \varsigma_6 \nabla_1 \tilde{\mu}_{23}^{\prime\pi}, \tag{5.134}$$

$$\pi_{232} = \pi_{223} = -\varsigma_4 \nabla_1 \tilde{\mu}_{13}^{\prime\pi} - (\varsigma_4 + \varsigma_5 + \varsigma_6) \nabla_2 \tilde{\mu}_{23}^{\prime\pi}, \tag{5.135}$$

$$\pi_{321} = \pi_{312} = -\varsigma_6 \left(\nabla_2 \tilde{\mu}_{13}^{\prime\pi} + \nabla_1 \tilde{\mu}_{23}^{\prime\pi} \right), \tag{5.136}$$

$$\pi_{311} = -(\varsigma_3 + 2\varsigma_6) \nabla_1 \tilde{\mu}_{13}^{\prime\pi} - \varsigma_3 \nabla_2 \tilde{\mu}_{23}^{\prime\pi}, \tag{5.137}$$

$$\pi_{322} = -\varsigma_3 \nabla_1 \tilde{\mu}_{13}^{\prime\pi} - (\varsigma_3 + 2\varsigma_6) \nabla_2 \tilde{\mu}_{23}^{\prime\pi}, \tag{5.138}$$

$$\pi_{333} = -(\varsigma_3 + 2\varsigma_4) \left(\nabla_1 \tilde{\mu}_{13}^{\prime\pi} + \nabla_2 \tilde{\mu}_{23}^{\prime\pi} \right). \tag{5.139}$$

The remaining components of the tensor $\hat{\boldsymbol{\pi}}^{m(3)}$ are equal to zero, namely:

$$\pi_{111} = \pi_{122} = \pi_{133} = \pi_{112} = \pi_{121} = \pi_{211} = \pi_{222} = \pi_{233}$$

$$= \pi_{212} = \pi_{221} = \pi_{331} = \pi_{313} = \pi_{323} = \pi_{332} = 0.$$

The equation of motion (3.17), formulae (3.11) and (3.13), the strain–displacement relations (5.127), and the constitutive equations (5.128)–(5.139) yield the following governing set of equations for the anti-plane shear state in the elastic half-space:

$$G\left(1 - \overline{\overline{\gamma}}\right)\left(\frac{\partial^2 u_3}{\partial x_1^2} + \frac{\partial^2 u_3}{\partial x_2^2} \right) - \frac{G_{\rho e}}{G_\rho}\left(\frac{\partial \tilde{\mu}_{13}^{\prime\pi}}{\partial x_1} + \frac{\partial \tilde{\mu}_{23}^{\prime\pi}}{\partial x_2} \right) = \rho_0 \frac{\partial^2 u_3}{\partial t^2}, \tag{5.140}$$

$$\left(\frac{\partial^2 \tilde{\mu}_{13}^{\prime\pi}}{\partial x_1^2} + \frac{\partial^2 \tilde{\mu}_{13}^{\prime\pi}}{\partial x_2^2} \right) - \underline{\lambda}^2 \tilde{\mu}_{13}^{\prime\pi} + \kappa' \frac{\partial}{\partial x_1}\left(\frac{\partial \tilde{\mu}_{13}^{\prime\pi}}{\partial x_1} + \frac{\partial \tilde{\mu}_{23}^{\prime\pi}}{\partial x_2} \right) = b_u \frac{\partial u_3}{\partial x_1}, \tag{5.141}$$

$$\left(\frac{\partial^2 \tilde{\mu}_{23}^{\prime\pi}}{\partial x_1^2} + \frac{\partial^2 \tilde{\mu}_{23}^{\prime\pi}}{\partial x_2^2} \right) - \underline{\lambda}^2 \tilde{\mu}_{23}^{\prime\pi} + \kappa' \frac{\partial}{\partial x_2}\left(\frac{\partial \tilde{\mu}_{13}^{\prime\pi}}{\partial x_1} + \frac{\partial \tilde{\mu}_{23}^{\prime\pi}}{\partial x_2} \right) = b_u \frac{\partial u_3}{\partial x_2}. \tag{5.142}$$

Here,

$$\underline{\lambda}^2 = \frac{1}{2G_\rho \varsigma_5}, \quad \kappa' = \frac{\varsigma_4 + \varsigma_6}{\varsigma_5}, \quad b_u = \frac{G_{\rho e}}{2\rho_0 G_\rho \varsigma_5}.$$

5.5.2 Problem Solution and Its Analysis

Let us introduce the function $m(x_1, x_2, t)$ through the formula

$$m(x_1, x_2, t) = \frac{\partial \tilde{\mu}_{13}^{\prime\pi}}{\partial x_1} + \frac{\partial \tilde{\mu}_{23}^{\prime\pi}}{\partial x_2}. \tag{5.143}$$

To find this function, using Eqs. (5.141) and (5.142), we get

$$\Delta m - \Lambda^2 m = b_u \frac{\Lambda^2}{\underline{\lambda}^2} \Delta u_3. \tag{5.144}$$

Here, $\Delta = \frac{\partial^2}{\partial x_1^2} + \frac{\partial^2}{\partial x_2^2}$ is the 2D Laplace operator in the Cartesian coordinates, and

$$\Lambda^2 = \frac{\underline{\lambda}^2}{1+\kappa'} = \frac{1}{2G_\rho(\varsigma_4 + \varsigma_5 + \varsigma_6)}. \tag{5.145}$$

We complete the set of equations (5.143) and (5.144) by the equation of motion, which, taking the formula (5.143) into account, is written as follows:

$$G(1-\bar{\bar{\gamma}})\Delta u_3 - \frac{G_{\rho e}}{G_\rho} m = \rho_0 \frac{\partial^2 u_3}{\partial t^2}. \tag{5.146}$$

Based on the coupled set of equations (5.144) and (5.146), we obtain a differential equation of the fourth order to determine the component $u_3(x_1, x_2, t)$ of the displacement vector

$$G(1-\bar{\bar{\gamma}})\Delta\Delta u_3 - \Lambda^2\left[G(1-\bar{\bar{\gamma}}) + \frac{G_{\rho e}}{G_\rho}\frac{b_u}{\underline{\lambda}^2}\right]\Delta u_3 = \rho_0 \frac{\partial^2}{\partial t^2}(\Delta u_3 - \Lambda^2 u_3). \tag{5.147}$$

In the sequel, we consider a steady state. In this case, we can represent the displacement field in the following general form:

$$u_3(x_1, x_2, t) = w(x_1, x_2)e^{-i\omega t}, \tag{5.148}$$

where ω is the frequency.

Upon substituting (5.148) into Eq. (5.147), we get the following equation to find the unknown function $w(x_1, x_2)$

$$b'\Delta\Delta w - a(\omega)\Delta w - k^2(\omega)w = 0, \tag{5.149}$$

where

$$k(\omega) = \frac{\omega}{c_2}, \quad c_2 = \sqrt{\frac{G}{\rho_0}}, \quad b' = \frac{1-\bar{\bar{\gamma}}}{\Lambda^2}, \quad a(\omega) = 1 - \frac{k^2}{\Lambda^2}. \tag{5.150}$$

Equation (5.149) describes the dispersion properties of the investigated medium. We see that the coefficients a and k in this equation are functions of a frequency. Note that depending on the frequency ω and parameters Λ and c_2, the coefficient $a(\omega)$ can assume positive and negative values, as well as can also be equal to zero. Here, c_2 is the velocity of the transverse wave within the classical theory of elasticity. The sign of the parameter $a(\omega)$ affects the solution to the differential equation (5.149). Frequencies $a < 0$ correspond to the condition $\omega > \Lambda\sqrt{G/\rho_0}$. For crystalline bodies $G = O(10^9)\mathrm{N/m^2}$, $\rho_0 = O(10^3)$ kg/m^3; thus, we get $\omega > 3.162\cdot 10^2\,\Lambda$ [m/s]. Assuming $\Lambda = O(10^9)$ m^{-1}, which is typical for crystalline materials, $\omega > 3.162\cdot 10^{11}$ (Hz) (i.e., ω belongs to a hypersonic range). It should be noted that for materials having a larger characteristic length, much lower frequencies correspond to the condition $a < 0$. For example, if $l_* = 10^{-4}$ (m), then $\omega > 3.162\cdot 10^6$ (Hz). Since the parameter Λ attains quite large values, and based on the formulae (5.150), it may be assumed that for low frequencies $a \approx 1$.

Since the wave is a surface wave, we can present the unknown function $w(x_1, x_2)$ in the following form:

$$w(x_1,x_2) = w_*(q,x_2)e^{iqx_1}, \tag{5.151}$$

where q is the wave number to be determined.

Using Eq. (5.149) and taking the expression (5.151) into account, we get an ordinary differential equation of the fourth order to find the function $w_*(q, x_2)$:

$$b'\frac{d^4w_*}{dx_2^4} - (a+2q^2b')\frac{d^2w_*}{dx_2^2} + (aq^2 + b'q^4 - k^2)w_* = 0. \tag{5.152}$$

We represent the solution to Eq. (5.152) as a linear combination of exponential functions, that is, $w_*(q,x_2) \sim e^{\alpha_i x_2}$, where α_i are the roots of the biquadratic equation:

$$b'\alpha^4 - (a+2q^2b')\alpha^2 + aq^2 + b'q^4 - k^2 = 0.$$

This equation has four roots. From these roots, we take only those satisfying the boundedness condition of the problem solution at $x_2 \to +\infty$. Thus, we can write

$$w_*(q,x_2) = B'(q)\,e^{-\alpha_1(q)x_2} + C'(q)\,e^{-\alpha_2(q)x_2}, \tag{5.153}$$

where $B'(q)$ and $C'(q)$ are constants depending on the wave number q, and

$$\alpha_1 = \sqrt{q^2 + g_1^2}\,,\quad \alpha_2 = \sqrt{q^2 - g_2^2}\,,\quad (\alpha_1 > 0,\, \alpha_2 > 0), \tag{5.154}$$

$$g_1 = \sqrt{\frac{\sqrt{D} + a}{2b'}}\,,\quad g_2 = \sqrt{\frac{\sqrt{D} - a}{2b'}}\,, \tag{5.155}$$

$$D = \left(1 + \frac{k^2}{\Lambda^2}\right)^2 - 4\bar{\bar{\gamma}}\frac{k^2}{\Lambda^2} = \left(1 + \frac{k^2}{\Lambda^2}\right)^2 \left[1 - \frac{4\bar{\bar{\gamma}}k^2}{\Lambda^2\left(1 + k^2/\Lambda^2\right)^2}\right]. \tag{5.156}$$

The frequency at which the parameters α_1 and α_2 change from real to complex (imaginary) values is the wave cut-off frequency. Parameters α_1 and α_2 will be real and positive quantities if q is a real number whose module is greater than g_2: $|q| > g_2$. Hence, we have the following restrictions on the wave number q: $-\infty < q < -g_2$ and $g_2 < q < +\infty$.

Using the formulae (5.148), (5.151), and (5.153), we obtain the displacement component u_3 as follows:

$$u_3(x_1, x_2, t) = \left[B'(q)e^{-\alpha_1(q)x_2} + C'(q)e^{-\alpha_2(q)x_2}\right]e^{i(qx_1 - \omega t)}, \tag{5.157}$$

where $x_2 \geq 0$, $\alpha_1 > 0$, $\alpha_2 > 0$, and $B'(q)$, $C'(q)$ are the unknown amplitudes.

Based on Eq. (5.146), we define the function $m(x_1, x_2, t)$ through the component u_3 of the displacement vector:

$$m = \frac{G_\rho}{G_{\rho e}}\left[G\left(1 - \bar{\bar{\gamma}}\right)\Delta u_3 - \rho_0\frac{\partial^2 u_3}{\partial t^2}\right]. \tag{5.158}$$

Using the formulae (5.157) and (5.158), we obtain the following expression for the function $m(x_1, x_2, t)$:

$$m(x_1, x_2, t) = \frac{GG_\rho}{G_{\rho e}}\left\{\left[\left(1 - \bar{\bar{\gamma}}\right)\left(\alpha_1^2 - q^2\right) + k^2\right]B'e^{-\alpha_1 x_2}\right.$$

$$\left. + \left[\left(1 - \bar{\bar{\gamma}}\right)\left(\alpha_2^2 - q^2\right) + k^2\right]C'e^{-\alpha_2 x_2}\right\}e^{i(qx_1 - \omega t)}. \tag{5.159}$$

In view of Eqs. (5.141)–(5.143), as well as the formulae (5.157) and (5.159), in order to find the potentials $\tilde{\mu}_{13}'^{\pi}$ and $\tilde{\mu}_{23}'^{\pi}$, we obtain two inhomogeneous equations

$$\frac{\partial^2 \tilde{\mu}_{13}^{\prime\pi}}{\partial x_1^2} + \frac{\partial^2 \tilde{\mu}_{13}^{\prime\pi}}{\partial x_2^2} - \underline{\lambda}^2 \tilde{\mu}_{13}^{\prime\pi} = iqb_u \left\langle \left(1 - \frac{\kappa' d_1}{\overline{\overline{\gamma}}\underline{\lambda}^2}\right) B' e^{-\alpha_1 x_2} \right.$$
$$\left. + \left(1 - \frac{\kappa' d_2}{\overline{\overline{\gamma}}\underline{\lambda}^2}\right) C' e^{-\alpha_2 x_2} \right\rangle e^{i(qx_1 - \omega t)}, \tag{5.160}$$

$$\frac{\partial^2 \tilde{\mu}_{23}^{\prime\pi}}{\partial x_1^2} + \frac{\partial^2 \tilde{\mu}_{23}^{\prime\pi}}{\partial x_2^2} - \underline{\lambda}^2 \tilde{\mu}_{23}^{\prime\pi} = -b_u \left\langle \alpha_1 B' \left(1 - \frac{\kappa' d_1}{\overline{\overline{\gamma}}\underline{\lambda}^2}\right) e^{-\alpha_1 x_2} \right.$$
$$\left. + \alpha_2 C' \left(1 - \frac{\kappa' d_2}{\overline{\overline{\gamma}}\underline{\lambda}^2}\right) e^{-\alpha_2 x_2} \right\rangle e^{i(qx_1 - \omega t)}. \tag{5.161}$$

Here,

$$d_1 = k^2 + \left(1 - \overline{\overline{\gamma}}\right) g_1^2 = \frac{1}{2}\Lambda^2 \left(1 + \frac{k^2}{\Lambda^2} + \sqrt{D}\right),$$

$$d_2 = k^2 - \left(1 - \overline{\overline{\gamma}}\right) g_2^2 = \frac{1}{2}\Lambda^2 \left(1 + \frac{k^2}{\Lambda^2} - \sqrt{D}\right).$$

The solution to Eqs. (5.160) and (5.161) that satisfies the boundedness conditions of the functions at $x_2 \to +\infty$ is

$$\tilde{\mu}_{13}^{\prime\pi}(x_1, x_2, t) = \left(A_1 e^{-\lambda x_2} + iqB' N_1 e^{-\alpha_1 x_2} - iqC' N_2 e^{-\alpha_2 x_2}\right) e^{i(qx_1 - \omega t)}, \tag{5.162}$$

$$\tilde{\mu}_{23}^{\prime\pi}(x_1, x_2, t) = \left(iA_1 \frac{q}{\lambda} e^{-\lambda x_2} - B' \alpha_1 N_1 e^{-\alpha_1 x_2} + C' \alpha_2 N_2 e^{-\alpha_2 x_2}\right) e^{i(qx_1 - \omega t)}, \tag{5.163}$$

where,

$$\lambda = \left|\sqrt{\underline{\lambda}^2 + q^2}\right|, \tag{5.164}$$

$$N_1 = \frac{b_u}{g_1^2 - \underline{\lambda}^2}\left(1 - \frac{\kappa' d_1}{\overline{\overline{\gamma}}\,\underline{\lambda}^2}\right) = \frac{b_u d_1}{\overline{\overline{\gamma}}\,\underline{\lambda}^2 g_1^2},$$

$$N_2 = \frac{b_u}{g_2^2 + \underline{\lambda}^2}\left(1 - \frac{\kappa' d_2}{\overline{\overline{\gamma}}\,\underline{\lambda}^2}\right) = \frac{b_u d_2}{\overline{\overline{\gamma}}\,\underline{\lambda}^2 g_2^2}.$$

Thus, the solution to the formulated problem, which satisfies the boundedness condition of the fields at infinity, is given by the formulae (5.157), (5.162), and (5.163), where the constants A_1, B', C' are determined from the boundary conditions on the half-space surface $x_2 = 0$:

$$\sigma_{23}(x_1, 0) = 0,\ \pi_{213}(x_1, 0) = 0,\ \pi_{223}(x_1, 0) = 0 \text{ for } -\infty < x_1 < +\infty. \quad (5.165)$$

Using the constitutive equations (5.129), (5.134), (5.135) and the formulae (5.157), (5.162), (5.163), we obtain

$$\sigma_{23}\left(x_1,x_2,t\right)=-G\left[A_1\frac{iq\lambda^2\overline{\overline{\gamma}}}{\lambda b_u}e^{-\lambda x_2}+B'\alpha_1\left(1-\overline{\overline{\gamma}}-\frac{d_1}{g_1^2}\right)e^{-\alpha_1 x_2}\right.$$
$$\left.+C'\alpha_2\left(1-\overline{\overline{\gamma}}+\frac{d_2}{g_2^2}\right)e^{-\alpha_2 x_2}\right]e^{i(qx_1-\omega t)}, \quad (5.166)$$

$$\pi_{213}\left(x_1,x_2,t\right)=\left[A_1\left(\varsigma_5\lambda+\frac{\varsigma_6 q^2}{\lambda}\right)e^{-\lambda x_2}\right.$$
$$\left.+iq\left(\varsigma_5+\varsigma_6\right)\left(B'\alpha_1 N_1 e^{-\alpha_1 x_2}-C'\alpha_2 N_2 e^{-\alpha_2 x_2}\right)\right]e^{i(qx_1-\omega t)}, \quad (5.167)$$

$$\pi_{223}\left(x_1,x_2,t\right)=\left\{iq\left(\varsigma_5+\varsigma_6\right)A_1 e^{-\lambda x_2}-B'N_1\left[\varsigma_4 g_1^2+\left(\varsigma_5+\varsigma_6\right)\alpha_1^2\right]e^{-\alpha_1 x_2}\right.$$
$$\left.-C'N_2\left[\varsigma_4 g_2^2-\left(\varsigma_5+\varsigma_6\right)\alpha_2^2\right]e^{-\alpha_2 x_2}\right\}e^{i(qx_1-\omega t)}. \quad (5.168)$$

Taking the boundary conditions (5.165) and the formulae (5.166)–(5.168) into account, we get a homogeneous linear algebraic set of equations to find the unknown constants A_1, B', and C':

$$A_1\frac{iq\lambda^2\overline{\overline{\gamma}}}{\lambda b_u}+B'\alpha_1\left(1-\overline{\overline{\gamma}}-\frac{d_1}{g_1^2}\right)+C'\alpha_2\left(1-\overline{\overline{\gamma}}+\frac{d_2}{g_2^2}\right)=0, \quad (5.169)$$

$$A_1\left(\varsigma_5\lambda+\frac{\varsigma_6 q^2}{\lambda}\right)+iq\left(\varsigma_5+\varsigma_6\right)\left(B'\alpha_1 N_1-C'\alpha_2 N_2\right)=0, \quad (5.170)$$

$$A_1 iq(\varsigma_5+\varsigma_6)-B'N_1\left[\varsigma_4 g_1^2+(\varsigma_5+\varsigma_6)\alpha_1^2\right]$$
$$-C'N_2\left[\varsigma_4 g_2^2-(\varsigma_5+\varsigma_6)\alpha_2^2\right]=0. \quad (5.171)$$

The boundary-value problem has a nontrivial solution when the determinant of the set of equations (5.169)–(5.171) is equal to zero, that is,

$$\begin{vmatrix} 1 & \lambda\alpha_1 k^2/d_1 & \lambda\alpha_2 k^2/d_2 \\ 1+\nu_2\lambda^2/q^2 & \lambda\alpha_1 & \lambda\alpha_2 \\ 1 & \alpha_1^2+\nu_1 g_1^2 & \alpha_2^2-\nu_1 g_2^2 \end{vmatrix} = 0, \qquad (5.172)$$

where

$$\nu_1 = \frac{\varsigma_4}{\varsigma_5+\varsigma_6}, \quad \nu_2 = \frac{\varsigma_5}{\varsigma_5+\varsigma_6}.$$

Note that here $d_i = d_i(k)$, $g_i = g_i(k)$, $\alpha_i = \alpha_i(q, k)$, and $\lambda = \lambda(q)$. For a surface SH wave from Eq. (5.172), we obtain the following dispersion equation:

$$\alpha_1\left(\nu_1 g_2^2 - \alpha_2^2\right)\left[\frac{k^2}{d_1}\left(1+\nu_2\frac{\lambda^2}{q^2}\right)-1\right]$$
$$+\alpha_2\left(\nu_1 g_1^2 + \alpha_1^2\right)\left[\frac{k^2}{d_2}\left(1+\nu_2\frac{\lambda^2}{q^2}\right)-1\right]+\alpha_1\alpha_2\lambda k^2\left(\frac{1}{d_1}-\frac{1}{d_2}\right)=0. \qquad (5.173)$$

From this equation, we can determine the wave number q as the function of k, and, hence, we can find the phase velocity $v = \omega/q$ at which the surface transverse wave of a specified frequency ω may propagate in a homogeneous, isotropic half-space. If one knows the characteristics of the material, the solutions to the algebraic equation (5.173) may be found using numerical methods. Having eliminated the irrationality of the dispersive equation (5.173), there appear external roots. Real roots should be taken from the found roots, so that they should satisfy the equation and correspond to the above-mentioned constraints, particularly to the condition $|g_2| < |q|$. The complex roots correspond to the wave cut-off frequency.

Equation (5.173) shows that the phase velocity of the surface SH wave depends on the frequency, that is, the medium is dispersive. This equation can be written as follows:

$$\Lambda^2 \left\langle \left[\left(1 + \nu_2 \frac{\underline{\lambda}^2}{q^2}\right)\left(1 + \frac{k^2}{\Lambda^2}\right) + 2\overline{\overline{\gamma}} \right] \left[\alpha_1 - \alpha_2 + \nu_1 \left(\frac{g_1^2}{\alpha_1} + \frac{g_2^2}{\alpha_2} \right) \right] \right.$$

$$\left. + \left(1 + \nu_2 \frac{\underline{\lambda}^2}{q^2}\right)\left(1 + \frac{k^2}{\Lambda^2}\right) \left[\alpha_1 + \alpha_2 + \nu_1 \left(\frac{g_1^2}{\alpha_1} - \frac{g_2^2}{\alpha_2} \right) \right] \sqrt{1 - \frac{4\overline{\overline{\gamma}} k^2}{\Lambda^2 \left(1 + k^2/\Lambda^2\right)^2}} \right\rangle$$

$$-2\lambda\left(1 - \overline{\overline{\gamma}}\right)\left(g_1^2 + g_2^2\right) = 0 . \quad (5.174)$$

Since $g_1^2 + g_2^2 \geq 0$ and $\lambda = \left|\sqrt{\underline{\lambda}^2 + q^2}\right| > 0$, the SH wave can propagate in a homogeneous isotropic half-space if and only if the parameter Λ is nonzero (i.e., $\Lambda \neq 0$), and the following inequalities are satisfied:

$$\overline{\overline{\gamma}} < 1 , \quad (5.175)$$

$$\left[\left(1 + \nu_2 \frac{\underline{\lambda}^2}{q^2}\right)\left(1 + \frac{k^2}{\Lambda^2}\right) + 2\overline{\overline{\gamma}} \right] \left[\alpha_1 - \alpha_2 + \nu_1 \left(\frac{g_1^2}{\alpha_1} + \frac{g_2^2}{\alpha_2} \right) \right]$$

$$+ \left(1 + \nu_2 \frac{\underline{\lambda}^2}{q^2}\right)\left(1 + \frac{k^2}{\Lambda^2}\right) \left[\alpha_1 + \alpha_2 + \nu_1 \left(\frac{g_1^2}{\alpha_1} - \frac{g_2^2}{\alpha_2} \right) \right]$$

$$\times \sqrt{1 - \frac{4\overline{\overline{\gamma}} k^2}{\Lambda^2 \left(1 + k^2/\Lambda^2\right)^2}} > 0. \quad (5.176)$$

For materials having the parameter $\overline{\overline{\gamma}}$ much less than unity ($\overline{\overline{\gamma}} \ll 1$), and for the frequencies $\omega << c_2\Lambda$, inequality (5.176) may be presented in a much simpler form. At such frequencies $k^2/\Lambda^2 << 1$, and the consequence of Eq. (5.176) is an inequality: $\alpha_1^2 + \nu_1 g_1^2 > 0$. Hence, we get $q^2 + g_1^2(1 + \nu_1) > 0$, which imposes constraints on the material characteristics:

$$\frac{\varsigma_4}{\varsigma_5 + \varsigma_6} > -1 .$$

For such frequencies, the dispersion equation (5.174) looks as follows:

$$\Lambda^2 \left(1 + \nu_2 \frac{\underline{\lambda}^2}{q^2}\right)\left(\alpha_1^2 + \nu_1 g_1^2\right) - \lambda\alpha_1\left(g_1^2 + g_2^2\right) = 0 .$$

Taking into account the formulae (5.145), (5.150), (5.154)–(5.156), and (5.162), as well as the relation $1 + \kappa' = (1 + \nu_1)/\nu_2$, the last equation can be written as:

$$q^4\left(q^2+\Lambda^2\right)\left[q^2+\frac{\left(1+\nu_1\right)}{\nu_2}\Lambda^2\right]=\left[q^2+\left(1+\nu_1\right)\Lambda^2\right]^4. \quad (5.177)$$

Equation (5.177) is a bicubic one (i.e., the coefficient of the leading term is equal to zero). From this equation, we determine the square of the wave number for a low-wave approximation. From the tree solutions to the bicubic equation, we choose the one(s) that correspond(s) to the following criteria: the solution is a real positive number that satisfies the condition: $q > \omega/c_2$. The frequencies $\omega < qc_2$ are the wave cut-off frequencies.

Problems

5.1 Using the relations for electroelastic media with an electric quadrupole, determine the dispersion relations for the time-harmonic plane wave that propagates in an isotropic medium of ideal dielectrics along the axis Ox of the Cartesian coordinate system (x, y, z).

5.2 Following a step similar to Section 5.1, study the dispersive properties of a time-harmonic spherical elastic wave propagating in an isotropic dielectric medium.

5.3 Determine the dispersion relations for isotropic polarized solids based on the relations of the local gradient theory of dielectrics that takes the inertia of the electric polarization and local mass displacement into account (see Section 3.2).

5.4 Study the effect of polarization inertia on the propagation of a time-harmonic spherical elastic wave.

Bibliography

Abazari, A. M., Safavi, S. M., Rezazadeh, G., and Villanueva, L. G. (2015). Modelling the size effects on the mechanical properties of micro/nano structures, *Sensors*, **15**, 28543–28562.

Angadi, M. A. and Thanigaimani, V. (1994). Size effect in thickness dependence of Young's modulus of MnTe and MnSe films, *J. Mat. Sci. Lett.*, **13**(10), 703–704.

Askar, A. (1972). Molecular crystals and the polar theories of the continua. Experimental values of material coefficients for KNO_3, *Int. J. Eng. Sci.*, **10**(3), 293–300.

Askar, A. and Lee, P. C. Y. (1974). Lattice dynamics approach to the theory of diatomic elastic dielectrics, *Phys. Rev. B*, **9**, 5291–5299.

Askar, A., Lee, P. C. Y., and Cakmak, A. S. (1970). Lattice dynamics approach to the theory of elastic dielectrics with polarization gradient, *Phys. Rev. B*, **1**, 3525–3537.

Askar, A., Lee, P. C. Y., and Cakmak, A. S. (1971). The effect of surface curvature and discontinuity on the surface energy density and other induced fields in elastic dielectrics with polarization gradient, *Int. J. Solids Struct.*, **7**(5), 523–537.

Askar, A., Pouget, J., and Maugin, G. A. (1984). Lattice model for elastic ferroelectrics and related continuum theories, in: *Mechanical Behavior of Electromagnetic Solid Continua* (G. A. Maugin, Ed.), North-Holland, Amsterdam: Elsevier, 151–156.

Askes, H. and Aifantis, E. C. (2011). Gradient elasticity in statics and dynamics: an overview of formulations, length scale identification procedures, finite element implementations and new results, *Int. J. Solids Struct.*, **48**, 1962–1990.

Axe, J. D., Harada, J., and Shirane, G. (1970). Anomalous acoustic dispersion in centrosymmetric crystals with soft optic phonons, *Phys. Rev. B.*, **1**, 1227–1234.

Bampi, F. and Morro, A. (1986). A variational approach to deformable electromagnetic solids, *Acta Physica Polonica*, **B17**(11), 937–949.

Benson, G. G. and Yun, K. S. 1967. *The Solid–Gas Interface* (E. A. Flood, ed.), New York: Dekker Inc., **1**, 203–269.

Beran, M. J. and McCoy, J. J. (1970). The use of strain gradient theory for analysis of random media, *Int. J. Solids Struct.*, **6**, 1267–1275.

Boukai, A. I., Bunimovich, Y., Tahir-Kheli, J., Yu, J. K., Goddaed, W. A., and Heath, J. R. (2008). Silicon nanowires as efficient thermoelectric materials, *Nature*, **451**, 168–171.

Bredov, M. M., Rumyantsev, V. V., and Toptyhin, I. N. (1985). *Classic Electrodynamics*, Moscow: Nauka. In Russian.

Bullen, K. E. and Bolt, B. A. (1985). *An Introduction to the Theory of Seismology*, London: Cambridge University Press.

Burak, Ya. (1966). The equations of electroelasticity of isotropic dielectrics in electrostatic field, *Fiz. Khim. Mech. Materialov* (*Phis. Chim. Mech. Materials*), **2**(1), 51–57. In Russian.

Burak, Y. (1987). Constitutive equations of locally gradient thermomechanics, *Dopovidi Akad. Nauk URSR (Proc. Acad. Sci. Ukraine. SSR)*, **12**, 19–23. In Ukrainian.

Burak, Ya. and Hrytsyna, O. (2011). The constitutive equations of local gradient theory of anisotropic dielectrics, *Roczniki Inżynierii Budowlanej*, **11**, 41–48.

Burak, Ya. I., Kondrat, V. F., and Hrytsyna, O. R. (2007). Subsurface mechanoelectromagnetic phenomena in thermoelastic polarized bodies in the case of local displacements of mass, *Materials Science*, **43**(4), 449–463.

Burak, Ya., Kondrat, V., and Hrytsyna, O. (2008). An introduction of the local displacements of mass and electric charge phenomena into the model of the mechanics of polarized electromagnetic solids, *J. Mech. Mat. Struct.*, **3**(6), 1037–1046.

Burak, Ya., Kondrat, V., and Hrytsyna, O. (2011). *Fundamentals of the Local Gradient Theory of Dielectrics*, Lira, Uzhgorod. In Ukrainian.

Bursian, E. V. and Trunov, N. N. (1974). Nonlocal piezoelectric effect, *Sov. Phys.-Solid State.*, **16**, 760–762.

Cady, W. G. (1922). Piezoelectric resonator, *Proc. Inst. Rad. Eng.*, **10**, 83–114.

Cady, W. (1946). Piezoelectricity. *An Introduction to the Theory and Application of Electromechanical Phenomena in Crystals*, New York: Dover.

Cao, W.-Z., Yang, X.-H., and Tian, X.-B. (2014). Numerical evaluation of size effect in piezoelectric micro-beam with linear micromorphic electroelastic theory, *J. Mech.*, **30**(5P), 467–476.

Catalan, G., Sinnamon, L. J., and Gregg, J. M. (2004). The effect of flexoelectricity on the dielectric properties of inhomogeneously strained ferroelectric thin films, *J. Phys. Condens. Matter*, **16**(13), 2253–2264.

Catalan, G., Noheda, B., McAneney, J., Sinnamon, L. J., and Gregg, J. M. (2005). Strain gradients in epitaxial ferroelectrics, *Phys. Rev. B*, **72**(2), 020102.

Catalan, G., Lubk, A., Vlooswijk, A. H. G., Snoeck, E., Magen, C., Janssens, A., Rispens, G., Rijnders, G., Blank, D. H. A., and Noheda, B. (2011). Flexoelectric rotation of polarization in ferroelectric thin films, *Nat. Mater.*, **10**(12), 963–967.

Chapla, Ye., Kondrat, S., Hrytsyna, O., and Kondrat, V. (2009). On electromechanical phenomena in thin dielectric films, *Task Quarterly*, **13**(1–2), 145–154.

Chen, J. (2013). Micropolar theory of flexoelectricity, *J. Adv. Math. Appl.*, **1**, 269–274(6).

Chen, Y. and Lee, J. D. (2003). Determining material constants in micromorphic theory through phonon dispersion relations, *Int. J. Eng. Sci.*, **41**, 871–886.

Chen, Y., Lee, J. D., and Eskandarian, A. (2004). Atomistic viewpoint of the applicability of microcontinuum theories, *Int. J. Solids Struct.*, **41**, 2085–2097.

Cheney, E. W. (2010). *Analysis for Applied Mathematics*, New York: Springer.

Chowdhury, K. L. (1982). On an axisymmetric boundary value problem for an elastic dielectric half-space, *Int. J. Solids Struct.*, **18**(3), 263–271.

Chowdhury, K. L. and Glockner, P. G. (1974). Representations in elastic dielectrics, *Int. J. Eng. Sci.*, **12**, 597–606.

Chowdhury, K. L. and Glockner, P. G. (1976). Constitutive equations for elastic dielectrics, *Int. J. Non-Linear Mech.*, **11**, 315–324.

Chowdhury, K. L. and Glockner, P. G. (1977a). On thermoelastic dielectrics, *Int. J. Solids Struct.*, **13**(11), 1173–1182.

Chowdhury, K. L. and Glockner, P. G. (1977b). Point charge in the interior of an elastic dielectric half space, *Int. J. Eng. Sci.*, **15**(8), 481–493.

Chowdhury, K. L. and Glockner, P. G. (1980). On a boundary value problem for an elastic dielectric half-plane, *Acta Mech.*, **37**, 65–74.

Chowdhury, K. L. and Glockner, P. G. (1981). On a similarity solution of the Boussinesq problem of elastic dielectrics, *Arch. Mech.*, **32**, 429–442.

Chowdhury, K. L., Epstein, M., and Glockner, P. G. (1979). On the thermodynamics of non-linear elastic dielectrics, *Int. J. Non-Linear Mech.*, **13**, 311–322.

Collet, B. (1981). One–dimensional acceleration waves in deformable dielectrics with polarization gradients, *Int. J. Eng. Sci.*, **19**(3), 389–407.

Collet, B. (1982). Shock waves in deformable dielectrics with polarization gradients, *Int. J. Eng. Sci.*, **20**(10), 1145–1160.

Collet, B. (1984). Shock waves in deformable ferroelectric materials, in: *Mechanical Behavior of Electromagnetic Solid Continua* (G. A. Maugin, Ed.), North–Holland, Amsterdam: Elsevier, 157–163.

Cosserat, E. and Cosserat, F. (1909). *Théorie des corps déformable*, A. Hermann et Fils, Paris.

Cross, L. E. (2006). Flexoelectric effects: Charge separation in insulating solids subjected to elastic strain gradients, *J. Mater. Sci.*, **41**, 53–63.

Curie, J. and Curie, P. (1880). Développement, par pression, de 1'électricité polaire dans les cristaux hémièdres a faces inclinées, *Compt. Rend.*, **91**, 294–295.

Curie, J. and Curie, P. (1881). Contractions et dilatations produits par des tensions électriques dans les cristaux hémièdres a faces inclinées, *Compt. Rend.*, **93**, 1137–1140.

De Groot, S. R. and Mazur, P. (1962). *Non-Equilibrium Thermodynamics*, Amsterdam: North Holland Publishing Company.

De Gennes, P. G. (1974). *The Physics of Liquid Crystals*, Oxford: Clarendon Press.

Dell'Isola, F., Della Corte, A., and Giorgio, I. (2017). Higher-gradient continua: The legacy of Piola, Mindlin, Sedov and Toupin and some future research perspectives, *Math. Mech. of Solids*, **22** (4), 852–872.

Demiray, H. (1977). A continuum theory of diatomic solids: Viewed as directed media, *J. Eng. Math.*, **11**(3), 257–271.

Demiray, H. and Dost, S. (1988). A variational formulation of diatomic elastic dielectrics, *Int. J. Eng. Sci.*, **26**(8), 865–871.

Demiray, H. and Eringen, A. C. (1973). On the constitutive relations of polar elastic dielectrics, *Lett. Appl. Eng. Sci.*, **1**, 517–527.

Derjaguin, B. V. (1980). Analytical calculation of repulsion forces arising when the non-ionic diffuse adsorption layers are overlapped, *Colloid and Polymer Science*, **258**(4), 433–438.

Derjaguin, B. V. and Churaev, N. V. (1984). *Wetting Films*, Nauka, Moscow. In Russian.

Dixon, R. C. and Eringen, A. C. (1965a). A dynamical theory of polar elastic dielectric – I, *Int. J. Eng. Sci.*, **3**, 359–377.

Dixon, R. C. and Eringen, A. C. (1965b). A dynamical theory of polar elastic dielectric – II, *Int. J. Eng. Sci.*, **3**, 379–398.

Dost, S. (1983). Acceleration waves in elastic dielectrics with polarization gradient effects, *Int. J. Eng. Sci.*, **21**(11), 1305–1311.

Dost, S. and Sahin, E. (1986). Wave propagation in rigid dielectrics with polarization inertia, *Int. J. Eng. Sci.*, **24**(8), 1445–1451.

Dost, S., Epstein, M., and Gödze, S. (1984). Propagation of acceleration waves in generalized thermoelastic dielectrics, in: *Mechanical Behaviour of Electromagnetic Solid Continua* (G. A. Maugin, Ed.), North–Holland, Amsterdam: Elsevier, 211–216.

Dumitrică, T., Landis, C. M., and Yakobson, B. I. (2002). Curvature-induced polarization in carbon nanoshells, *Chem. Phys. Lett.*, **360**(1–2), 182–188.

Edelen, D. G. B. (1969). Protoelastic bodies with large deformation, *Arch. Rat. Mech. Anal.*, **34**, 283–300.

Edelen, D. G. B. and Laws, N. (1971). On the thermodynamics of systems with nonlocality, *Arch. Rat. Mech. Anal.*, **43**, 24–35.

Enakoutsa, K., Della Corte, A., and Giorgio, I. (2016). A model for elastic flexoelectric materials including strain gradient effects, *Math. Mech. Solids*, **21**(2), 242–254.

Eringen, A. C. (1964). Mechanics of micromorphic materials, in: *Proc. 11th Int. Congress Appl. Mech.*, Berlin: Springer-Verlag, 131–138.

Eringen, A. C. (1966a). Linear theory of micropolar elasticity, *J. Math. Mech.*, **16**, 909–924.

Eringen, A. C. (1966b). Theory of micropolar fluids, *J. Math. Mech.*, **16**, 1–18.

Eringen, A. C. (1967). *Mechanics of Continua*, New York: John Wiley & Sons.

Eringen, A. C. (1968). Mechanics of micromorphic continua, in: *Mechanics of Generalized Continua, IUTAM Symposium* (E. Kröner, Ed.), Berlin: Springer-Verlag, 18–35.

Eringen, A. C. (1969). Micropolar fluids with stretch, *Int. J. Eng. Sci.*, **7**, 115–127.

Eringen, A. C. (1970). Mechanics of micropolar continua, in: *Contributions to Mechanics* (Dabir, Ed.), Oxford: Pergamon Press, 23–40.

Eringen, A. C. (1971). Micropolar elastic solids with stretch, *Ari Kitabevi Matbassi*, **24**, 1–18.

Eringen, A. C. (1976). *Continuum Physics. IV: Polar and Nonlocal Field Theories,* New York: Academic Press.

Eringen, A. C. (1984). Theory of nonlocal piezoelectricity, *J. Math. Phys.*, **25**(3), 717–727.

Eringen, A. C. (1999). *Microcontinuum Field Theories I: Foundation and Solids*, New York: Springer Verlag.

Eringen, A. C. (2002). *Nonlocal Continuum Field Theories,* New York: Springer-Verlag.

Eringen, A. C. (2003). Continuum theory of micromorphic electromagnetic thermoelastic solids, *Int. J. Eng. Sci.*, **41**, 653–665.

Eringen, A. C. (2004). Electromagnetic theory of microstretch elasticity and bone modeling, *Int. J. Eng. Sci.*, **42**, 23–242.

Eringen, A. C. (2006). Micromorphic electromagnetic theory and waves, *Found. Phys.*, **36**, 902–919.

Eringen, A. C. and Edelen, D. (1972). On nonlocal elasticity, *Int. J. Eng. Sci.*, **10**, 233–248.

Eringen, A. C. and Kim, B. S. (1977). Relation between non-local elasticity and lattice dynamics, *Cryst. Lattice Defects*, **7**, 51–57.

Eringen, A. C. and Maugin, G. A. (1990). *Electrodynamics of Continua I.* Springer.

Eringen, A. C. and Suhubi, E. S. (1964). Nonlinear theory of simple microelastic solids I, *Int. J. Eng. Sci.*, **2**, 189–203.

Erofeev, V. I. (2003). *Wave Processes in Solids with Microstructure,* Singapore: World Scientific.

Fedorchenko, A. M. (1988). *Theoretical Physics. Electrodynamics: Textbook,* Vyshcha shkola, Kyiv. In Russian.

Fousek, J., Cross, L. E., and Litvin, D. B. (1999). Possible piezoelectric composites based on flexoelectric effect, *Mater. Lett.*, **39**, 289–291.

Fu, J. Y., Zhu, W., Li, N., and Cross, L. E. (2006). Experimental studies of the converse flexoelectric effect induced by inhomogeneous electric field in a barium strontium titanate composition, *J. Appl. Phys.*, **100**(2), 024112–024112-6.

Galeş, C. (2011). Spatial behavior in the electromagnetic theory of microstretch elasticity, *Int. J. Solids Struct.*, **48**, 2755–2763.

Geguzin, Ya. E. and Goncharenko, N. N. (1962). Surface energy and processes on the surface of solids, *Uspekhi Fizicheskikh Nauk (Advances in Physical Sciences)*, **LXXVI**(2), 283–328. In Russian.

Germer, L. N., Mac Rae, A. U., and Hartman, C. D. (1961). (110) Nickel surface, *J. Appl. Phys.*, **32**, 2432–2439.

Gharbi, M., Sun, Z. H., Sharma, P., and White, K. (2009). The origins of electromechanical indentation size effect in ferroelectrics, *Appl. Phys. Lett.*, **95**, 142901(3).

Gou, P. F. (1971). Effects of gradients of polarization on stress-concentration at a cylindrical hole in an elastic dielectric, *Int. J. Solids Struct.*, **7**(11), 1467–1476.

Grad, H. (1952). Statistical mechanics, thermodynamics and fluid dynamics of systems with an arbitrary number of integrals, *Comm. Pure Appl. Math.*, **5**, 455–494.

Green, A. E. and Rivlin, R. S. (1964). On Cauchy's equations of motion, *Z. Angew. Math. Phys.*, **15**, 290–293.

Günther, W. (1958). Zur Static und Kinematic des Cosseratschen Kontinuums, *Abh. Braunschweig. Wiss. Ges.*, **10**, 195–213.

Gurevich, V. L. and Tagantsev, A. K. (1982). Theory for the thermopolarization effect in dielectrics having a center of inversion, *JETR Lett.*, **35**(3), 106–108.

Gyarmati, I. (1970). *Non-equilibrium Thermodynamics. Field Theory and Variational Principles*, New York: Springer, Berlin, Heidelberg.

Hadjesfandiari, A. R. (2013). Size-dependent piezoelectricity, *Int. J. Solids Struct.*, **50**, 2781–2791.

Hadjigeorgiou, E. P., Kalpakides, V. K., and Massalas, C. V. (1999a). A general theory for elastic dielectrics. I. The vectorial approach, *Int. J. Non-Linear Mech.*, **34**, 831–841.

Hadjigeorgiou, E. P., Kalpakides, V. K., and Massalas, C. V. (1999b). A general theory for elastic dielectrics. II. The variational approach, *Int. J. Non-Linear Mech.*, **34**, 967–980.

Han, L. and Vlassak, J. J. (2009). Determining the elastic modulus and hardness of an ultrathin film on a substrate using nanoindentation, *J. Mat. Res.*, **24**(3), 1114–1126.

Hong, J., Catalan, G., Scott, J. F., and Artacho, E. (2010). The flexoelectricity of barium and strontium titanates from first principles, *J. Phys. Condens. Matter.*, **22**, 112201(6).

Hrytsyna, O. (2008). Constitutive equations of thermomechanics of rheological media with fading memory taking into account the local displacements of mass, *Fiz.-mat. modelyuvannya ta inform. tekhnologii (Physico-Mathematical Modelling and Informational Technologies)*, **7**, 52–57. In Ukrainian.

Hrytsyna, O. (2010). The theorem of reciprocity of work for local gradient linear thermo-electro-magnetoelasticity, *Fiz.-mat. modelyuvannya ta inform. tekhnologii (Physico-Mathematical Modelling and Informational Technologies)*, **12**, 69–77. In Ukrainian.

Hrytsyna, O. (2011). Dynamic equations of local gradient electro-magneto-thermomechanics of dielectrics, *Fiz.-mat. modelyuvannya ta inform. tekhnologii (Physico-Mathematical Modelling and Informational Technologies)*, **13**, 44–59. In Ukrainian.

Hrytsyna, O. (2013a). Oscillations of a layer of a crystal with cubic symmetry under the action of harmonic electric field, *Mater. Sci.*, **48**(5), 653- 663.

Hrytsyna, O. (2013b). The stress–strain state and disjoining pressure in a thin elastic layer of ideal dielectric. *Fiz.-mat. modelyuvannya ta inform. tekhnologii (Physico-Mathematical Modelling and Informational Technologies)*, **18**, 85–90. In Ukrainian.

Hrytsyna, O. (2013c). Surface energy of deformable solids, *Fiz.-mat. modelyuvannya ta inform. tekhnologii (Physico-Mathematical Modelling and Informational Technologies)*, **17**, 43–54. In Ukrainian.

Hrytsyna, O. (2014). Effects of local mass displacement on electromechanical fields in dielectric medium with cylindrical hole, *Fiz.-mat. modelyuvannya ta inform. tekhnologii (Physico-Mathematical Modelling and Informational Technologies)*, (20), 97–106. In Ukrainian.

Hrytsyna, O. (2015). The propagation of Rayleigh waves in nonferromagnetic dielectrics, *Fiz.-mat. modelyuvannya ta inform. tekhnologii (Physico-Mathematical Modelling and Informational Technologies)*, **22**, 31–40. In Ukrainian.

Hrytsyna, O. R. (2016). Effect of heating on the subsurface inhomogeneity of the electric and mechanical fields in dielectrics, *J. Math. Sci.*, **212**(2), 167–181.

Hrytsyna, O. R. (2017a). Electrothermomechanics of nonferromagnetic polarizable solid bodies with regard for the tensor nature of the local displacements of mass, *J. Math. Sci.*, **226**(2), 139–151.

Hrytsyna, O. R. (2017b). Influence of subsurface inhomogeneity on the propagation of SH waves in isotropic materials, *Mater. Sci.*, **53**(2), 273–281.

Hrytsyna, O. R. and Kondrat, V. F. (2018). Gradient-type theory for electro-thermoelastic non-ferromagnetic dielectrics: Accounting for quadrupole polarization and irreversibility of local mass displacement, in: *Nanooptics, Nanophotonics, Nanostructures, and Their Applications* (O. Fesenko and L. Yatsenko, Eds.), Springer Proceedings in Physics, **210**, 147–160.

Hrytsyna, O. and Moroz, H. (2019). Some general theorems for local gradient theory of electrothermoelastic dielectrics, *J. Mech. Mat. and Struct,* **14**(1), 25–41.

Hu, S. L. and Shen, S. P. (2009). Electric field gradient theory with surface effect for nano-dielectrics, *CMC Comput. Mater. Contin.*, **13**, 63–87.

Ieşan, D. (2006). On the microstretch piezoelectricity, *Int. J. Eng. Sci.*, **44**, 819–829.

Ieşan, D. (2008). Thermopiezoelectricity without energy dissipation, *Proc. Royal Soc. A*, **464**, 631–657.

Ieşan, D. and Quintanilla, R. (2007). Some theorems in the theory of microstretch thermopiezoelectricity, *Int. J. Eng. Sci.*, **45**, 1–16.

Indenbom, V. L., Loginov, E. B., and Osipov, M. A. (1981). Flexoelectric effect and crystal structure, *Kristalografija*, **26**, 1157–1162. In Russian.

Jagnoux, P. and Vincent, A. (1989). Ultrasonic imaging by leaky Rayleigh waves, *NDT Int.*, **22**(6), 339–346.

Jaramillo, T. J. (1929). *A Generalization of the Energy Function of Elasticity Theory.* Dissertation, Department of Mathematics, University of Chicago.

Jiang, X. N., Huang, W. B., and Zhang, S. J. (2013). Flexoelectric nano-generator: Materials, structures and devices, *Nano Energy*, **2**, 1079–1092.

Jirásek, M. (2004). Nonlocal theories in continuum mechanics, *Acta Polytechnica*, **44**(5–6), 16–34.

Joffrin, J. and Levelut, A. (1970). Mise en evidence et mesure du pouvoir rotatoire acoustique naturel du quartza, *Solid State Commun.*, **8**, 1573–1575.

Jurov, V. M., Portnov, V. S., Ibraev, N. H., and Guchenko, S. A. (2011). Superficial tension of solid state, small particles and thin films, *Uspekhi Sovremennogo Estestvoznaniya. Fiz.-Mat. Nauki.* (*Adv. Curr. Nat. Sci. Phys. Math. Sci.*), **11**, 55–58. In Russian.

Kafadar, C. B. (1971). Theory of multipoles in classical electromagnetism, *Int. J. Eng. Sci.*, **9**, 831–853.

Kalinin, S. V. and Meunier, V. (2008). Electronic flexoelectricity in low-dimensional systems, *Phys. Rev. B.*, **77**(3), 033403(4).

Kalpakides, V. K. (1996). On the dynamical theory of thermoelastic dielectrics with polarization inertia, *Mech. Res. Commun.*, **23**(3), 247–256.

Kalpakides, V. K. and Agiasofitou, E. K. (2002). On material equations in second order gradient electroelasticity, *J. Elast.*, **67**, 205–227.

Kalpakidis, V. K. and Massalas, C. V. (1993). Tiersten's theory of thermoelectroelasticity: An extension, *Int. J. Eng. Sci.*, **31**, 157–164.

Ke, L. L. and Wang, Y. S. (2012). Thermo-electric-mechanical vibration of the piezoelectric nanobeams based on the nonlocal theory, *Smart Mater. Struct.*, **21**, 025018.

Ke, L. L. and Wang, Y. S. (2014). Free vibration of size-dependent magneto-electro-elastic nanobeams based on the nonlocal theory, *Physica E*, **63**, 52–61.

Khoroshun, L. (2006). Constructing the dynamical equations of electromagnetomechanics of dielectrics and piezoelectrics on basis of two-continuum mechanics, *Fiz.-mat. modelyuvannya ta inform. tekhnologii* (*Phys. Math. Modelling Inform. Technol.*), **3**, 177–198. In Russian.

Kim, J. J., Marzouk, H. A., Eloi, C. C., and Robertson, J. D. (1995). Effect of water vapor on the nucleation and growth of chemical vapor deposited copper films on spin–coated polyamide, *J. Appl. Phys.*, **78**(1), 245–250.

Kogan, S. M. (1964). Piezoelectric effect during inhomogeneous deformation and acoustic scattering of carriers in crystals, *Sov. Phys. Solid State*, **5**, 2069–2070.

Kondrat, V. and Hrytsyna, O. (2009a). Equations of thermomechanics of deformable bodies with regard for irreversibility of local displacement of mass, *J. Math. Sci.*, **168**(5), 688–698.

Kondrat, V. F. and Hrytsyna, O. R. (2009b). On the description of the Mead anomaly in thin dielectric films, *Dopovidi Akad. Nauk Ukr.* (*Reports Nat. Acad. Sci. Ukraine*), **3**, 84–89. In Ukrainian.

Kondrat, V. and Hrytsyna, O. (2009c). Linear theories of electro-magnetomechanics of dielectrics, *Fiz.-mat. modelyuvannya ta inform. tekhnologii* (*Physico-Mathematical Modelling and Informational Technologies*), **9**, 7–46. In Ukrainian.

Kondrat, V. and Hrytsyna, O. (2010a). Mechanoelectromagnetic interaction in isotropic dielectrics with regard for the local displacement of mass, *J. Math. Sci.*, **168**(5), 688–698.

Kondrat, V. and Hrytsyna, O. (2010b). Mechanical and electromagnetic wave interaction in linear isotropic dielectrics with local mass displacement and polarization inertia, in: *Vibrations in Physical Systems* (C. Cempel and M. W. Dobry, Eds.), **XXIV**, Poznan, 227–232.

Kondrat, V. and Hrytsyna, O. (2010c). Electromagnetomechanical waves dispersion in local gradient dielectrics with polarization inertia, *Fiz.-*

mat. modelyuvannya ta inform. tekhnologii (*Physico-Mathematical Modelling and Informational Technologies*), **11**, 81–90. In Ukrainian.

Kondrat, V. and Hrytsyna, O. (2011). Electromagnetic solids with irreversible process of local mass displacement, *Arch. Mech.*, **63**(3), 255–266.

Kondrat, V. and Hrytsyna, O. (2012a). Local gradient theory of dielectrics with polarization inertia and irreversibility of local mass displacement, *J. Mech. Mat. Struct.*, **7**(3), 285–296.

Kondrat, V. and Hrytsyna, O. (2012b). Equation of local gradient electromagnetothermomechanics of dielectrics with regard for polarization inertia, *Mater. Sci.*, **47**(4), 535–544.

Kondrat, V. and Hrytsyna, O. (2012c). Relations of gradient thermomechanics taking into account the irreversibility and inertia of local mass displacement, *J. Math. Sci.*, **183**(1), 100–112.

Kondrat, V. F. and Hrytsyna, O. R. (2018). Equations of the local gradient electromagnetothermomechanics of polarizable nonferromagnetic bodies with regard for electric quadrupole moments, *J. Math. Sci.*, **224**(4).

Korn, G. A. and Korn, T. M. (1968). *Mathematical Handbook for Scientists and Engineers*, New York: McGraw-Hill Book Company.

Kovalenko, A. D. (1969). *Thermoelasticity: Basic Theory and Application*, The Netherlands: Wolters-Noordhoff Groningen.

Kraut, E. (1971). Surface elastic waves: A review, in: *Acoustic Surface Wave and Acousto-optic Devises*, New York: Optosonic Press, 18–20.

Krichen, S. and Sharma, P. (2016). Flexoelectricity: A perspective on an unusual electromechanical coupling, *J. Appl. Mech.*, **83**, 030801.

Kröner, E. (1963). On the physical reality of torque stresses in continuum mechanics, *Int. J. Eng. Sci.*, **1**, 261.

Kunin, I. A. (1975). *Theory of Elastic Media with Microstructure. Nonlocal Theory of Elasticity*, Moscow, Nauka. In Russian.

Lakes, R. (1988). Cosserat micromechanics of structured media experimental methods, in: *Third Technical Conference Proceedings of the American Society for Composites*, September 25–29, 1988, Seattle, Washington, 505–516.

Lakes, R. (1995). Experimental methods for study of Cosserat elastic solids and other generalized elastic continua, in: *Continuum Models for Materials with Microstructure* (H. Muhlaus and J. Wiley, Eds.), New York: Wiley, 1–22.

Landau, L. D. and Lifshitz, E. M. (1984). *Electrodynamics of Continuous Media*, Oxford: Pergamon.

Langevin, P. (1918). Précédé et appareil d'émission et de réception des ondes élastiques sousmarines a 1'aide des propriétés piézoélectriques du quartz, *Fr. Pat.*, (505703).

Lee, J. D. and Chen, Y. (2004). Electromagnetic wave propagation in micromorphic elastic solids, *Int. J. Eng. Sci.*, **42**, 841–848.

Lee, J. D., Chen, Y., and Eskandarian, A. (2004). A micromorphic electromagnetic theory, *Int. J. Solids Struct.*, **41**, 2099–2110.

Li, X. X., Ono, T., Wang, Y. L., and Esashi, M. (2003). Ultrathin single-crystalline-silicon cantilever resonators: Fabrication technology and significant specimen size effect on Young's modulus, *Appl. Phys. Lett.*, **83**, 3081–3083.

Li, X. F., Yang, J. S., and Jiang, Q. (2005). Spatial dispersion of short surface acoustic waves in piezoelectric ceramics, *Acta Mech.*, **180**(1–4), 11–20.

Liang, X., Zhang, R., Hu, S., and Shen, S. (2017). Flexoelectric energy harvesters based on Timoshenko laminated beam theory, *J. Intel. Mat. Syst. Str.*, **28**(15), 2064–2073.

Liebold, C. and Müller, W. H. (2015). Applications of strain gradient theories to the size effect in submicro-structures incl. Experimental analysis of elastic material parameters, *Bull. TICMI*, **19**(1), 45–55.

Lippmann, G. (1881). Principe de la conversation de 1'électricité, *Ann. de Chim. et de Phys. (IV)*, **24**, 145–178.

Liu, C., Ke, L. L., and Wang, Y. S. (2013). Thermo-electro-mechanical vibration of piezoelectric nanoplates based on the nonlocal theory, *Comput. Struct.*, **106**, 167–174.

Lurie, A. I. (1990). *Nonlinear Theory of Elasticity*, Amsterdam, New York, North Holland: Elsevier Science Pub. Co.

Ma, W. H. and Cross, L. E. (2001a). Large flexoelectric polarization in ceramic lead magnesium niobate, *Appl. Phys. Lett.*, **79**(19), 4420–4422.

Ma, W. H. and Cross, L. E. (2001b). Observation of the flexoelectric effect in relaxor $Pb(Mg_{1/3}Nb_{2/3})O_3$ ceramics, *Appl. Phys. Lett.*, **78**(19), 2920–2921.

Ma, W. H. and Cross, L. E. (2003). Strain-gradient-induced electric polarization in lead zirconate titanate ceramics, *Appl. Phys. Lett.*, **82**(19), 3293–3295.

Ma, W. H. and Cross, L. E. (2006). Flexoelectricity of barium titanate, *Appl. Phys. Lett.*, **88**(23), 232902–232904.

Majdoub, M. S. (2010). Flexoelectricity and piezoelectricity in nanostructures and consequences for energy harvesting and storage, *Theses of Ph.D.*, University of Houston.

Majdoub, M. S., Sharma, P., and Çagin, T. (2008a). Dramatic enhancement in energy harvesting for a narrow range of dimensions in piezoelectric nanostructures, *Phys. Rev. B*, **78**, 121407(R).

Majdoub, M. S., Sharma, P., and Çagin, T. (2008b). Enhanced size-dependent piezoelectricity and elasticity in nanostructures due to the flexoelectric effect, *Phys. Rev. B*, **77**, 125424(9).

Majdoub, M. S., Maranganti, R., and Sharma, P. (2009). Understanding the origins of the intrinsic dead layer effect in nanocapacitors, *Phys. Rev. B*, **79**, 115412(8).

Majorkowska-Knap, K. and Lenz, J. (1989). Piezoelectric love waves in non-classical elastic dielectrics, *Int. J. Eng. Sci.*, **27**(8), 879–893.

Mao, S. (2016). *Continuum and Computational Modeling of Flexoelectricity*. Publicly Accessible Penn Dissertations, 1878. http://repository.upenn.edu/edissertations/1878.

Maranganti, R. and Sharma, P. (2009). Atomistic determination of flexoelectric properties of crystalline dielectrics, *Phys. Rev. B*, **80**, 054109(10).

Maranganti, R., Sharma, N. D., and Sharma, P. (2006). Electromechanical coupling in nonpiezoelectric materials due to nanoscale nonlocal size effects: Green's functions and embedded inclusions, *Phys. Rev. B*, **74**, 014110(14).

Marchenko, I. G., Neklyudov, I. M., and Marchenko, I. I. (2009). Collective atomic ordering processes during the low-temperature film deposition, *Dopovidi NAN Ukrainy* (*Proc. Nat. Acad. Sci. Ukraine*), **10**, 97–103. In Russian.

Marsden, J. E. and Tromba, A. (2003). *Vector Calculus*, 5th ed., New York: W. H. Freeman & Company.

Maskevich, V. S. and Tolpygo, K. V. (1957). Investigation of long-wavelength vibrations of diamond-type crystals with an allowance for long-range forces, *Sov. Phys. JETP*, **5**, 435–437.

Massalas, C. V., Kalpakides, V. K., and Foutsitzi, G. (1994). Some comments on the extended Tiersten's theory of thermoelectroelasticity, *Mech. Res. Commun.*, **21**(4), 343–351.

Maugin, G. A. (1977). Deformable dielectrics II. Voigt's intramolecular force balance in elastic dielectrics, *Arch. Mech.*, **29**, 143–151.

Maugin, G. A. (1979). Nonlocal theories or gradient-type theories: A matter of convenience? *Arch. Mech.*, **31**, 15–26.

Maugin, G. A. (1980). The method of virtual power in continuum mechanics: Applications to coupled fields, *Acta Mechanica*, **35**, 1–80.

Maugin, G. A. (1988). *Continuum Mechanics of Electromagnetic Solids*, Amsterdam: North-Holland Publishing Company.

Maugin, G. A. (1999). *Nonlinear Waves in Elastic Crystals*, Oxford: Oxford University Press.

Maugin, G. A. and Pouget, J. 1980. Electroacoustic equations for one-domain ferroelectric bodies, *J. Acoust. Soc. Am.*, **68**, 575–587.

Maugin, G .A., Pouget, J., Drouot, R., and Collet, B. (1992). *Non-Linear Electromechanical Couplings*, Wiley, New York.

Mead, C. A. (1961). Anomalous capacitance of thin dielectric structures, *Phys. Rev. Lett.*, **6**, 545–546.

Miller, R. E. and Shenoy, V. B. (2000). Size-dependent elastic properties of nanosized structural elements, *Nanotechnology*, **11**, 139–147.

Mindlin, R. D. (1964). Micro-structure in linear elasticity, *Arch. Ration. Mech. Anal.*, **16**, 51–78.

Mindlin, R. D. (1965). Second gradient of strain and surface-tension in linear elasticity, *Int. J. Solids Struct.*, **1**, 417–438.

Mindlin, R. D. (1967). Theories of elastic continua and crystal lattice theories, in: *IUTAM Symposium Mechanics of Generalized Continua*, (E. Kröner, Ed.), Berlin: Springer-Verlag, 312–320.

Mindlin, R. D. (1968). Polarization gradient in elastic dielectrics, *Int. J. Solids Struct.*, **4**, 637–642.

Mindlin, R. D. (1969). Continuum and lattice theories of influence of electromechanical coupling on capacitance of thin dielectric films, *Int. J. Solids Struct.*, **5**, 1197–1208.

Mindlin, R. D. (1971). Electromechanical vibrations of centrosymmetric cubic crystal plates, *Q. J. Mech. Appl. Math.*, **35**(4), 404–408.

Mindlin, R. D. (1972a). Elasticity, piezoelectricity and crystal lattice dynamics, *J. Elast.*, **2**(4), 217–282.

Mindlin, R. D. (1972b). A continuum theory of a diatomic, elastic dielectric, *Int. J. Solids Struct.*, **8**, 369–383.

Mindlin, R. D. (1972c). Coupled elastic and electromagnetic fields in a diatomic, electric continuum, *Int. J. Solids Struct.*, **8**, 401–408.

Mindlin, R. D. (1972d). Electromagnetic radiation from a vibrating quartz plate, *Int. J. Solids Struct.*, **9**, 697–702.

Mindlin, R. D. (1973). On the electrostatic potential of a point charge in a dielectric solid, *Int. J. Solids Struct.*, **9**, 233–235.

Mindlin, R. D. (1974). Electromagnetic radiation from a vibrating, elastic sphere, *Int. J. Solids Struct.*, **10**(11), 1307–1314.

Mindlin, R. D. and Toupin, R. A. (1971). Acoustical and optical activity in alpha quartz, *Int. J. Solids Struct.*, **7**(9), 1219–1227.

Naumov, I., Bratkovsky, A. M., and Ranjan, V. (2009). Unusual flexoelectric effect in two-dimensional noncentrosymmetric sp^2-bonded crystals, *Phys. Rev. Lett.*, **102**, 217601.

Nguyen, T. D., Mao, S., Yeh, Y. W., Purohit, P. K., and McAlpine, M. C. (2013). Nanoscale flexoelectricity, *Adv. Mater.*, **25**, 946–974.

Nowacki, W. (1970). *Teoria sprężystości,* Państwowe Wydawnictwo Naukowe, Warszawa. In Polish.

Nowacki, W. (1983). *Efekty elektromagnetyczne w stałych ciałach odkształcalnych*, Państwowe Wydawnictwo Naukowe, Warszawa. In Polish.

Nowacki, W. (1986). *Theory of Asymmetric Elasticity,* Polish Scientific Publishers, Warszawa.

Nowacki, J. P. (2004). Electro-elastic fields of a plane thermal inclusion in isotropic dielectrics with polarization gradient, *Arch. Mech.*, **56**(1), 33–57.

Nowacki, J. P. (2006). *Static and Dynamic Coupled Fields in Bodies with Piezoeffects or Polarization Gradient,* Lecture Notes in Applied and Computational Mechanics, Springer, **26**.

Nowacki, J. P. and Glockner, P. G. (1981). Propagation of waves in the interior of a thermoelastic dielectrics half-space, *Int. J. Eng. Sci.*, **19**(3), 603–613.

Nowacki, J. P. and Hsieh, R. K. T. (1986). Lattice defects in linear isotropic dielectrics, *Int. J. Eng. Sci.*, **24**(10), 1655–1666.

Ojaghnezhad, F. and Shodja, H. M. (2013). A combined first principles and analytical determination of the modulus of cohesion, surface energy, and the additional constants in the second strain gradient elasticity, *Int. J. Solids Struct.*, **50**. 3967–3974.

Pidstryhach, Ya. S. (1965). Diffusion theory of inelasticity of metals, *Zhurnal Pricl. Mekh. i Tekhn. Fiziki (J. Appl. Mech. Tech. Phys.),* **2**, 67–72. In Russian.

Pidstryhach, Ya. (1967). One nonlocal theory of deformation for solid bodies, *Pricladnaya mekhanika (Applied Mechanics)*, **3**(2), 71–76. In Russian.

Pierce, G. W. (1925). Piezoelectric oscillators applied to the precision measurement of the velocity of sound in air and CO_2 at high frequencies, *Proc. Amer. Acad.*, **60**, 271–302.

Pine, A. S. (1970). Direct observation of acoustical activity in α–Quartz, *Phys. Rev.*, **B2**, 2049–2054.

Polyzos, D. and Fotiadis, D. I. (2012). Derivation of Mindlin's first and second strain gradient elastic theory via simple lattice and continuum models, *Int. J. Solids Struct.*, **49**, 470–480.

Pouget, J. and Maugin, G. A. (1980). Coupled acoustic-optic modes in deformable ferroelectrics, *J. Acoust. Soc. Am.*, **68**, 588–601.

Pouget, J. and Maugin, G. A. (1981a). Bleustein–Gulyaev surface modes in elastic ferroelectrics, *J. Acoust. Soc. Am.*, **69**, 1304–1318.

Pouget, J. and Maugin, G. A. (1981b). Piezoelectric Rayleigh waves in elastic ferroelectrics, *J. Acoust. Soc. Am.*, **69**, 1319–1325.

Pouget, J., Askar, A., and Maugin, G. A. (1986a). Lattice model for elastic ferroelectric crystals: Microscopic approximation, *Phys. Rev. B.*, **33**, 6304–6319.

Pouget, J., Askar, A., and Maugin, G. A. (1986b). Lattice model for elastic ferroelectric crystals: Continuum approximation, *Phys. Rev. B.*, **33**, 6320–6325.

Prechtl, A. (1980). Deformable bodies with electric and magnetic quadrupoles, *Int. J. Eng. Sci.*, **18**, 665–680.

Rafikov, E. and Savinov, A. (1994). Frequency and temperature dependence of the thermopolarization response in dynamic investigations, *Physica Status Solidi (A)*, **144**(2), 471–477.

Robinson, C. R., White, K. W., and Sharma, P. (2012). Elucidating the mechanism for indentation size-effect in dielectrics, *Appl. Phys. Lett.*, **101**(12), 122901.

Romeo, M. (2010). Dynamic eigenvector problem in thermoelectroelasticity of dissipative ionic crystals, *European J. Mech. A Solids.*, **29**, 308–316.

Romeo, M. (2011). Micromorphic continuum model for electromagnetoelastic solids, *Z. Angew. Math. Phys.*, **62**, 513–527.

Sahin, E. and Dost, S. (1988). A strain-gradient theory of elastic dielectrics with spatial dispersion, *Int. J. Eng. Sci.*, **26**(12), 1231–1245.

Schaëfer, H. (1967). Das Cosserat-Kontinuum, *ZAMM*, **47**(8), 485–498.

Schiff, J. L. 1999. *The Laplace Transform: Theory and Applications*, New York: Springer-Verlag.

Schwartz, J. (1969). Solutions of the equations of equilibrium of elastic dielectrics: Stress functions, concentrated force, surface energy, *Int. J. Solids Struct.*, **5**(11), 1209–1220.

Sdobnyakov, N. Yu., Samsonov, V. M., Bazulev, A. N., and Kulpin, A. N. (2007). On the surface tension of nanocrystals of the different nature, *Kondensirovannye sredy i mezhfaznue granitsy* (*Condensed Matter and Interphases*), **9**(3), 255–260. In Russian.

Sharma, N. D., Maranganti, R., and Sharma, P. (2007). On the possibility of piezoelectricity nanocomposites without using piezoelectric materials, *J. Mech. Phys. Solids*, **55**, 2338–2350.

Shen, S. and Hu, S. (2010). A theory of flexoelectricity with surface effect for elastic dielectrics, *J. Mech. Phys. Solids*, **58**(5), 665–677.

Shuttleworth, R. (1950). The surface tension of solids, *Proc. Phys. Soc. A.*, **63**, 444–457.

Sirdeshmukh, D. B., Sirdeshmukh, L., and Subhadra, K. G. (2001). *Alkali Halides: A Handbook of Physical Properties.* Springer.

Sladek, J., Sladek, V., Kasala, J., and Pan, E. (2017). Nonlocal and gradient theories of piezoelectric nanoplates, *Procedia Engineering*, **190**, 178–185.

Sladek, J., Sladek, V., Wünsche, M., and Zhang, C. (2018). Effects of electric field and strain gradients on cracks in piezoelectric solids, *Eur. J. Mechanics A Solids*, **71**, 187–198.

Suhubi, E. S. (1969). Elastic dielectrics with polarization gradients, *Int. J. Eng. Sci.*, **7**, 993–997.

Suhubi, E. S. and Eringen, A. C. (1964). Nonlinear theory of simple microelastic solids II, *Int. J. Eng. Sci.*, **2**, 389–404.

Tagantsev, A. K. (1986). Piezoelectricity and flexoelectricity in crystalline dielectrics, *Phys. Rev. B.*, **34**, 5883(7).

Tagantsev, A. K. (1987). Pyroelectric, piezoelectric, flexoelectric, and thermal polarization effects in ionic crystals, *Soviet Physics Uspekhi*, **30**, 588–603.

Tagantsev, A. K. (1991). Electric polarization in crystals and its response to thermal and elastic perturbations, *Phase Transit.*, **35**(3–4), 119–203.

Tang, C. and Alici, G. (2011a). Evaluation of length-scale effects for mechanical behaviour of micro- and nanocantilevers: I. Experimental determination of length-scale factors, *J. Phys. D. Appl. Phys.*, **44**, 335501.

Tang, C. and Alici, G. (2011b). Evaluation of length-scale effects for mechanical behaviour of micro- and nanocantilevers: II. Experimental verification of deflection models using atomic force microscopy, *J. Phys. D. Appl. Phys.*, **44**, 335502.

Taucher, T. R. and Guzelsu, A. N. (1972). An experimental study of dispersion of stress waves in a fiber-rein-forced composite, *J. Appl. Mech.*, **39**, 98–102.

Tekoğlu, C. and Onck, P. R. (2008). Size effects in two-dimensional Voronoi foams: A comparison between generalized continua and discrete models, *J. Mech. Phys. Solids*, **56**, 3541–3564.

Tiersten, H. F. and Tsai, C. F. (1972). On the interaction of the electromagnetic field with heat conducting deformable insulators, *J. Math. Phys.*, **13**, 361–378.

Tolman, R. C. (1949). The effect of droplet size on surface tension, *J. Chem. Phys.*, **17**(3), 333–337.

Toupin, R. A. (1956). The elastic dielectric, *J. Rat. Mech. Anal.*, **5**, 849–915.

Toupin, R. A. (1962). Elastic materials with couple-stresses, *Arch. Ration. Mech. Anal.*, **11**, 385–414.

Toupin, R. A. (1963). A dynamical theory of elastic dielectrics, *Int. J. Eng. Sci.*, **1**, 101–126.

Truesdell, C. and Noll, W. (1992). *The Non-Linear Field Theories of Mechanics*, 2nd ed., New York: Springer-Verlag.

Ván, P. 2003. Weakly nonlocal irreversible thermodynamics, *Annalen der Physik.*, **12**(3), 146–173.

Vardoulakis, I. and Georgiadis, H. G. (1997). SH surface waves in a homogeneous gradient-elastic half-space with surface energy, *J. Elasticity*, **47**, 147–165.

Vladimirov, V. S. (1971). *Equations of Mathematical Physics*, 3. *Pure and Applied Mathematics Ser.* New York: M. Dekker.

Voigt, W. (1910). *Lehrbuch der Kristall-Physik*, Leipzig: B. G. Teubner.

Wang, X. (2016). *Modelling and Simulation of the Flexoelectric Effect on a Cantilevered Piezoelectric Nanoplate*, Electronic Thesis and Dissertation Repository. 4134. https://ir.lib.uwo.ca/etd/4134.

Wang, X. and Lee, J. D. (2010). Micromorphic theory: A gateway to nano world, *Int. J. Smart Nano Mater.*, **1**(2), 115–135.

Wang, G.-F., Yu, S.-W., and Feng, X.-Q. (2004). A piezoelectric constitutive theory with rotation gradient effects, *Eur. J. Mech. A Solid.*, **23**, 455–466.

Wang, X., Pan, E., and Feng, W. J. (2008). Anti-plane Green's functions and cracks for piezoelectric material with couple stress and electric field gradient effects, *Eur. J. Mech. A Solids*, **27**(3), 478–486.

Wood, R. W. and Loomis, A. L. (1927). The physical and biological effects of high-frequency sound waves of great intensity, *Philos. Mag.*, **7**(4), 417–436.

Yan, Z. and Jiang, L. Y. (2013a). Flexoelectric effect on the electroelastic responses of bending piezoelectric nanobeams, *J. Appl. Phys.*, **113**, 194102.

Yan, Z. and Jiang, L. Y. (2013b). Size-dependent bending and vibration behaviour of piezoelectric nanobeams due to flexoelectricity, *J. Phys. D*, **46**, 355502.

Yan, Z. and Jiang, L. (2017). Modified continuum mechanics modeling on size-dependent properties of piezoelectric nanomaterials: A review, *Nanomaterials*, **7**, 27.

Yang, J. S. (1997). Thin film capacitance in case of a nonlocal polarization law, *Int. J. Appl. Electromagn. Mech.*, **8**, 307–314.

Yang, J. (2004). Effects of electric field gradient on an anti-plane crack in piezoelectric ceramics, *Int. J. Fract.*, **127**, L111–L116.

Yang, J. (2006). Review of a few topics in piezoelectricity, *Appl. Mech. Rev.*, **59**, 335–345.

Yang, J. S. and Yang, X. M. (2004). Electric field gradient effect and thin film capacitance, *World J. Eng.*, **2**, 41–45.

Yang, Z. and Yang, J. (2009). Effect of electric field gradient on the propagation of short piezoelectric interface waves, *Int. J. Appl. Electromagn. Mech.*, **29**(2), 101–108.

Yang, X. M., Hu, Y. T., and Yang, J. S. (2004). Electric field gradient effects in antiplane problems of polarized ceramics, *Int. J. Solids Struct.*, **41**, 6801–6811.

Yang, X. M., Hu, Y. T., and Yang, J. S. (2005). Electric field gradient effects in anti-plane problems of a circular cylindrical hole in piezoelectric materials of 6 mm symmetry, *Acta Mech.*, **18**, 29–36.

Yang, X. M., Zhou, H. G., and Li, J. Y. (2006). Electric field gradient effects in anti-plane circular inclusion in polarized ceramics, *Proc. Roy. Soc. A*, **462**, 3511–3522.

Yang, W., Liang, X., and Shen, S. (2015). Electromechanical responses of piezoelectric nanoplates with flexoelectricity, *Acta Mechanica*, **226**(9), 3097–3110.

Yariv, A. and Yeh, P. (1984). *Optical Waves in* Crystals, New York: Wiley–Interscience.

Yudin, P. V. and Tagantsev, A. K. (2013). Fundamentals of flexoelectricity in solids, *Nanotechnology*, **24**, 432001.

Yue, Y. M., Xu, K. Y., and Aifantis, E. C. (2014). Microscale size effects on the electromechanical coupling in piezoelectric material for anti-plane problem, *Smart Materials and Structures*, **23**(12), 125043.

Zeng, Y., Hu, Y. T., and Yang, J. S. (2005). Electric field gradient effects in piezoelectric anti-plane crack problems, *J. Huazhong Univ. Sci. Technol.*, **22**, 31–35.

Zeng, X., Lee, J. D., and Chen, Y. (2006). Determining material constants in nonlocal micromorphic theory through phonon dispersion relations, *Int. J. Eng. Sci.*, **44**, 1334–1345.

Zhang, Z. R., Yan, Z., and Jiang, L. Y. (2014). Flexoelectric effect on the electroelastic responses and vibrational behaviors of a piezoelectric nanoplate, *J. Appl. Phys.*, **116**, 014307.

Zhang, R., Liang, X., and Shen, S. (2016). A Timoshenko dielectric beam model with flexoelectric effect, *Meccanica*, **51**, 1181–1188.

Zholudev, E. S. (1966). Symmetry and piezoelectric properties of crystals, *Czech. J. Phys.*, **16**(5), 368–381. In Russian.

Zhu, W., Fu, J. Y., Li, N., and Cross, L. (2006). Piezoelectric composite based on the enhanced flexoelectric effects, *Appl. Phys. Lett.*, **89**(19), 192904.

Zubko, P., Catalan, G., Buckley, A., Welche, P. R. L., and Scott, J. F. (2007). Strain-gradient-induced polarization in $SrTiO_3$ single crystals, *Phys. Rev. Lett.*, **99**(16), 167601(4).

Zubko, P., Catalan, G., and Tagantsev, A. K. (2013). Flexoelectric effect in solids, *Annu. Rev. Mater. Res.*, **43**, 387–421.

Index